Study Guide

Biology

NINTH EDITION

Eldra P. Solomon

Linda R. Berg

Diana W. Martin

Prepared by

Jennifer Metzler
Ball State University

Robert Yost
Indiana University-Purdue University Indianapolis

BROOKS/COLE
CENGAGE Learning™

Australia • Brazil • Japan • Korea • Mexico • Singapore • Spain • United Kingdom • United States

ISBN-13: 978-0-538-73167-6
ISBN-10: 0-538-73167-2

Brooks/Cole
20 Davis Drive
Belmont, CA 94002-3098
USA

Cengage Learning is a leading provider of customized learning solutions with office locations around the globe, including Singapore, the United Kingdom, Australia, Mexico, Brazil, and Japan. Locate your local office at: **www.cengage.com/global**

Cengage Learning products are represented in Canada by Nelson Education, Ltd.

To learn more about Brooks/Cole, visit **www.cengage.com/brookscole**

Purchase any of our products at your local college store or at our preferred online store **www.cengagebrain.com**

Printed in the United States of America
1 2 3 4 5 6 7 14 13 12 11 10

CONTENTS

□

iii

TO THE STUDENT

□

This study guide has been prepared to help you learn biology. We have included the ways of studying and learning that successful students find helpful. If you study the textbook, use this study guide regularly, and follow the guidelines provided here, you will learn more about biology and, in the process, you are likely to earn a better grade. It is our sincere desire to provide you with enjoyable experiences during your excursion into the world of life.

Study guide chapters contain the following sections: *Chapter Overview, Reviewing Concepts, Building Words, Matching, Making Comparisons, Making Choices,* and *Visual Foundations.* Although these sections provide a fairly comprehensive review of the material in each chapter, they may not cover all of the material that your instructor expects you to know. The ways that you can use these sections are described below.

CHAPTER OVERVIEW

Each chapter in this study guide begins with an overview of the chapter. It focuses on the major concepts of the chapter, and, when appropriate, relates them contextually to what you have learned in previous chapters. We believe you will find it beneficial to read this section carefully *before* you undertake any of the exercises in the chapter, and again *after* you have completed them. Review this material any time you find yourself losing sight of "the big picture."

REVIEWING CONCEPTS

This section outlines the textbook chapter. It contains important concepts and fill-in questions. Be sure that you understand them. Examine the textbook material to be sure that you have correctly filled in all of the blank spaces. Once completed, check your work in the *Answers* section at the back of the study guide.

BUILDING WORDS

The language of biology is fundamental to an understanding of the subject. Many terms used in biology have common word roots. If you know word roots, you can frequently figure out the meaning of a new term. This section provides you with a list of some of the prefixes and suffixes (word roots) that are used in the chapter.

Learn the meanings of these prefixes and suffixes, then use them to construct terms in the blank spaces next to the definitions provided. Once you have filled in the blanks for all of the definitions, check your work in the *Answers* section at the back of the study guide.

Learning the language of biology also means that you can communicate in its language, and effective communication in any language means that you can read, write, and speak the language. Use the glossary in your textbook to learn how to pronounce new terms. Say the words out loud. You will discover that pronouncing words aloud will not only teach you verbal communication, it will also help you remember them.

MATCHING

This section tests your knowledge of key terms used in the chapter. Check you answers in the *Answers* section.

MAKING COMPARISONS

This section uses tables to test your ability to recognize relationships between key elements in the chapter. Check your answers in the *Answers* section

MAKING CHOICES

The items in this section were selected for two purposes. First, and perhaps most importantly, they help you develop study skills. They provide a means to re-examine, integrate, and analyze selected textbook material. To answer many of them, you must understand how material presented in one part of a chapter relates to material presented in other parts of the chapter. This section provides a means to evaluate your mastery of the subject matter and to identify material that needs additional study.

Once you have studied the textbook material and your notes, close your textbook, put away your notes, and try answering the multiple choice items. Check your answers in the *Answers* section. Don't be discouraged if you miss some answers. Take each item that you missed one at a time, go back to the textbook and your notes, and determine what you missed and why. Do not merely copy down the correct answers — that defeats the purpose. If you dig the answers out of the textbook, you will find that your understanding of the material will improve. Some points that you thought you already knew will become clearer, and you will develop a better understanding of the relationships between different aspects of biology.

VISUAL FOUNDATIONS

Selected visual elements from the textbook are presented in this section. Coloring the parts of biological illustrations causes one to focus on the subject more carefully. Coloring also helps to clearly delineate the parts and reveals the relationship of parts to one another and to the whole.

Begin by coloring the open boxes (❑). This will provide you with a quick reference guide to the illustration when you study it later. Close your textbook while you are doing this section. After completing the *Visual Foundations* section, use your textbook to check your work. Make notes about anything you missed for future reference. Come back to these figures periodically for review. If an item to be colored is listed in the singular form, and there is more than one of them in the illustration, be sure to color all of them. For example, if you are asked to color the "mitochondrion," be sure to color all of the mitochondria shown in the illustration.

OTHER STUDY SUGGESTIONS

When possible, read material in the textbook *before* it is covered in class. Don't be concerned if you don't understand all of the material at first, and don't try to memorize every detail in this first reading. Reading before class will contribute greatly to your understanding of the material as it is presented in class, and it will help you take better notes. As soon as possible after the lecture, recopy all of your class notes. You should do this while the classroom presentation is still fresh in your mind. If you are prompt and disciplined about it, you will be able to recall additional details and include them in your notes. Use material in this study guide and your textbook to fill in the gaps.

What is the best way for you to study? That is something you have to figure out for yourself. Reading all of the chapters assigned by your instructor is basic, but it is not all there is to studying.

A technique that many successful students have found beneficial is to organize a study group of 4-6 students. Meet periodically, especially before examinations, and go over the material together. Bring along your textbooks, study guides, and expanded notes. Have each member of the group ask every other member several questions about the material you are reviewing. Discuss the answers as a group until the principles and details are clear to everyone. Use the combined notes of all members to fill in any gaps in your notes. Consider the reasons for correct and incorrect answers to questions and items in the *Reviewing Concepts* and *Making Choices* sections. This study technique can do wonders. There is probably no better way to find out how well you know the material than to verbalize it in front of a critical audience. Just ask any biology instructor!

If you follow the guidelines presented here and carefully complete the work provided in this study guide, you will make giant strides toward understanding the principles of biology. If you decide to specialize in some area of biology, you will find the general background you obtained in this course to be invaluable.

CHAPTER 1

A View of Life

Biology is the science of life. Knowledge about biological principles can help individuals make informed decisions about aspects of their lives. The basic themes of biology include evolution, information transfer, and energy transfer. Living things share a common set of characteristics that distinguishes them from nonliving things. They are organized at several levels, each more complex than the previous, from the chemical level up through the biosphere. Organisms transmit information within themselves and to other organisms in order to survive using chemical, electrical, and behavioral signals. They require a continuous input of energy to maintain the chemical transactions and cellular organization essential to their organization. Evolution shows the relationships that exist between diverse life forms on Earth and provides a framework for study. To make sense of the millions of organisms that have evolved on our planet, scientists have developed a binomial and hierarchical system of nomenclature. The tree of life indicates evolutionary relationships between organisms. Organisms are studied by a system of observation, question, hypothesis, experiment, data analysis, and revised hypothesis known as the scientific method. Scientific endeavors have many ethical implications as scientific techniques and technology continue to advance.

REVIEWING CONCEPTS

Fill in the blanks.

INTRODUCTION

1.
Knowledge of biological concepts is a vital tool for understanding the challenges that confront modern society. Some of the more important challenges are: (a)_____,
(b)_____,
(c)_____,
(d)_____,
(e)_____.

THREE BASIC THEMES

2. The three basic themes of biology are (a)_____,
(b)_____, and (c)_____.

CHARACTERISTICS OF LIFE

3. The characteristics of life include:

_____

_____

_____.

Organisms are composed of cells

4. Organisms can be _____ or _____.

5. The two major cell types are _____ and
   _____.

Organisms grow and develop

6. Living things grow by increasing _____ of
   cells or both.

7. _____ is a term that encompasses all the changes that
   occur during the life of an organism.

Organisms regulate their metabolic processes

8. Metabolism can be described as the sum of all the _____
   that take place in the organism.

9. _____ is the tendency of organisms to maintain a
   balanced internal environment.

Organisms respond to stimuli

10. Physical or chemical changes in an internal or external environment that
    evoke a response from all life forms are called _____.

Organisms reproduce

11. We know that organisms come from previously existing organisms as a result
    of scientists like Francesco Redi in the 17th century and _____
    in the 19th century.

12. In general, living things reproduce in one of two ways:
    _____.

Populations evolve and become adapted to the environment

13. Inherited characteristics that enhance an organism's ability to survive in a
    particular environment are called _____.

LEVELS OF BIOLOGICAL ORGANIZATION

14. Biologists learn about living things using _____ to study their parts.

15. _____ also help biologists see what characteristics are new at a
    higher level of organization.

Organisms have several levels of organization

16. The simplest component of life is the _____.

Several levels of ecological organization can be identified

17. All of the ecosystems on Earth are collectively referred to as the
    _____.

18. All of the members of one species that live in the same geographic area make
    up a _____.

INFORMATION TRANSFER

DNA transmits information from one generation to the next

19. _____ are the units of hereditary material.

20. The two scientists credited with working out the structure of DNA are
    _____.

Information is transmitted by chemical and electrical signals

21. _____ are large molecules important in determining the
    structure and function of cells and tissues.

THE ENERGY OF LIFE

22. All the energy transformations and chemical processes that occur in organisms are referred to as _____.

23. _____ is the process by which molecular energy is released to do cellular work.

24. A self-sufficient ecosystem contains three major categories of organisms: _____. It must also have a continuous input of energy.

25. Organisms that can produce their own food from simple raw materials are called _____ or _____.

EVOLUTION: THE BASIC UNIFYING CONCEPT OF BIOLOGY

26. Populations change over time via _____.

Biologists use a binomial system for naming organisms

27. The science of classifying and naming organisms is _____.

28. The simplest category of classification is the _____.

29. The scientific name (binomial) of an organism consists of the _____ names for that organism.

Taxonomic classification is hierarchical

30. Families are grouped into _____.

31. The broadest of the taxonomic groups is the _____.

The tree of life includes three domains and six kingdoms

32. The three domains of life are (a)_____, (b)_____, and (c)_____, which includes plants, protists, animals, and fungi.

33. The _____ have been broken up into five supergroups based on molecular analysis.

Species adapt in response to changes in their environment

34. Adaptations to environmental change occur due to _____.

Natural selection is an important mechanism by which evolution proceeds

35. Charles Darwin's famous book, "_____ _____," published in 1859, had and still has a profound influence on the biological sciences, as well as impacting on the thinking of society in general.

36. Darwin's theory of natural selection was based on the observations that individuals within species show (a)_____, and more individuals are produced than can possibly (b)_____, this causes (c)_____, and individuals with the most advantageous characteristics are the most likely to (d)_____ and pass those adaptations on to their offspring.

Populations evolve as a result of selective pressures from changes in the environment

37. The source of variation is (a)_____ which occurs in (b)_____.

38. All the genes in a population make up its _____.

39. Natural selection favors organisms with traits that enable them to effectively respond to _____.

THE PROCESS OF SCIENCE

40. The _____ is a system of observation, question, hypothesis, experiment, data analysis, and revised hypothesis.

Science requires systematic thought processes

41. With _____ reasoning, one begins with supplied information or premises and draws conclusions based on those premises.

42. With _____ reasoning, one draws a conclusion from specific observations.

Scientists make careful observations and ask critical questions

43. In 1928, bacteriologist _____ discovered the antibiotic, penicillin, but could not grow the antibiotic producing organism.

Chance often plays a role in scientific discovery

44. Significant discoveries are usually made by those who are in the habit of looking critically at nature and recognizing a _____.

A hypothesis is a testable statement

45. The characteristics of a good hypothesis include being (a) _____, (b) _____, and (c) _____.

46. A hypothesis cannot be _____.

47. _____ are becoming very useful in hypothesis development.

Many predictions can be tested by experiment

48. An experimental group differs from a control group only with respect to the _____ being studied.

Researchers must avoid bias

49. In a _____, neither the patient nor the physician knows who is getting the experimental drug and who is getting the placebo.

Scientists interpret the results of experiments and make conclusions

50. Sampling errors can lead to inaccurate conclusions because _____.

51. _____ are strengthened when others repeat a scientist's work.

A theory is supported by tested hypotheses

52. A _____ is an integrated explanation of a number of hypotheses, each supported by consistent results from many observations or experiments.

Many hypotheses cannot be tested by direct experiment

53. Two examples of hypotheses that cannot be tested by direct experiment include _____.

Paradigm shifts allow new discoveries

54. A _____ is a set of assumptions or concepts that constitute a way of thinking about reality.

Systems biology integrates different levels of information

55. Biologists that integrate data from various levels of complexity with the goal of understanding the big picture are known as _____.

Science has ethical dimensions

56. Some of the more controversial areas of scientific inquiry with important
    ethical dimensions include _____

    _____.

BUILDING WORDS

Use combinations of prefixes and suffixes to build words for the definitions that follow.

Prefixes	The Meaning		Suffixes	The Meaning
a-	without, not, lacking		-logy	"study of"
auto-	self, same		-stasis	equilibrium caused by opposing forces
bio-	life		-troph	nutrition, growth, "eat"
eco-	home		-zoa	animal
hetero-	different, other			
homeo-	similar, "constant"			
hypo-	under			
photo-	light			
pro-	first, earliest form of			

Prefix	Suffix	Definition
_____	_____	1. The study of life.
_____	_____	2. The study of groups of organisms interacting with one another and their nonliving environment.
_____	-sexual	3. Reproduction without sex.
_____	_____	4. Maintaining a constant internal environment.
_____	-system	5. A community along with its nonliving environment.
_____	-sphere	6. The planet earth including all living things:
_____	_____	7. An organism that produces its own food.
_____	_____	8. An organism that is dependent upon other organisms for food, energy, and oxygen.
_____	-karyote	9. Simplest cell type.
_____	-synthesis	10. The conversion of solar (light) energy into stored chemical energy by plants, blue-green algae, and certain bacteria.
_____	-thesis	11. A tentative explanation.

MATCHING

Terms:

a. Adaptation   f. Domain   k. Species
b. Cell   g. Kingdom   l. Stimulus
c. Cilia   h. Order   m. Taxonomy
d. Consumer   i. Phylum   n. Organ
e. Decomposer   j. Population   o. Community

For each of these definitions, select the correct matching term from the list above.

_____ 1. Populations of different species that live in the same geographical area at the same time.

_____ 2. The broadest category of classification used.

_____ 3. The science of naming, describing, and classifying organisms.

_____ 4. Hairlike structures that project from the surface of some cells and are used by some organisms for locomotion.

_____ 5. A physical or chemical change in the internal or external environment of an organism potentially capable of provoking a response.

_____ 6. A group of tissues that work together to carry out specific functions.

_____ 7. An evolutionary modification that improves the chances of survival and reproductive success in a given environment.

_____ 8. A microorganism that breaks down organic material.

_____ 9. The basic structural and functional unit of life, which consists of living material bounded by a membrane.

_____10. A group of organisms with similar structural and functional characteristics that in nature breed only with each other and have a close common ancestry; a group of organisms with a common gene pool.

MAKING COMPARISONS

Fill in the blanks.

Level of Ecological Organization	Examples
#1	Group of elephants in Kalahari desert
Community	#2
Ecosystem	#3
#4	All ecosystems on Earth

MAKING CHOICES

Place your answer(s) in the space provided. Some questions may have more than one correct answer.

_____ 1. In science, a tentative explanation for an observation or phenomenon is known as a/an
    a. supposition   d. theory
    b. theory   e. hypothesis
    c. belief

_____ 2. Organisms that rely on others for complex food molecules are known as
    a. heterotrophs.   d. consumers.
    b. cyanobacteria.   e. producers.
    c. autotrophs.

_____ 3. The kingdom(s) containing organisms that have a nucleus surrounded by a membrane ("true nucleus") and are unicellular is/are
    a. Bacteria.   d. Plantae.
    b. Protista.   e. Animalia.
    c. Fungi.

_____ 4. A snake is a member of the kingdom/phylum
    a. Prokaryotae/vertebrate.   d. Animalia/chordata.
    b. Fungi/ascomycete.   e. Plantae/angiosperm.
    c. Protista/Aves.

_____ 5. The domain(s) containing organisms that are both single-celled and have a chromosome that is not separated from the cytoplasm by a membrane is/are
    a. Eukarya.   d. Plantae.
    b. Animalia.   e. Archae.
    c. Bacteria.

_____ 6. Maintenance of an internal balanced internal environment within an organism is known as
    a. homeostasis.   d. adaptation.
    b. metabolism.   e. cyclosis.
    c. response.

_____ 7. Prokaryotic organisms
    a. lack a nuclear membrane.   d. are simple plants.
    b. include bacteria.   e. comprise two domains.
    c. contain membrane-bound organelles.

_____ 8. To comply with the scientific method, a hypothesis must
    a. be true.   d. withstand rigorous investigations.
    b. be testable.   e. eventually reach the status of a principle.
    c. be popular and fundable.

_____ 9. The characteristic(s) common to all living things include(s)
    a. reproduction.   d. growth.
    b. presence of a true nucleus.   e. multicellularity.
    c. adaptation.

The Federal Center for Disease Control tested 600 female prostitutes who regularly use condoms. They found that none of the women had the AIDS virus. Use this study and data to answer questions 10-12.

_____10. In terms of determining a relationship between use of condoms by women and the kind of sexual behavior that might lead to contraction of AIDS, this experiment

a. could be called a theory.

b. is well-designed.

c. should be repeated with more subjects.

d. is useless.

e. should include male subjects.

_____11. The experiment

a. proves that condoms are effective in preventing AIDS.

b. indicates that AIDS is not transmitted through sexual activity.

c. suggests that condoms may be effective in preventing AIDS.

d. probably means that none of the partners of the female prostitutes had AIDS.

e. shows that women are less prone to contract AIDS than men.

_____12. To determine the effectiveness of using condoms in preventing AIDS among women, this study was lacking in not having

a. enough subjects.

b. male subjects.

c. controls.

d. subjects who do not use condoms during sex.

e. women who are not prostitutes.

_____13. The number of domains in the tree of life is

a. one.

b. two.

c. three.

d. four

e. none.

_____14. Evolution by natural selection depends upon:

a. mutations.

b. adaptations.

c. differential reproduction

d. change in individuals.

e. environmental change.

_____15. Which of the following is true concerning the levels of organization in multicellular organisms?

a. tissues are comprised of cells

b. organs are made up of tissues

c. a group of tissues and organs make up an organ system

d. molecules are comprised of atoms

e. cells are made up of atoms, molecules, and organelles.

VISUAL FOUNDATIONS

Questions 1-3 refer to the diagram below. Some questions may have more than one correct answer.

_____ 1. The diagram represents:

    a. developmental theory.

    b. a controlled experiment.

    c. deductive reasoning.

    d. scientific method.

    e. all of the above.

_____ 2. Which of the following statements is/are supported by the diagram?

    a. A theory and a hypothesis are identical.

    b. Conclusions are based on results obtained from carefully designed experiments.

    c. Using results from a single sample can result in sampling error.

    d. A tested hypothesis may lead to new observations.

    e. Science is infallible.

_____ 3. Which of the following statements best describes the activity depicted by the box labeled #4 in the diagram?

    a. Make a prediction that can be tested.

    b. Generate a related hypothesis.

    c. Form generalized inductive conclusions.

    d. Develop a scientific principle.

    e. Hypothesis not supported.

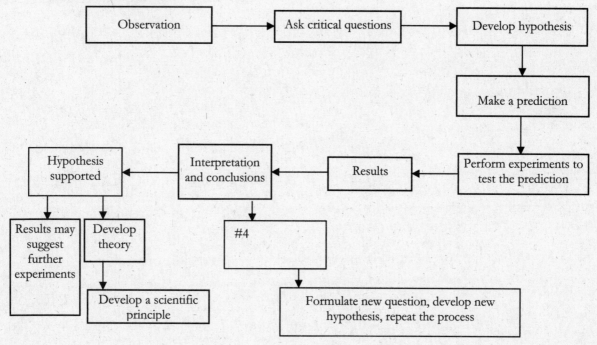

CHAPTER 2

❑

Atoms and Molecules:
The Chemical Basis of Life

Living things, like everything else on our planet, are made of atoms and molecules. Therefore, to understand life you must first understand the interaction and activity of atoms and molecules. Living things share remarkably similar chemical compositions. About 96% of an organism's mass is made up of just four elements — oxygen, carbon, hydrogen, and nitrogen. The smallest portion of an element that retains its chemical properties is the atom. Atoms, in turn, consist of subatomic particles. Those most important to an understanding of life are the protons, neutrons, and electrons. The number and arrangement of electrons are most important biologically as they determine the chemical behavior of atoms. Electrons transferred during redox reactions are also very important to energy conversions. Molecules and compounds are formed by the joining of atoms. The interaction between atoms determines chemical bonds and attractive forces that produce compounds. The focus is on small, simple substances, known as inorganic compounds, which play critical roles in life's processes. One of these, water, is considered to have been essential to the origin of life, as well as to the continued survival and evolution of life on Earth. Acids, bases, and salts and how they generate electrolytes are also important for proper cellular function.

REVIEWING CONCEPTS
Fill in the blanks.

INTRODUCTION

1. Among the biologically important groups of inorganic compounds are _____.

ELEMENTS AND ATOMS

2. An element is a substance that cannot be separated into simpler parts by _____.

3. Over 96% of an organism's weight is made up of the four elements _____ _____ _____.

4. The three main parts of an atom are the
(a)_____
_____. The
(b)_____ compose the atomic nucleus, and the (c)_____ occupy the mostly empty space surrounding the nucleus.

5. (a)_____ have one unit of positive charge,
   (b)_____ are neutral, and (c)_____
   have a negative charge.

An atom is uniquely identified by its number of protons

6. The number of protons in an atom is called the
   _____.

7. The symbol $_8O$ indicates that oxygen has _____
   protons.

Protons plus neutrons determine atomic mass

8. An atomic mass unit (amu) is also called a
   _____.

9. The total number of protons and neutrons in an atom is referred to
   as the _____.

Isotopes of an element differ in number of neutrons

10. The (a)_____ of
    isotopes vary, but the (b)_____ remains constant.

11. Some isotopes are unstable and emit radiation when they decay,
    these are termed _____.

Electrons move in orbitals corresponding to energy levels

12. Electrons with the same energy comprise an
    _____.

13. The electrons in the outermost energy level of an atom are called
    _____ electrons.

CHEMICAL REACTIONS

14. All elements in the same group on the periodic table have similar
    _____.

Atoms form compounds and molecules

15. A chemical compound consists of
    _____
    _____ that are combined in a
    fixed ratio.

16. A _____ forms when two or more atoms are
    joined very strongly.

Simplest, molecular, and structural chemical formulas give different information

17. A _____ formula simply uses chemical
    symbols to describe the chemical composition of a compound.

18. A _____ formula shows not only the
    types and numbers of atoms in a molecule but also their
    arrangement.

One mole of any substance contains the same number of units

19. The molecular mass of a compound is the sum of the
    _____
    _____.

20. The amount of a compound in grams that is equal to its atomic mass is called one _____.

21. A 1 molar solution contains one (a)_____ of substance (b)_____ in one (c)_____.

Chemical equations describe chemical reactions

22. (a)_____ are substances that participate in the reaction, (b)_____ are substances formed by the reaction.

23. Arrows are used to indicate the direction of a reaction; reactants are placed to the (a)_____ of arrows, products are placed to the (b)_____ of arrows.

CHEMICAL BONDS

24. (a)_____ can hold atoms together and involve (b)_____.

In covalent bonds electrons are shared

25. Depending on the number of electrons shared they can be_____.

26. Covalently bonded atoms with similar electronegativities are called (a)_____ , and covalently bonded atoms with differing electronegativities are called (b)_____.

Ionic bonds form between cations and anions

27. Ionic compounds are comprised of positively charged ions called (a)_____, and negatively charged ions called (b)_____.

28. In the presence of a (a)_____ an ionic compound will dissolve to produce a (b)_____.

29. The process known as _____ has occurred when cations and anions of an ionic compound are surrounded by the charged ends of water molecules.

Hydrogen bonds are weak attractions

30. These bonds tend to form between an (a)_____ covalently bonded to an (b)_____ and an atom with a partial (c)_____ charge.

31. Hydrogen bonds are individually (a)_____, but are collectively (b)_____ when present in large numbers.

Van der Waals interactions are weak forces

32. Van der Waals interactions are weak forces based on fluctuating _____.

REDOX REACTIONS

33. Reactions involving oxidation *and* reduction are called _____ reactions.

34. When iron rusts, it loses electrons and is said to be (a)_____, while at the same time oxygen gains electrons and is (b)_____.

WATER

35. In humans, water makes up about _____% of total body weight.

Hydrogen bonds form between water molecules

36. Cohesion of water molecules is due to (a)_____ between the molecules and contributes to (b)_____.

37. _____ is the ability of water molecules to stick to other substances.

38. The tendency of water to move through narrow tubes as the result of adhesion and cohesion is known as

_____.

Water molecules interact with hydrophilic substances by hydrogen bonding

39. Substances that interact readily with water are said to be (a)_____, whereas those that don't are called (b)_____.

40. When nonpolar molecules interact with each other instead of water it is called an _____.

Water helps maintain a stable temperature

41. Evaporative cooling depends on the high_____ of water.

42. A _____ is a unit of heat energy.

43. Water has a high _____, which helps marine organisms maintain a relatively constant internal temperature.

ACIDS, BASES, AND SALTS

44. An acid is a proton donor that dissociates in solution to yield (a)_____. A base is a proton acceptor that generally dissociates in solution to yield (b)_____.

pH is a convenient measure of acidity

45. The pH scale covers the range (a)_____ and is a(b)_____scale. Neutrality is at pH (c)_____.

46. A solution with a pH above 7 is a(an) (a)_____. A solution with a pH below 7 is a(an) (b)_____.

Buffers minimize pH change

47. A _____ is a weak acid or a weak base.

An acid and a base react to form a salt

48. Dissociated ions called _____ can conduct an electric current when a salt, an acid, or a base is dissolved in water.

BUILDING WORDS

Use combinations of prefixes and suffixes to build words for the definitions that follow.

Prefixes	The Meaning		Suffixes	The Meaning
equi-	equal		-hedron	face
hydr(o)-	water		-philic	loving, friendly, lover
neutro-	neutral		-phobic	fearing
non-	not			
tetra-	four			

Prefix	Suffix	Definition
_____	-n	1. An uncharged subatomic particle.
_____	-librium	2. The condition of a chemical reaction when the forward and reverse reaction rates are equal.
_____	_____	3. The shape of a molecule when it appears as a three-dimensional pyramid.
_____	-ation	4. The process whereby ionic compounds combine chemically with water and dissolve.
_____	_____	5. Not attracted to water; insoluble in water.
_____	-electrolyte	6. A substance that does not form ions when dissolved in water, and therefore does not conduct an electric current.
_____	_____	7. Having a strong affinity for water; soluble in water.

MATCHING

Terms:

a. Anion
b. Atom
c. Atomic mass
d. Calorie
e. Cation
f. Covalent bond
g. Electrolyte
h. Electron
i. Hydrogen bond
j. Polar
k. Ionic bond
l. Oxidation
m. Proton
n. Reduction
o. Valence electrons

For each of these definitions, select the correct matching term from the list above.

_____ 1. The loss of electrons, or the loss of hydrogen atoms from a compound.

_____ 2. The total number of protons and neutrons in the atomic nucleus.

_____ 3. Substances dissolved in water that can conduct electrical current.

_____ 4. The smallest subdivision of an element that retains its chemical properties.

_____ 5. Chemical bond that involves one or more shared pairs of electrons.

_____ 6. The name of the outermost electrons.

_____ 7. An ion with a negative electric charge.

_____ 8. A molecule with partial opposite charges.

_____ 9. An ion with a positive electric charge.

_____10. The amount of heat required to raise 1 gram of a substance 1 degree Celsius.

_____11. Type of interaction key to water's unique properties.

MAKING COMPARISONS

Fill in the blanks.

Name	Formula
Sodium ion	Na^+
Carbon dioxide	#1
Calcium	#2
#3	Fe_2O_3
Methane	#4
Chloride ion	#5
#6	K^+
Hydrogen ion	#7
#8	Mg
Ammonium ion	#9
Water	#10
Glucose	#11
Bicarbonate ion	#12
#13	H_2CO_3
#14	$NaCl$
#15	OH^-
Sodium hydroxide	#16

MAKING CHOICES

Place your answer(s) in the space provided. Some questions may have more than one correct answer.

_____ 1. Substances that cannot be broken down into simpler substances by ordinary chemical means are

a. compounds.

b. covalently bonded atoms.

c. mixtures.

d. molecules.

e. elements.

____ 2. Adhesion of water molecules results from
   a. attraction between water and other substances.
   b. hydrogen bonding between water molecules.
   c. arrested adhesion.
   d. covalent bonding.
   e. capillary action.

____ 3. The symbol $_6C$ stands for
   a. calcium with six electrons.
   b. carbon with six protons.
   c. carbon with three protons and three neutrons.
   d. calcium with three protons and three electrons.
   e. an isotope of carbon.

____ 4. Anions are
   a. negative ions.
   b. positive ions.
   c. isotopes.
   d. incomplete products.
   e. negative molecules.

Use the following chemical equation to answer questions 5-10.

$$2 H_2 + O_2 \rightarrow 2 H_2O \text{ (water)}$$

____ 5. The product has a molecular mass of
   a. one.
   b. twelve.
   c. larger than either reactant alone.
   d. sixteen.
   e. eighteen.

____ 6. The reactant(s) in the formula is/are
   a. hydrogen.
   b. oxygen.
   c. H_2O.
   d. water.
   e. the arrow.

____ 7. How many hydrogen atoms are involved in this reaction?
   a. one
   b. two
   c. three
   d. four
   e. eight

____ 8. The product(s) is/are
   a. hydrogen.
   b. oxygen.
   c. H_2O.
   d. water.
   e. the arrow.

____ 9. The lightest reactant has a mass number of
   a. one.
   b. two.
   c. larger than the product.
   d. eight.
   e. ten.

____ 10. The direction in which this reaction tends to occur is
   a. right to left.
   b. from products to reactants.
   c. from hydrogen and oxygen to water.
   d. left to right.
   e. from reactants to products.

____ 11. Reduction has occurred when
   a. an atom gains an electron.
   b. a molecule loses an electron.
   c. hydrogen bonds break.
   d. an atom loses an electron.
   e. an ion gains an electron.

_____12. When two atoms combine by sharing two pairs of their electrons, the bond is a
    a. divalent, single.
    b. covalent, single.
    c. divalent, double.
    d. covalent, double.
    e. covalent, polar.

_____13. A form of an element that is progressing toward a more stable state by emitting radiation is called a(n)
    a. Dalton atom.
    b. radioisotope.
    c. enerton.
    d. quantum unitron.
    e. radionuclide.

_____14. The outer electron energy level of an atom
    a. may contain two or more orbitals.
    b. determines the element.
    c. contains the same number of electrons as protons.
    d. often contains at least one neutron.
    e. determines chemical properties.

_____15. If the number of neutrons in an atom is changed, it will also change
    a. atomic mass.
    b. atomic number.
    c. the element.
    d. the number of electrons.
    e. the elements chemical properties.

_____16. Dissociation of which of the following into ions may result in the formation of electrolytes?
    a. salts
    b. acids
    c. hydrogen
    d. bases
    e. alcohols

_____17. A substance containing more hydrogen (H^+) and less hydroxyl (OH^-) ions would be
    a. acidic, pH below 7.
    b. alkaline, pH below 7.
    c. acidic, pH above 7.
    d. alkaline, pH above 7.
    e. neutral, pH 7.

_____18. H−O−H is/are
    a. a structural formula.
    b. water.
    c. a chemical formula.
    d. a molecule.
    e. a structure with the molecular mass 18.

_____19. The atomic nucleus contains
    a. protons.
    b. neutrons.
    c. electrons.
    d. positrons.
    e. isotopes.

VISUAL FOUNDATIONS

Color the parts of the illustration below as indicated.

RED ☐ electron

YELLOW ☐ neutron

BLUE ☐ proton

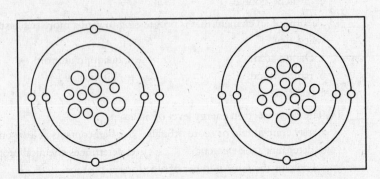

Questions 1-2 refer to the illustration above. Two atoms are represented in this illustration. Some questions may have more than one correct answer.

1. The atom on the left is called (a)_____. The atom on the right is called (b)_____.

2. These atoms:
   a. are ions.
   b. have the same atomic mass.
   c. have the same mass.
   d. are from different elements.
   e. have the same atomic number.

CHAPTER 3

The Chemistry of Life: Organic Compounds

In the previous chapter, the focus was on the smaller, simpler substances known as inorganic compounds and the bonds that produce them. The focus in this chapter is on the larger and more complex organic compounds. Most of the chemical compounds in organisms are organic compounds. Carbon is the central component of organic compounds, probably because it forms bonds with a greater number of different elements than any other type of atom. Adding functional groups allows large organic compounds to interact better with each other. The major groups of organic compounds are the carbohydrates, lipids, proteins, and nucleic acids. Carbohydrates, built from monosaccharides, serve as energy sources for cells, and as structural components of fungi, plants, and some animals. Lipids, which are not characterized by structure, function as energy storage molecules, as structural components of cell membranes, and as hormones. Proteins, made from amino acids, serve a multitude of functions within the cell important to structure and basic cellular function. They also exhibit several levels of organization, allowing the production of a fully functional protein. Nucleic acids (made from nucleotides) transmit hereditary information and determine what proteins a cell manufactures.

REVIEWING CONCEPTS
Fill in the blanks.

INTRODUCTION
1. Macromolecules are built from_____.
2. _____are the four major "classes" of macromolecules in living systems.

CARBON ATOMS AND MOLECULES
3. Carbon has_____ (#?) valence electrons.
4. Carbon atoms can form
_____(#?) bonds with a large variety of other atoms.
5. (a)_____ made of carbon and (b)_____,
can be branched or (c)_____ chains or
(d)_____.

Isomers have the same molecular formula but different structures

6. _____ isomers are compounds that differ in the covalent arrangements of their atoms.

7. _____ isomers are compounds that are different in the spatial arrangement of atoms or groups of atoms.

8. _____ are isomers are molecules that are mirror images of one another.

Functional groups change the properties of organic molecules

9. The (a) _____ of the major organic compounds are a consequence of the types and arrangements of (b) _____ groups.

10. R — OH is the symbol for a (a) _____ group, and its chemical nature is (b) _____.

11. R — COOH is the symbol for a _____ group, and its chemical nature is (b) _____.

12. R-NH$_2$ is the symbol for a (a) _____ group, and its chemical nature is (b) _____.

Many biological molecules are polymers

13. Polymers are macromolecules produced by linking _____ together.

14. The process by which polymers are degraded into their subunit components is _____.

15. (a) _____ reactions are used to link subunits to build (b) _____.

CARBOHYDRATES

16. Carbohydrates contain the elements _____ _____ in a ratio of about 1:2:1.

Monosaccharides are simple sugars

17. The simple sugar (a) _____ is abundant fuel molecule in most organisms. In cells its structure is usually in a (b) _____ from.

Disaccharides consist of two monosaccharide units

18. Disaccharides form when two monosaccharides are bonded by a _____ linkage.

19. Two examples would be common table sugar, (a) _____, and the sugar in milk, (b) _____.

Polysaccharides can store energy or provide structure

20. Long chains of _____ link together to form a polysaccharide.

21. _____ is the main storage carbohydrate of plants.

22. Starch occurs in two forms. The simpler form, _____, is unbranched.

23. Plant cells store starch in organelles called _____.

24. _____ is the main storage carbohydrate of animals.

25. (a) _____ is the most abundant carbohydrate on earth and is used to make plant (b)_____.

Some modified and complex carbohydrates have special roles

26. _____ is a complex carbohydrate that forms the external skeleton of arthropods.

27. A carbohydrate combined with a protein forms (a)_____; carbohydrates combined with lipids are called (b)_____.

LIPIDS

28. Lipids are not defined by (a)_____ but by being (b)_____ in water.

29. Lipids are composed of the elements _____.

Triacylglycerol is formed from glycerol and three fatty acids

30. A triacylglycerol consists of one (a)_____ molecule combined with three (b)_____ molecules.

31. These molecules are commonly called (a)_____ and are used for energy (b)_____.

Saturated and unsaturated fatty acids differ in physical properties

32. Saturated fatty acids are solid due to _____.

33. Unsaturated fats are (a)_____ due to the presence of carbon (b)_____.

34. Trans fats are produced by (a)_____ and mimic (b)_____.

Phospholipids are components of cell membranes

35 A phospholipid molecule assumes a distinctive configuration in water because of its _____ property, which means that one end of the molecule is hydrophilic and the other end is hydrophobic.

36. They are made up of a (a)_____ attached at one end to two (b)_____ and at the other end to a (c)_____ group attached to an (d)_____compound.

Carotenoids and many other pigments are derived from isoprene units

37. Carotenoids consist of five-carbon hydrocarbon monomers known as (a)_____. Animals convert these pigments to (b)_____.

Steroids contain four rings of carbon atoms

38. A steroid consists of (a)_____ atoms arranged in four attached rings, three of which contain(b) _____ carbon atoms and the fourth contains (c)_____.

39. _____is an important example in animal cell membranes.

Some chemical mediators are lipids

40. Some chemical mediators are produced by the modification of _____ that have been removed from phospholipids.

PROTEINS

Amino acids are the subunits of proteins

41. All amino acids contain (a)_____ groups but vary in the (b)_____ group.

42. _____amino acids are ones animals cannot synthesize.

Peptide bonds join amino acids

43. The bonds that join the subunits of proteins are called _____ bonds.

44. Proteins are large, complex molecules made of subunits called _____.

45. This bond occurs between the (a)_____ group of one amino acid and the (b)_____group of the other.

Proteins have four levels of organization

46. The (a)_____of a protein is its 3-D shape, which is key to its (b)_____.

47. The levels of organization distinguishable in protein molecules are _____.

48. The amino acid sequence is the primary structure of a _____ and is what other levels are derived from.

49. Secondary structure results from (a)_____ bonding involving the backbone. The two most common types are (b)_____

50. Tertiary structure depends on interactions among (a) _____ and generates the (b)_____of the overall molecule.

51. Quaternary structure results from interactions among _____to generate a functional protein.

The amino acid sequence of a protein determines its conformation

52. _____help proteins fold.

53. The conformation of a protein determines its _____.

54. When a protein loses its shape and function it is called _____.

55. Three examples of human disease caused by misfolded proteins are_____.

NUCLEIC ACIDS

56. Nucleic acids consist of _____subunits.

57. Each nucleotide subunit in a nucleic acid is composed of a
    (a)_____ _____
    nitrogenous base, a five-carbon (b)_____, and an
    inorganic (c)_____ group.

58. Nucelotides are connected
    by_____.

Some nucleotides are important in energy transfers and other cell functions

59. Energy for life functions is supplied mainly by the nucleotide
    _____.

IDENTIFYING BIOLOGICAL MOLECULES

BUILDING WORDS

Use combinations of prefixes and suffixes to build words for the definitions that follow.

Prefixes	The Meaning		Suffixes	The Meaning
amphi- decomposition	two, both, both sides		-lysis	breaking down,
amylo-	starch		-mer	part
di-	two, twice, double		-plast	formed, molded, "body"
hex-	six		-saccharide	sugar
hydr(o)-	water			
iso-	equal, "same"			
macro-	large, long, great, excessive			
mono-	alone, single, one			
poly-	much, many			
tri-	three			
pent-	five			

Prefix	Suffix	Definition
_____	_____	1. A compound that has the same molecular formula as another compound but a different structure and different properties.
_____	-molecule	2. A large molecule consisting of thousands of atoms.
_____	_____	3. A single organic compound that links with similar compounds in the formation of a polymer.
_____	_____	4. The degradation of a compound by the addition of water.
_____	_____	5. A simple sugar.
_____	-ose	6. A sugar that consists of six carbons.
_____	_____	7. Two monosaccharides covalently bonded to one another.
_____	_____	8. A macromolecule consisting of repeating units of simple sugars.
_____	_____	9. A starch-forming granule.

_____ -acylglycerol 10. A compound formed of one fatty acid and one glycerol molecule.

_____ - acylglycerol 11. A compound formed of two fatty acids and one glycerol molecule.

_____ - acylglycerol 12. A compound formed of three fatty acids and one glycerol molecule.

_____ -pathic 13. Having one hydrophilic end and one hydrophobic end.

_____ -peptide 14. A compound consisting of two amino acids.

_____ -peptide 15. A compound consisting of a long chain of amino acids.

_____ -ose 16. A sugar that consists of five carbons.

MATCHING

Terms:

a. Amino acid
b. Amino group
c. Carotenoid
d. Condensation
e. Glycogen
f. Glucose
g. Hydrocarbon
h. Lipid
i. Nucleotide
j. Phosphate group
k. Phospholipid
l. Protein
m. Starch
n. Steroid
o. Tertiary structure

For each of these definitions, select the correct matching term from the list above.

_____ 1. A molecule composed of one or more phosphate groups, a 5-carbon sugar, and a nitrogenous base.

_____ 2. A group of yellow to orange pigments in plants.

_____ 3. A large complex organic compound composed of chemically linked amino acid subunits.

_____ 4. Complex lipid molecules containing carbon atoms arranged in four interlocking rings.

_____ 5. An organic compound containing an amino group and a carboxyl group.

_____ 6. The three-dimensional structure of a protein molecule.

_____ 7. An organic compound that is made up of only carbon and hydrogen atoms.

_____ 8. Fatlike substance found in cell membranes.

_____ 9. Polysaccharide used by plants to store energy.

_____10. The principle carbohydrate stored in animal cells.

MAKING COMPARISONS

Fill in the blanks.

Functional Group	Abbreviated Formula	Class of Compounds Characterized by Group	Description
Carboxyl	R — COOH	Carboxylic acids (organic acids)	Weakly acidic; can release a H^+ ion
#1	R — PO_4H_2	#2	Weakly acidic; one or two H^+ ions can be released
Methyl	#3	Component of many organic compounds	#4
#5	R---NH_2	#6	Weakly basic; can accept an H^+ ion
Sulfhydryl	R — SH	#7	Helps stabilize internal structure of proteins
#8	R---OH	#9	#10

MAKING CHOICES

Place your answer(s) in the space provided. Some questions may have more than one answer.

_____ 1. The primary structure of a protein results from
    a. ionic bonds.
    b. hydrogen bonds.
    c. covalent bonds.
    d. polymerization.
    e. peptide bonds.

_____ 2. Condensation reactions result in
    a. smaller molecules.
    b. larger molecules.
    c. reduced ambient water.
    d. increase in polymers.
    e. increase in monomers.

_____ 3. An isomer can be described as one of a group of molecules with the same molecular formulas but different
    a. elements.
    b. compounds.
    c. structures.
    d. physical properties.
    e. chemical properties.

_____ 4. The complex carbohydrate storage molecule in plants is
    a. glycogen.
    b. starch.
    c. usually in an extracellular position.
    d. composed of simple glucose subunits.
    e. more soluble than the animal storage molecule.

_____ 5. Nucleotides
    a. are polypeptide polymers.
    b. are polymers of simple carbohydrates.
    c. contain a base, phosphate, and sugar.
    d. comprise complex proteins.
    e. contain genetic information.

_____ 6. Proteins contain

    a. glycerol.   d. amino acids.

    b. polypeptides.   e. nucleotides.

    c. peptide bonds.

_____ 7. The main structural components of cells and tissues are

    a. carbon-based molecules.   d. organic compounds.

    b. carbohydrates, lipids, and DNA.   e. DNA, RNA, and ATP.

    c. basically strings of covalently bonded carbons.

_____ 8. Compared to saturated fatty acids, unsaturated fatty acids

    a. contain more hydrogens.   d. are more prone to be solids.

    b. contain more double bonds.   e. contain more carbohydrates.

    c. are more like oils.

_____ 9. Glycogen is

    a. a molecule in animals.   d. composed of simple carbohydrate subunits.

    b. a molecule in plants.   e. more soluble than starch.

    c. stored mainly in the liver and muscle of vertebrates.

_____ 10. A monomer is to a polymer as a

    a. chain is to a link.   d. necklace is to a bead.

    b. link is to a chain.   e. bead is to a necklace.

    c. monosaccharide is to a polysaccharide.

_____ 11. $C_6H_{12}O_6$ is the formula for

    a. glucose.   d. fructose.

    b. a type of monosaccharide.   e. a type of disaccharide.

    c. cellulose.

_____ 12. Proteins

    a. are a major source of energy.   d. are found in both plants and

    animals.

    b. contain carbon, hydrogen, and oxygen.   e. are polypeptide polymers.

    c. can combine with carbohydrates.

_____ 13. Phospholipids

    a. are found in cell membranes.   d. are polar.

    b. are in the amphipathic lipid group.   e. are polypeptide polymers.

    c. contain glycerol, fatty acids, and phosphate groups.

_____ 14. Polymers are

    a. made from monomers.   d. macromolecules.

    b. formed by condensation reactions.   e. smaller than monomers.

    c. found in monomers.

_____ 15. –COOH

    a. is a carboxyl group.   d. coverts to $–COO^+$ .after accepting a proton.

    b. is an amino group.   e. is a polymer.

    c. becomes ionized $–COO^-$ .after donating a proton

VISUAL FOUNDATIONS

Color the parts of the illustration below as indicated.

RED ☐ saturated fatty acid
GREEN ☐ glycerol
YELLOW ☐ phosphate group
BLUE ☐ choline
ORANGE ☐ unsaturated fatty ac

Questions 1-4 refer to the illustration above. Some questions may have more than one correct answer.

_____ 1. The illustration represents

    a. a polypeptide.   d. lecithin.

    b. a phospholipid.   e. starch.

    c. glycogen.

_____ 2. The hydrophobic end of this molecule is composed of

    a. amino acids.   d. phosphate groups.

    b. fatty acids.   e. steroids.

    c. glucose.

_____ 3. This kind of molecule contributes to the formation of

    a. neutral fat.   d. cellulose.

    b. cell membranes.   e. complex proteins.

    c. RNA.

_____ 4. This molecule is amphipathic, meaning that

    a. it contains nonpolar groups.   d. it is found in all life forms.

    b. it is very large.   e. it can act as an acid.

    c. it contains hydrophobic and hydrophilic regions.

CHAPTER 4

Organization of the Cell

Chapters 2 and 3 introduced the inorganic and organic materials critical to an understanding of the cell, the basic unit of life. In this chapter and those that follow, you will see how cells utilize these chemical materials. Because all cells come from preexisting cells, they have similar needs and therefore share many fundamental features. Most cells are microscopically small because of limitations on the movement of materials across the plasma membrane as well as the functions they perform. Cells are studied by a combination of methods. These include various types of light and electron microscopy and cell fractionation. Prokaryotic cells do not have membrane-bound organelles, but they do share several similarities with eukaryotes. Eukaryotic cells are more complex, containing a variety of membranous organelles. Although each organelle has its own particular structure and function, all of the organelles of a cell work together in an integrated fashion, many as part of the endomembrane system. The cytoskeleton provides mechanical support to the cell and functions in cell movement, the transport of materials within the cell, and cell division. Most cells are surrounded by some type of covering, which could be a glycocalyx, an extracellular matrix, or a cell wall.

REVIEWING CONCEPTS
Fill in the blanks.

INTRODUCTION
1. (a)_____ are the building blocks of complex multicellular organisms, but an organism can also be a (b)_____cell.

THE CELL: BASIC UNIT OF LIFE
The cell theory is a unifying concept in biology
2. The two components of the cell theory are
_____
_____
_____.

The organization of all cells is basically similar
3. Cellular organization allows cells to maintain _____ in a constantly changing environment.
4. All cells are enclosed by a _____.
5. The internal structures of cells that carry on life's activities are called _____.

Cell size is limited
6. A micrometer is 1/1,000,000 of a meter and therefore _____ of a millimeter.

7. In general, the smaller the cell, the larger the
_____.

Cell size and shape are adapted to function

8. The _____ of cells are adapted to the functions they perform.

METHODS FOR STUDYING CELLS

Light microscopes are used to study stained or living cells

9. (a)_____ is the ratio of microscopic size to actual size of an object, and (b)_____ is the capacity to distinguish fine detail in an image.

10. Light microscopes have limited resolution due to the (a)_____ of (b)_____ used.

11. Lack of contrast can be remedied by _____.

Electron microscopes provide a high-resolution image that can be greatly magnified

12. Powerful electron microscopes are used to study the _____ of cells.

13. In SEM, the bean of (a)_____ does not (b)_____ through the sample while in (c)_____ it does.

Biologists us biochemical techniques to study cell components

14. Cell components are separated by a method known as _____.

PROKARYOTIC AND EUKARYOTIC CELLS

Organelles of prokaryotic cells are not surrounded by membranes

15. (a)_____ cells are relatively lacking in complexity, and their genetic material is not enclosed by membranes but is located in a (b)_____.

16. Outside of their cell membrane, prokaryotic cells have a (a)_____ and may have (b)_____ to aid in movement.

17. The bacterial (a)_____, used for polypeptide synthesis, is (b)_____ than the eukaryotic version.

Membranes divide the eukaryotic cell into compartments

18. _____ cells are relatively complex and possess both membrane-bound organelles and a "true" nucleus.

19. In eukaryotic cells, DNA is contained in the _____.

20. Compartmentalization allows cellular activity to be (a)_____ and occur (b)_____.

21. The internal membrane system in the eukaryotic cell is known as the _____.

THE CELL NUCLEUS

22. The (a)_____ consists of two concentric membranes that separate the nuclear contents from the surrounding cytoplasm and contains (b)_____that regulate material passage.

23. The _____provides support and helps organize nuclear contents.

24. DNA is associated with proteins and RNA, forming a complex known as _____.

25. The (a)_____ helps to make (b)_____ used in ribosomes.

ORGANELLES IN THE CYTOPLASM

Ribosomes manufacture proteins

26. Ribosomes are tiny particles found free in the cytoplasm or attached to certain membranes; they consist of _____.

The endoplasmic reticulum is a network of internal membranes

27. The _____ is a complex of intracytoplasmic membranes that compartmentalize the cytoplasm.

28. The smooth ER synthesizes lipids including
    (a)_____. It also stores
    (b)_____ and can be a major site of
    (c)_____.

29. Rough RER is studded with _____ that are involved in protein synthesis.

30. Some proteins constructed on RER are transported by
    (a)_____ for secretion to the outside or insertion in other membranes. Misfolded proteins will be degraded by the
    (b)_____.

The Golgi complex processes, sorts, and modifies proteins

31. In many cells the Golgi consists of flattened membranous sacs called _____.

32. Cells that make large quantities of _____ will have many Golgi stacks.

Lysosomes are compartments for digestion

33. Lysosomes are small sacs containing _____ that can break down (lyse) complex molecules, foreign substances, and "dead" organelles.

34. Primary lysosomes are formed by budding from the _____.

Vacuoles are large, fluid-filled sacs with a variety of functions

35. The membrane of a vacuole is called the _____.

36. The three main types of vacuoles are

_____.

Peroxisomes metabolize small organic compounds

37. Peroxisomes get their name from the fact that they produce

_____ during oxidation

reactions.

38. Their major functions are (a)_____ and

(b)_____.

Mitochondria and chloroplasts are energy-converting organelles

39. These organelles are thought to have evolved from

_____.

Mitochondria make ATP through cellular respiration.

40. This organelle is the site of (a)_____ which

primarily occurs on folds of the innermembrane called

(b)_____.

41. Mitochondria are also involved in _____, programmed

cell death.

42. Mitochondria negatively affect health and aging through the

release of _____.

Chloroplasts convert light energy to chemical energy through photosynthesis

43. _____ are organelles that contain green

pigments that trap light energy for photosynthesis.

44. _____ membranes contain chlorophyll that traps

sunlight energy and converts it to chemical energy in ATP.

45. The matrix within chloroplasts, where carbohydrates are

synthesized, is called the (a)_____, while the ATP

that supplies energy to drive the process is synthesized on

membranes called (b)_____.

THE CYTOSKELETON

46. The cytoskeleton is a dense network of protein fibers responsible

for the _____ of cells.

47. _____ and

_____ are both made up of globular

protein subunits that can rapidly assemble and disassemble.

Microtubules are hollow cylinders

48. Microtubules consist of the two proteins

_____ and _____.

49. (a)_____ help microtubules

function. Dynein and kinesin are examples of

(b)_____ proteins.

Centrosomes and centrioles function in cell division

50. In animal cells, the (a)_____ is the location of the
(b)_____ where minus ends of microtubules
are anchored.

51. (a)_____ found in animal cells within the
(b)_____ are important in (c)_____.

Cilia and flagella are composed of microtubules

52. The cellular appendages that are long and few in number are called
(a)_____ while numerous, short ones are
(b)_____. Both are used for (c)_____.

53. These structure are anchored to the cell by the
_____.

Microfilaments consist of intertwined strings of actin

54. The fibers generate the _____ used for cell shape
and movement.

55. In muscle cells actin is associated with the protein
_____ to form fibers associated with muscle
contraction.

Intermediate filaments help stabilize cell shape

56. Keratin and neurofilaments are examples of tough, flexible fibers
called _____ that stabilize cell
shape and provide mechanical strength.

CELL COVERINGS

57. In most eukaryotic cells polysaccharide side chains of proteins and
lipids form a _____, or cell coat, that is
part of the plasma membrane.

58. The animal cell extracellular matrix is composed
of_____.

59. Plant cell walls are composed primarily of _____
and smaller quantities of other polysaccharides.

BUILDING WORDS

Use combinations of prefixes and suffixes to build words for the definitions that
follow.

Prefixes	The Meaning	Suffixes	The Meaning
chloro-	green	-karyo(te)	nucleus
chromo-	color	-plast(id)	formed, molded, "body"
cyto-	cell	-some	body
eu-	good, well, "true"		
leuko-	white (without color)		
lyso-	loosening, decomposition		
micro-	small		
myo-	muscle		
pro-	"before"		

Prefix	Suffix	Definition
_____	-phyll	1. A green pigment that traps light for photosynthesis.
_____	_____	2. Organelles containing pigments that give fruits and flowers their characteristic colors.
_____	-plasm	3. Cell contents exclusive of the nucleus.
_____	-skeleton	4. A complex network of protein filaments within the cell.
glyoxy-	_____	5. A microbody containing enzymes used to convert stored fats in plant seeds to sugars.
_____	_____	6. An organelle that is not pigmented and is found primarily in roots and tubers, where it is used to store starch.
_____	_____	7. An organelle containing digestive enzymes.
_____	-filaments	8. Small, solid filaments, made of actin, that make up part of the cytoskeleton of eukaryotic cells.
_____	-tubules	9. Small, hollow filaments, made of tubulin, that make up part of the cytoskeleton of eukaryotic cells.
_____	-villi	10. Small, finger-like projections from cell surfaces that increase surface area.
_____	-sin	11. A muscle protein that, together with actin, is responsible for muscle contraction.
peroxi-	_____	12. An organelle containing enzymes that split hydrogen peroxide, rendering it harmless.
_____	_____	13. Precursor organelles.
_____	-karyotes	14. Organisms that evolved before organisms with nuclei.
_____	-sol	15. The fluid component of the cytoplasm.
_____	_____	16. An organism with a distinct nucleus surrounded by nuclear membranes.

MATCHING

Terms:

a. Actin
b. Centriole
c. Chloroplast
d. Endoplasmic reticulum
e. Granum
f. Mitochondrion
g. Nuclear envelope
h. Nucleolus
i. Plasma membrane
j. Ribosome
k. Secretory vesicle
l. Stroma
m. Vacuole
n. Vesicle
o. Cristae

For each of these definitions, select the correct matching term from the list above.

_____ 1. One of a pair of small, cylindrical organelles lying at right angles to each other near the nucleus.

_____ 2. Site of ribosome synthesis.

_____ 3. An intracellular organelle that is the site of aerobic respiration.

_____ 4. A fluid-filled, membrane-bounded sac found within the cytoplasm; may function in storage, digestion, or water elimination.

_____ 5. Any small sac, especially a small spherical membrane-bounded compartment, within the cytoplasm.

_____ 6. A chlorophyll containing intracellular organelle of some plant cells.

_____ 7. A stack of thylakoids within a chloroplast.

_____ 8. The fluid region of the chloroplast.

_____ 9. An interconnected network of intracellular membranes.

_____10. An organelle that is part of the protein synthesis machinery.

_____11. Folds of the innermembrane in the mitochondria.

MAKING COMPARISONS

Fill in the blanks.

Cell Structure	Location of Structure	Function of Structure	Kind of Cell Containing This Structure
Nuclear area containing a single strand of circular DNA	Cytoplasm	Inheritance, control center	Prokaryotic
#1	Cytoplasm and RER	Synthesis of polypeptides	#2
Golgi complex	#3	#4	Most eukaryotic
Mitochondria	#5	#6	#7
Complex chromosomes	#8	Inheritance	#9
#10	Centrosome	#11	#12
#13	#14	Protection, support	Prokaryotic and eukaryotic
#15	#16	Encloses cellular contents, regulates passage of materials into and out of cell	#17
Chloroplasts	#18	#19	Eukaryotic (primarily plants)

MAKING CHOICES

Place your answer(s) in the space provided. Some questions may have more than one correct answer.

_____ 1. A micrometer (μm) is
    a. one billionth of a meter.
    b. one millionth of a centimeter.
    c. one thousandth of a millimeter.
    d. one millionth of a micrometer.
    e. one thousandth of a micrometer.

_____ 2. Which of the following is in the nucleolus, but not normally found in the rest of chromatin?
    a. DNA
    b. protein
    c. chromosomes
    d. RNA
    e. ribosomes

_____ 3. Cell membrane functions include
    a. isolation of chemical reactions.
    b. selective permeability.
    c. moderating between a cell's internal and external environments.
    d. maintaining cell shape.
    e. concentration of reactants.

_____ 4. The low resolution attained by the light microscope is attributed to its use of
    a. fixed specimens.
    b. electromagnetic lenses.
    c. short wavelengths.
    d. grid patterns.
    e. electrons instead of light.

_____ 5. Cells are small at least in part because as size increases the surface-to-volume ratio
     a. doubles.   d. decreases.
     b. decreases to half.   e. reduces efficiency of cell activities.
     c. increases.

_____ 6. Inner membranous folds in mitochondrion
     a. are called cristae.   d. contain enzymes.
     b. increase membrane surface area.   e. contain structural proteins.
     c. extend into the intermembrane space.

_____ 7. The Golgi complex functions to
     a. modify proteins.   d. sort molecules.
     b. process proteins.   e. break down large carbohydrates.
     c. packages glycoproteins.

_____ 8. The membranes that comprise the endomembrane system include
     a. Golgi complex.   d. transport vesicles.
     b. lysosomes.   e. plasma membrane.
     c. endoplasmic reticulum.

_____ 9. Lysosomes
     a. contain digestive enzymes.   d. break down complex molecules.
     b. contain nucleic acids.   e. break down organelles.
     c. possess a membrane.

_____ 10. Which of the following cells contain plastids?
     a. animal   d. some prokaryotic
     b. plant   e. algae
     c. some eukaryotic

_____ 11. The "cytoskeleton" of eukaryotic cells
     a. changes constantly.   d. functions in cell movements.
     b. includes microfilaments.   e. includes protein.
     c. includes some DNA.

_____ 12. The cell theory states that
     a. new cells come from preexisting cells.   d. living things are composed of cells.
     b. all cells are descended from ancient cells.   e. cells contain genetic material.
     c. cells divide.

_____ 13. Chloroplasts and mitochondria both
     a. are found in plant cells.   d. are found in animal cells.
     b. have two membranes.   e. contain a matrix.
     c. contain DNA.

_____ 14. Ribosomal RNA is synthesized in the
     a. cytoplasm.   d. mitochondria.
     b. nuclear matrix.   e. nucleolus.
     c. cristae.

_____ 15. Actin, myosin, and tubulin are
     a. proteins.   d. components of filaments.
     b. in chromatin.   e. components of the plasma membrane.
     c. primarily in neurons.

_____ 16. Fluorescence microscopy
     a. uses stains bound to antibodies.   d. can detect location of specific molecules.
     b. uses stains that absorb visible light.   e. provide better resolution than electron microscopes.
     c. uses special filters.

VISUAL FOUNDATIONS

Color the parts of the illustration below as indicated.

RED ▢ plasma membrane
GREEN ▢ nuclear area
YELLOW ▢ cell wall
PINK ▢ capsule
BLUE ▢ flagellum
ORANGE ▢ pili

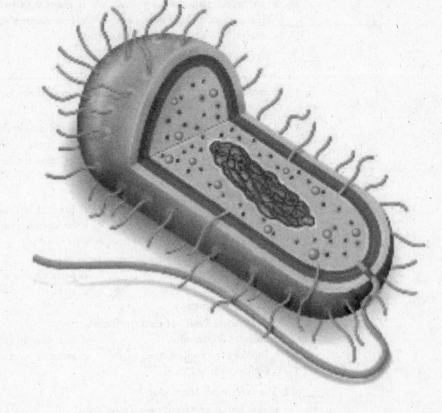

Questions 1 and 2 pertain to the illustration above.

1. Is this a prokaryotic cell or a eukaryotic cell?

   _____

2. Which of the colored parts is/are not unique to this kind of cell? _____

Color the parts of the illustration below as indicated.

RED ☐ plasma membrane

ORANGE ☐ nucleus

YELLOW ☐ cell wall

BLUE ☐ prominent vacuole

GREEN ☐ chloroplast

BROWN ☐ mitochondrion

TAN ☐ endomembrane system (Note: The plasma membrane and the outer portion of the nuclear envelop were assigned other colors. Use tan to identify the remaining structures in this system.)

Questions 3-5 pertain to the illustration below.

3. What kind of cell is illustrated by this generalized diagram? _____

4. Which of the colored parts is/are considered components of the cytoplasm of this cell?

_____

5. Which of the colored parts is/are characteristic of this kind of cell and not generally associated with other types of cells?

_____

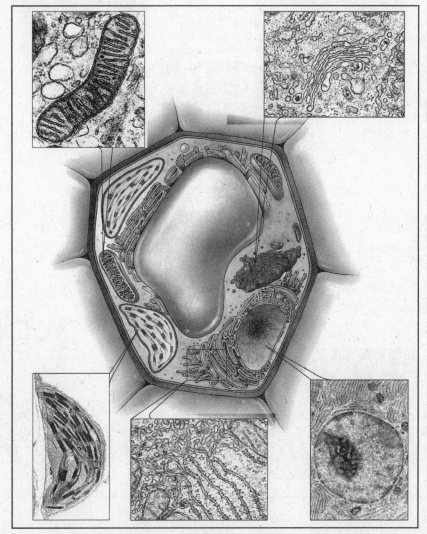

Color the parts of the illustration below as indicated.

RED   ☐ plasma membrane

ORANGE   ☐ nucleus

YELLOW   ☐ centriole

BROWN   ☐ mitochondrion

BROWN   ☐ endomembrane system (Note: The plasma membrane and the outer portion of the nuclear envelope have already been assigned colors. Use tan to identify the remaining structures in this system.)

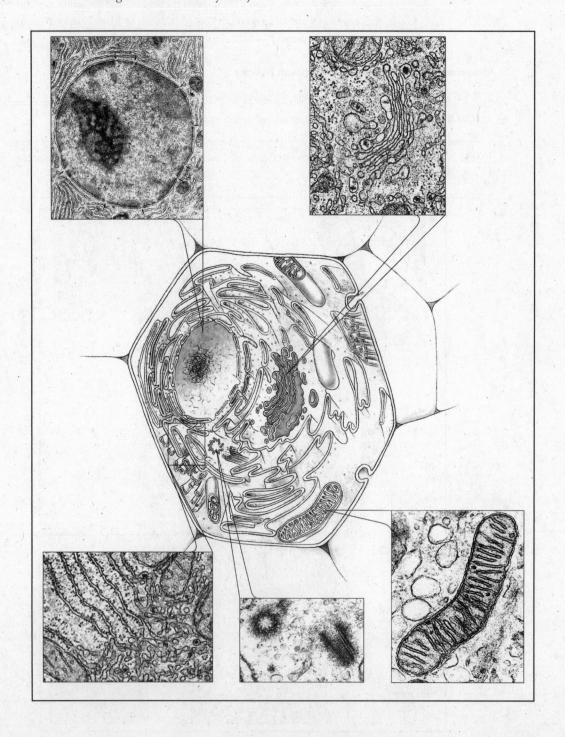

Questions 6-8 pertain to the illustration above.

6. What kind of cell is illustrated by this generalized diagram?

   _____

7. Which of the colored parts is/are considered components of the cytoplasm of this cell?

   _____

   _____

8. Which of the colored parts is/are characteristic of this kind of cell and not generally associated with other types of cells?

   _____

CHAPTER 5

Biological Membranes

Biological membranes serve to enclose cells. They separate the interior world of the cell from the exterior and are an essential step in the evolution of life. In eukaryotes, biological membranes also enclose a variety of internal structures and form extensive internal membrane systems. Membranes are phospholipid bilayers with both integral and peripheral proteins associated. The fluid mosaic model explains our current understanding of their structure. This structure allows membranes to have several important physical properties. Membranes and their associated proteins have varied functions including; providing work surfaces for many chemical reactions, regulating movement of materials in and out of the cell, transmitting signals and information, holding cells together, and being an essential part of energy transfer and storage systems. Molecules pass through membranes in a variety of ways because of their selective permeability. In multicellular organisms, cell membranes have specialized junctions associated with them that allow neighboring cells to form strong connections, prevent material passage, or to establish rapid communication.

REVIEWING CONCEPTS

Fill in the blanks.

INTRODUCTION

1. All cells are physically separated from the external environment by a _____.

2. Membrane proteins known as _____ are responsible for adhesion between cells in multicellular sheets.

THE STRUCTURE OF BIOLOGICAL MEMBRANES

Phospholipids form bilayers in water

3. The physical properties of biological membranes are due primarily to their _____.

4. The headgroups of phospholipid molecules are said to be (a)_____ because they readily associate with water. On the other hand, the (b)_____ ends turn away from water and associate with each other.

5. The cylindrical shape and strongly _____ character of phospholipid molecules are features responsible for the formation of the bilayer.

The fluid mosaic model explains membrane structure

6. The theory of membrane structure known as the (a)_____ model holds that membranes consist of a dynamic, fluid (b)_____ and embedded or otherwise associated globular (c)_____.

Biological membranes are two-dimensional fluids

7. The crystal-like properties of many phospholipid bilayers is due to the orderly arrangement of

   (a)_____ on the outside and

   (b)_____ on the inside.

8. The two-dimensional fluid property of the lipid bilayer is due to the constant motion of

   _____.

9. The presence of _____helps to prevent the membrane from solidifying.

10. _____functions as a "fluidity buffer" in animal cells. Plant cells have similar molecule types.

Biological membranes fuse and form closed vesicles

11. Products are secreted from cells by the fusion of vesicles with the

    _____.

Membrane proteins include integral and peripheral proteins

12. (a)_____ have hydrophobic regions that interact with the fatty acid tails of the membrane phospholipids. One type

    (b)_____pass through the entire membrane.

13. _____ are not embedded in the lipid bilayer.

Proteins are oriented asymmetrically across the bilayer

14. Membrane proteins that will become part of the inner surface of the plasma membrane are manufactured by

    (a)_____ and moved to the membrane through the (b)_____.

15. Proteins destined for the outer cell surface are manufactured on

    (a)_____ and then modified in the ER lumen to become (b)_____. They then move through the (c)_____.

OVERVIEW OF MEMBRANE PROTEIN FUNCTIONS

16. A variety of membrane functions is made possible by the diversity of _____ molecules in membranes.

17. The proteins called (a)_____help anchor the cell. Channel and pump proteins are important for

    (b)_____. (c)_____proteins transmit signals by (d)_____.

18. Antigens are important in (a)_____and other membrane proteins form (b)_____between cells.

CELL MEMBRANE STRUCTURE AND PERMEABILITY

19. Membranes are _____ because they let some, but not all, substances pass.

Biological membranes present a barrier to polar molecules

20. Small _____ molecules easily permeate plasma membranes.

21. The membrane is relatively impermeable to (a)_____ and most (b)_____ polar molecules.

Transport proteins transfer molecules across membranes

22. (a)_____ bind the molecule being transported and undergo a (b)_____.

23. (a)_____ from tunnels through the membrane and are often (b)_____.

24. _____ are channel proteins that facilitate the transport of water through the plasma membrane in response to osmotic gradients.

PASSIVE TRANSPORT

25. Passive transport does not require the cell to spend metabolic _____.

Diffusion occurs down a concentration gradient

26. Diffusion rate is affected by temperature and the _____ of the moving particles.

27. In diffusion, there is a net movement of particles from (a)_____ to (b)_____ areas of concentration.

Osmosis is diffusion of water across a selectively permeable membrane

28. The movement of water through a selectively permeable membrane from a region of higher concentration to a region of lower concentration is called _____.

29. The _____ of a solution is determined by the amount of dissolved substances in the solution.

30. When there is no net movement of water, a cell is in an _____ solution.

31. If a solution has a high solute concentration, it is (a)_____. If a cell is placed in this solution there is net movement of water (b)_____ of the cell, causing the cell to (c)_____.

32. A (a)_____ solution has a low solute concentration and a cell placed in this type of solution will (b)_____ as the net movement of water is (c)_____ the cell.

33. (a)_____ is generated in cells that have a cell wall when placed in a (b)_____ solution. This same cell will undergo (c)_____ when placed in a (d)_____ solution.

Facilitated diffusion occurs down a concentration gradient

34. In facilitated diffusion, the net movement is always from a region of (a)_____ solute concentration to a region of (b)_____ solute concentration.

35. In this type of diffusion, a (a)_____protein makes the membrane permeable. These proteins are either (b)_____or_____.

ACTIVE TRANSPORT

36. The energy for diffusion comes from the (a)_____, while the energy for active transport usually comes from (b)_____.

Active transport systems "pump" substances against their concentration gradients

37. Because of the sodium-potassium pump, there are _____ (fewer or more) potassium ions inside the cell relative to the sodium ions outside

38. The distribution of sodium and potassium across cell membranes causes the inside of the cell to be _____ charged relative to the outside.

39. The unequal ion distribution creates an (a)_____. When both a charge and concentration difference are present an (b)_____ will be established.

Carrier proteins can transport one or two solutes

40. (a)_____ transport one type of substance in one direction and (b)_____ move two types of substances in one direction, while (c)_____move two substances in opposite directions.

Co-transport systems indirectly provide energy for active transport

41. The process whereby the movement of one solute down its concentration gradient provides the energy to transport another solute up its concentration gradient is known as _____.

EXOCYTOSIS AND ENDOCYTOSIS

42. These mechanisms are similar to active transport because the cell must spend _____.

In exocytosis, vesicles export large molecules

43. Exocytosis is the exportation of waste materials or specific products of secretion by the fusion of a _____ with the plasma membrane of a cell.

In endocytosis, the cell imports materials

44. In (a)_____ a cell ingests large, solid particles. In (b)_____ a cell takes in dissolved materials.

45. _____is how cells take in macromolecules like cholesterol.

CELL JUNCTIONS

46. Three types of junctions, the
    (a)_____
    _____, are specialized structures
    associated with the plasma membrane of animal cells, while
    (b)_____ connect plant cells.

Anchoring junctions connect cells of an epithelial sheet

47. The two common types of anchoring junctions are
    (a)_____ that are anchored to
    intermediate filaments in the cell, and
    (b)_____ that connect with microfilaments of
    the cytoskeleton.

Tight junctions seal off intercellular spaces between some animal cells

48. The _____, composed of tight
    junctions, prevents many substances in the blood from passing into
    the brain.

Gap junctions allow the transfer of small molecules and ions

49. Gap junctions consist of a cluster of
    (a)_____ that form
    (b)_____ that connect the cytoplasm of
    adjacent cells to allow communication.

**Plasmodesmata allow certain molecules and ions to move between plant
  cells**

50. Plasmodesmata in plant cells are functionally equivalent to
    _____ in animal cells.

51. Most plasmodesmata contain a cylinder-shaped structure called the
    _____.

BUILDING WORDS

Use combinations of prefixes and suffixes to build words for the definitions that follow.

Prefixes	The Meaning		Suffixes	The Meaning
desm(o)-	bond		-cyto(sis)	cell
endo-	within		-desm(a)	bond
exo-	outside, outer, external		-some	body
hyper-	over			
hypo-	under			
iso-	equal, "same"			
phago-	eat, devour			
pino-	drink			

Prefix	Suffix	Definition
_____	_____	1. A process whereby materials are taken into the cell.
_____	_____	2. The process whereby waste or secretion products are ejected from a cell by fusion of a vesicle with the plasma membrane.
_____	-tonic (-osmotic)	3. Having an osmotic pressure or solute concentration that is greater than a standard solution.
_____	-tonic (-osmotic)	4. Having an osmotic pressure or solute concentration that is less than a standard solution.
_____	-tonic (-osmotic)	5. Having an osmotic pressure or solute concentration that is the same as a standard solution.
_____	_____	6. The ingestion of large solid particles, such as bacteria and food, by a cell.
_____	_____	7. A type of endocytosis whereby fluid is engulfed by vesicles originating at the cell surface.
_____	_____	8. A button-like plaque (body) present on two opposing cell surfaces, that holds (bonds) the cells together by means of protein filaments that span the intercellular space.
plasmo-	_____	9. A cytoplasmic channel connecting (bonding) adjacent plant cells and allowing for the movement of small molecules and ions between cells.

MATCHING

Terms:

a.   Active transport
b.   Concentration gradient
c.   Co-transport
d.   Dialysis
e.   Diffusion
f.   Facilitated diffusion
g.   Fluid-mosaic model
h.   Plasmodesmata
i.   Osmosis
j.   Signal transduction
k.   Selectively permeable
     membrane
l.   Tight junction
m.   Turgor pressure

For each of these definitions, select the correct matching term from the list above.

_____ 1. The modern picture of membranes in which protein molecules float in a phospholipid bilayer.

_____ 2. The diffusion of water across a selectively permeable membrane.

_____ 3. The transport of ions or molecules across a membrane and down a concentration gradient by a specific carrier protein.

_____ 4. Regions in a system of differing concentration, such as exist in a cell and its environment, that cause molecules to move from areas of higher concentration to lower concentration.

_____ 5. Energy-requiring transport of a molecule across a membrane from a region of low concentration to a region of high concentration.

_____ 6. A membrane that allows some substances to cross it more easily than others.

_____ 7. The random movement of molecules from a region of higher concentration to one of lower concentration of that substance.

_____ 8. The internal pressure in a plant cell caused by the diffusion of water into the cell.

_____ 9. A specialized structure between some animal cells, producing a tight seal that prevents materials from passing through the spaces between the cells.

_____10. Structure allowing passage of certain small molecules and ions between plant cells.

MAKING COMPARISONS

Fill in the blanks.

Transport Mechanism	Description of the Transport Mechanism
Diffusion	Net movement of a substance from an area of high concentration to an area of low concentration
Phagocytosis	#1
#2	The passive transport of solutes down a concentration gradient aided by a specific protein in a membrane
#3	The active transport of substances into the cell by the formation of invaginated regions or "folds" of the plasma membrane that pinch off and become cytoplasmic vesicles
#4	Transfer of solutes by proteins located within the membrane
Desmosomes	#5
Exocytosis	#6
Osmosis	#7
#8	Transport of substances across a membrane requiring the expenditure of energy by the cell
Pinocytosis	#9
#10	Points of attachment between cells that hold cells together

MAKING CHOICES

Place your answer(s) in the space provided. Some questions may have more than one correct answer.

_____ 1. Simple diffusion
    a. moves molecules with a gradient.
    b. does not occur in prokaryotes.
of cells.
    c. requires use of ATP.

    d. involves protein channels.
    e. moves substances both into and out

_____ 2. Active transport
    a. can move molecules against a gradient.
cells.
    b. does not occur in prokaryotes.
of cells.
    c. requires use of ATP.

    d. occurs in animal cells, not in plant
    e. moves substances both into and out

_____ 3. A molecule is called amphipathic when it
    a. prevents free passage of substances.
    b. is in a membrane.
    c. has hydrophobic and hydrophilic regions.

    d. is a lipid.
    e. is embedded in a bilipid layer.

_____ 4. If membranes were not fluid and dynamic, which of the following might still occur normally?

    a. active transport

    b. facilitated diffusion

    c. simple diffusion

    d. endo- and exocytosis

    e. osmosis

_____ 5. Membrane fusion enables

    a. diversity of proteins.

    b. exocytosis.

    c. endocytosis.

    d. fusion of vesicles and plasma membrane.

    e. formation of Golgi complexes.

Plants are placed in three beakers containing the following solutions: beaker A, distilled water; beaker B, isotonic solution; beaker C, 13% salt solution. Use this information to answer questions 7-9.

_____ 6. The cells in beaker C

    a. shriveled.

    b. swelled.

    c. plasmolyzed.

    d. were unaffected.

    e. probably eventually burst.

_____ 7. The cells in beaker A

    a. shriveled.

    b. swelled.

    c. plasmolyzed.

    d. were unaffected.

    e. probably eventually burst.

_____ 8. The cells in beaker B

    a. shriveled.

    b. swelled.

    c. plasmolyzed.

    d. were unaffected.

    e. probably eventually burst.

_____ 9. Plasma membranes of eukaryotic cells have a large amount of

    a. cholesterols.

    b. phospholipid.

    c. rigidity.

    d. fluidity.

    e. protein.

_____ 10. Diffusion rate depends on

    a. the flow of water.

    b. concentration gradient.

    c. energy from the cell.

    d. the plasma membrane.

    e. kinetic energy.

_____ 11. Endocytosis may include

    a. secretion vacuoles.

    b. pinocytosis.

    c. phagocytosis.

    d. a combination of inbound particles with proteins.

    e. receptor-mediation.

_____ 12. Plasmodesmata

    a. are channels in cytoplasm.

    b. connect plant cells.

    c. are plant cell structures equivalent to animal cell desmosomes.

    d. are the same as several plasmodesma.

    e. connect ER of adjacent cells.

_____ 13. The principal cell adhesion molecules in vertebrates are known as

    a. cadherins.

    b. integral proteins.

    c. glycoproteins.

    d. plasmodesma.

    e. β-pleated molecules.

_____14. Sugars may be added to proteins
    a. and porins.
    b. to form gated channels.
    c. resulting in a glycoprotein.
    d. in the lumen of ER.
    e. in the extracellular space.

VISUAL FOUNDATIONS

Color the parts of the illustration below as indicated. Label the interior and exterior of the cell, the carbohydrate chains, and the lipid bilayer.

RED   ☐ alpha helix
GREEN   ☐ hydrophilic region of transmembrane proteins
YELLOW   ☐ hydrophobic region of transmembrane proteins
BLUE   ☐ glycolipid
ORANGE   ☐ glycoprotein
BROWN   ☐ cholesterol
TAN   ☐ peripheral protein

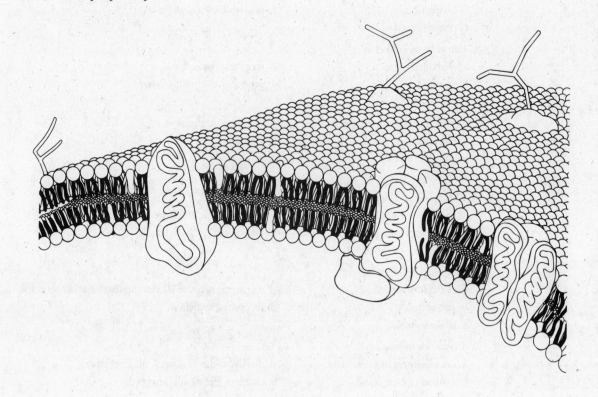

1. What is the name of the model for membrane structure illustrated above?_____

2. Which of the colored parts is most important in forming junctions between adjacent cells?
_____

3. Which of the colored parts is important in maintaining fluidity in animal cells?_____

CHAPTER 6

Cell Communication

The process of cell communication is vital to maintaining homeostasis in all types of cells. In multicellular organisms, cells must also communicate with one another, sometimes directly and sometimes indirectly. Organisms of different species, and even different kingdoms and domains, communicate with one another. This process involves cells sending and receiving signals, information crossing the plasma membrane and is transmitted through the cell, and finally, cells responding to the signals they receive. These signals come from the extracellular environment, many times from other cells and are typically chemical in nature. The signals are received by several known receptor proteins. Once received the signal is transmitted through a signaling pathway, which moves the signal inside the cell and amplifies it. Problems in cell communication can cause or contribute to a variety of diseases, including diabetes and cancer. Over billions of years, elaborate systems of cell communication have evolved, but many molecules involved are highly conserved across species.

REVIEWING CONCEPTS
Fill in the blanks.

INTRODUCTION
1. To maintain homeostasis, cells must continuously
   _____with each other.
2. Prokaryotes, protists, fungi, plants, and animals all communicate
   with other members of their species by
   _____.
3. A community of microorganisms attached to a solid surface is
   known as a _____.

CELL SIGNALING: AN OVERVIEW
4. The term _____ refers to the mechanisms
   by which cells communicate with one another.
5. The cells that can respond to a signaling molecule are called
   _____.
6. _____ is the process by which a cell
   converts an extracellular signal into an intracellular signal that
   results in a response.

SENDING SIGNALS
7. Most neurons signal one another by releasing chemical compounds
   called _____.

8. _____ is a process in which a signaling molecule diffuses through the interstitial fluid and acts on nearby cells.

9. _____ are paracrine regulators that modify cAMP levels and interact with other signaling molecules to regulate metabolic activities.

RECEPTION

10. A cell is programmed to receive signals based on the _____ it can synthesize.

11. A signaling molecule that binds to a specific receptor is called a (a)_____. Most are (b)_____ but some are (c)_____ and can pass through the membrane.

12. A receptor generally has at least three (a)_____. The external one is for (b)_____ of the ligand, while the internal (c)_____ transmits the signal.

13. Pigments in plants, some algae, and some animals that absorb blue light and play a role in biological rhythms are called _____.

Cells regulate reception

14. Receptor down-regulation occurs in response to (a)_____ hormone concentrations and involves (b)_____ the number of receptors.

15. Receptor up-regulation occurs in response to (a)_____ hormone concentrations and involves (b)_____ the number of receptors.

Three types of receptors occur on the cell surface

16. (a)_____ receptors convert (b)_____ signals to (c)_____ ones.

17. (a)_____ receptors couple signaling molecules to (b)_____ pathways and are major targets for (c)_____ development.

18. (a)_____ are transmembrane proteins with a binding site for a signaling molecule outside the cell and an enzyme component inside the cell. A (b)_____ is an example that performs (c)_____.

Some receptors are located inside the cell

19. Intracellular receptors that regulate the expression of specific genes are called (a)_____ and usually bind (b)_____ molecules.

SIGNAL TRANSDUCTION

20. Many regulatory molecules transmit information to the cell's interior without physically crossing the _____.

21. A chain of signaling molecules is called a
    (a)_____ and its job is to relay and
    (b)_____the original signal.

Signaling molecules can act as molecular switches

22. (a)_____typically activates kinases while
    (b)_____ typically inactivates them.

Ion channel-linked receptors open or close channels

23. Gamma-aminobutyric acid is a neurotransmitter that binds to
    _____ in neurons, thereby inhibiting
    neural signaling.

G-protein-linked receptors initiate signal transduction

24. G proteins are a group of regulatory proteins important in many
    _____ pathways.

25. G proteins usually relay a signal to a
    _____.

Second messengers are intracellular signaling agents

26. Second messengers are usually _____.

27. In many signaling cascades in prokaryotic and animal cells, the
    second messenger is _____.

28. Adenylyl cyclase is an enzyme on the _____ side of
    the plasma membrane.

29. When adenylyl cyclase is activated, it catalyzes the formation of
    (a)_____, which in turn activates (b)_____,
    leading to the phosphorylation of (c)_____, resulting in
    some response in the cell.

30. The enzyme (a)_____splits PIP_2 into
    (b)_____which act as second messengers to affect
    (c)_____activity and levels of (d)_____ions.

31. Ion pumps in the plasma membrane normally maintain a
    _____ calcium ion concentration in the cytosol compared to its
    concentration in the extracellular fluid.

Many enzyme-linked receptors activate protein kinase signaling pathways

32. Most enzyme-linked receptors are
    (a)_____ , enzymes that
    (b)_____ the amino acid tyrosine in proteins.

Many activated intracellular receptors are transcription factors

33. Some hydrophobic signaling molecules diffuse across the
    membranes of target cells and bind with
    _____ in the cytosol or in the nucleus.

Scaffold proteins increase efficiency

34. Scaffold proteins organize groups of
    (a)_____ into
    (b)_____ , ensuring that signals are
    relayed accurately, rapidly, and more efficiently.

Signals can be transmitted in more than one direction

35. Transmembrane proteins that connect the cell to the extracellular matrix are called (a)_____and can participate in (b)_____signaling.

RESPONSES TO SIGNALS

36. The three categories of responses are_____
_____.

Ras pathways involve tyrosine kinase receptors and G proteins

37. (a)_____are small G proteins active when bound to (b)_____. Mutations in these proteins are associated with (c)_____.

38. A well-studied ras pathway called (a)_____ is the main signaling pathway for cell (b)_____.

The response to a signal is amplified

39. The process of magnifying the strength of a signaling molecule is called _____.

Signals must be terminated

40. Signal termination returns the receptor and each of the components of the _____ pathway to their _____ states.

EVOLUTION OF CELL COMMUNICATION

41. Some signaling molecules are highly (a)_____ such as, (b)_____.

42. Evidence suggests that cell communication first evolved in _____ and continued to change over time as new types of organisms evolved.

BUILDING WORDS

Use combinations of prefixes and suffixes to build words for the definitions that follow.

Prefixes	**The Meaning**	**Suffixes**	**The Meaning**
di-	two, twice, double	-chrome	color
intra-	within	-crine	to secrete
neuro-	nerve		
para-	beside, near		
phyto-	plant		
tri-	three		
inter-	between		

Prefix	**Suffix**	**Definition**
_____	_____	1. Pertains to cell secretions (signaling molecules) that diffuse through the interstitial fluid and act on nearby cells.
_____	-transmitter	2. Substance used by neurons to transmit impulses across a synapse.
_____	_____	3. A blue-green, proteinaceous pigment involved in photoperiodism and a number of other light-initiated physiological responses of plants.
_____	-cellular	4. Pertains to the inside of a cell.
_____	-phosphate	5. A chemical compound with three phosphate groups.
_____	-phosphate	6. A chemical compound with two phosphate groups.
_____	-acylglycerol	7. A compound formed of two fatty acids and one glycerol molecule.
_____	-cellular	8. Signals that are sent between cells.

MATCHING

Terms:

a. Adenylyl cyclase
b. Calmodulin
c. Enzyme-linked receptor
d. G protein-linked receptor
e. Guanosine diphosphate
f. Hormone
g. Intracellular receptor
h. Ligand
i. Neurotransmitter
j. Nitric oxide
k. Ras protein
l. Signal amplification
m. Signal termination
n. Signal transduction
o. Transcription factor
p. Tyrosine kinase

For each of these definitions, select the correct matching term from the list above.

_____ 1. A signaling molecule that binds to a specific receptor.

_____ 2. The process in which a receptor converts an extracellular signal into an intracellular signal that causes some change in the cell.

_____ 3. A gaseous signaling molecule that passes into target cells.

_____ 4. A signaling molecule released by neurons to signal one another.

_____ 5. A chemical messenger secreted by endocrine glands.

_____ 6. A transmembrane protein composed of seven alpha helices connected by loops that extend into the cytosol or outside the cell.

_____ 7. A transmembrane protein with a binding site for a signaling molecule outside the cell and a binding site for an enzyme inside the cell.

_____ 8. A receptor located in the cytosol or in the nucleus.

_____ 9. An enzyme-linked receptor that phosphorylates specific tyrosines in certain signaling proteins inside the cell.

_____ 10. One of a group of small G proteins that, when activated, trigger a cascade of reactions.

_____ 11. A receptor located in the nucleus that activates or represses specific genes.

_____ 12. An enzyme on the cytoplasmic side of the plasma membrane that catalyzes the formation of cyclic AMP from ATP.

_____ 13. A protein that when combined with calcium ions affects the activity of protein kinases and protein phosphatases.

_____ 14. The process of enhancing signal strength as a signal is relayed through a signal transduction pathway.

_____ 15. The process of inactivating the receptor and each component of the signal transduction pathway once they have done their jobs.

MAKING COMPARISONS

Refer to this illustration of signal transduction to fill in blanks in the table below.

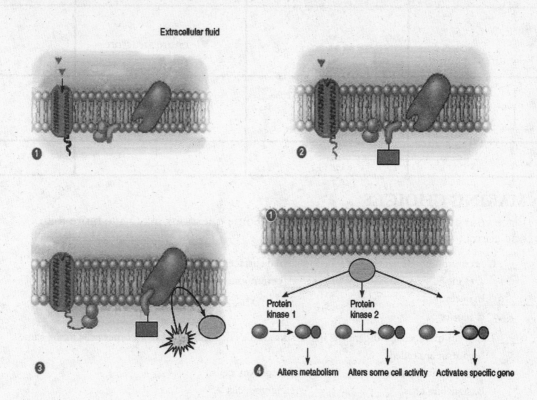

Identify the structures in the table and fill in the blanks.

Structure	Name of Structure	Structure	Name of Structure
entire structure	receptor molecule	at the arrow	#1
at the arrow	#2	at the arrow	#3
at the arrow	#4	entire structure	#5

Structure	Name of Structure	Structure	Name of Structure
entire structure	#6	▼ entire structure	#7
at the arrow	#8		

MAKING CHOICES

Place your answer(s) in the space provided. Some questions may have more than one correct answer.

_____ 1. A community of bacteria attached to a solid surface is a
a. clone.
b. quorum.
c. biofilm.
d. communication population.
e. colony.

_____ 2. Cells in multicellular organisms that respond to a chemical signal from other cells in the same organism are called
a. transmitter cells.
b. signaling cells.
c. responders.
d. secretory cells.
e.. target cells

_____ 3. The process in cells of converting an extracellular signal into an intracellular signal to which a response occurs is known as
a. optimization.
b. transduction.
c. internal regulation.
d. chemical transposition.
e. chemical signal reception.

_____ 4. The process whereby a signaling molecule diffuses through interstitial fluid and affects nearby cells is
a. paracrine regulation.
b. cellular reception.
c. interstitial reception.
d. interstitial modification.
e. endocrine facilitation.

_____ 5. The signaling chemicals transferred across the gap between neurons are called
a. neuroreceptors.
b. synaptic transmitters.
c. neurotransmitters.
d. neuronal enhancers.
e. neuron communicators.

_____ 6. Endocrine glands
a. have no ducts.
b. secrete hormones.
c. communicate directly with neurons.
d. secrete chemical messengers into interstitial fluid.
e. secrete hormones into capillaries.

_____ 7. Proteins and glycoproteins that bind with ligands are called
a. enzymes.
b. messengers.
c. signaling molecules.
d. bonding elements.
e. receptors.

_____ 8. Receptor down-regulation may involve
    a. destruction of receptors by lysosomes.   d. increased intake of insulin by cells.
    b. decreased numbers of receptors.   e. degradation of receptors.
    c. regulation of blood glucose levels.

_____ 9. Receptor up-regulation occurs in response to
    a. low hormone concentrations.   d. increase in protein membrane complexes.
    b. high hormone concentrations.   e. integration.
    c. receptor down-regulation.

_____ 10. Receptors on the cell surface that convert chemical signals into electrical signals are called
    a. G protein-linked receptors.   d. protein complex receptors.
    b. ion channel-linked receptors.   e. ligand-gated channels.
    c. enzyme-linked receptors.

_____ 11. Proteins that regulate the expression of genes are
    a. cell surface proteins.   d. intracellular receptors.
    b. cell membrane glycoproteins.   e. intranuclear integrators.
    c. transcription factors.

_____ 12. When gamma-aminobutyric acid (GABA) binds to a ligand-gated chloride ion channel in a neuron
    a. the channel opens.   d. neural signaling is inhibited.
    b. adjacent channels close.   e. the neuron's electrical potential changes.
    c. chloride ions flow into the neuron.

_____ 13. Second messengers
    a. are ions or small molecules.   d. are cAMP.
    b. may amplify a neuronal signal.   e. are hormones.
    c. are found mainly in interstitial fluids.

_____ 14. Calcium ion messengers are not involved in
    a. microtubule disassembly.   d. liver cell activation.
    b. muscle contraction.   e. initiation of development.
    c. blood clotting.

_____ 15. Calmodulin
    a. combines to only one specific enzyme.   d. is found in eukaryotic cells.
    b. produces the binding ezyme calmodulinase.   e. is a Ca^{2+} binding protein.
    c. changes the shape of the 4- Ca^{2+} binding complex

_____ 16. Ras proteins
    a. are activated when bound to GTP.   d. trigger a series of reactions called the Ras pathway.
    b. are small proteins.   e. phosphorylate glucosamine.
    c. are a group of G proteins.

_____ 17. Integrins
    a. are transmembrane proteins.   d. are enzymes that catalyze signaling reactions.
    b. are intracellular signaling transmitters.   e. may be involved in "inside-out" signaling.
    c. interface between the cell and the extracellular matrix.

_____ 18. The components of signal transduction pathways have been identified in
    a. protists.   d. plants.
    b. fungi.   e. yeasts.
    c. animals.

VISUAL FOUNDATIONS

Color the parts of this illustration of the mechanism of intracellular reception as indicated.

RED ▢ extracellular signaling molecules

GREEN ▢ signaling molecules moving through the cytosol

YELLOW ▢ intranuclear receptor

BLUE ▢ transcription factor (activated receptor)

ORANGE ▢ DNA

BROWN ▢ m-RNA

PINK ▢ ribosomes

TAN ▢ protein that alters cell activity

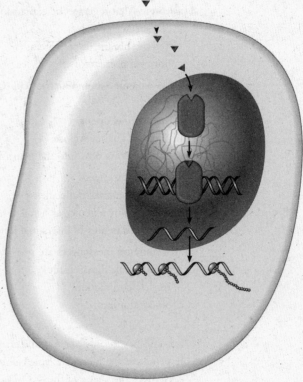

Color the parts of this illustration of types of cell signaling as indicated.

RED ☐ generalized signaling cells
GREEN ☐ generalized target cells
YELLOW ☐ receptors
BLUE ☐ signaling molecules
ORANGE ☐ signaling neuron
BROWN ☐ receptor neuron

CHAPTER 7

Energy and Metabolism

You should now have an understanding of the chemical materials that function in life's processes and of the structure of a cell. You will now examine one of the basic themes of biology – the flow of energy through organisms. Life depends on a continuous input of energy. The myriad chemical reactions of cells that enable them to function involve energy transformations between potential and kinetic energy. These transformations are governed by the laws of thermodynamics which explain why organisms cannot produce energy but must continuously capture it from somewhere else, and why in every energy transaction, some energy is dissipated as heat. In the metabolism of living things, some chemical reactions occur spontaneously, releasing free energy that is then available to perform work. Other reactions require an input of free energy before they can occur, and sometimes these reactions are coupled together. ATP, the energy currency of the cell, has a structure that easily allows it to perform its vital function. Energy transfers often occur in cells through redox reactions utilizing electron carrier molecules. Chemical reactions in organisms are regulated by enzymes, substances that lower the amount of energy needed to activate reactions. Enzymes have several unique characteristics and are highly regulated. The effectiveness of some drugs is due to their ability to inhibit enzymes that are critical to the normal functioning of certain pathogenic organisms.

REVIEWING CONCEPTS

Fill in the blanks.

INTRODUCTION

1. Life depends on a continuous supply of outside energy. Fortunately, _____ capture outside sources of energy and incorporate it in chemical bonds (food).

2. Plants convert radiant energy to _____ energy.

BIOLOGICAL WORK

3. Energy is the capacity to do _____.

Organisms carry out conversions between potential energy and kinetic energy

4. Energy is in one of two forms: (a)_____ is "stored energy" and (b)_____ is "energy of motion."

THE LAWS OF THERMODYNAMICS

5. The study of energy and its transformations is called _____.

The total energy in the universe does not change

6. The first law of thermodynamics states that energy can be neither
   (a)_____ nor _____, however, it
   can be (b)_____ and changed
   in form.

The entropy of the universe is increasing

7. In every energy conversion or transfer some energy is dissipated as
   _____.

8. The term entropy refers to the
   _____, in the universe.

ENERGY AND METABOLISM

Enthalpy is the total potential energy of a system

9. The energy required to break a chemical bond is referred to as
   _____.

10. The total potential energy of a chemical reactions, or enthalpy,
    equals the _____ of the
    reactants and products.

Free energy is available to do cell work

11. An increase in _____ leads to a decrease in the
    amount of free energy.

Chemical reactions involve changes in free energy

12. The change in free energy during a reaction is equal to the change
    in (a)_____ minus the product of the
    absolute temperature multiplied by the change in
    (b)_____.

Free energy decreases during an exergonic reaction

13. _____ are spontaneous and
    they release energy that can perform work.

14. The mathematical/chemical symbol for change in free energy is
    _____.

Free energy increases during an endergonic reaction

15. In an endergonic reaction, free energy has a _____
    value.

Diffusion is an exergonic process

16. In a concentration gradient, energy moves from a region of
    (a)_____ concentration to a region of (b)_____ concentration.

**Free-energy changes depend on the concentrations of reactants and
    products**

17. In a state of _____, the rate of the
    reverse reaction equals the rate of the forward reaction.

Cells drive endergonic reactions by coupling them to exergonic reactions

18. Exergonic reactions (a)_____[release or require input of?] free energy; endergonic reactions (b)_____[release or require input of?] free energy.

19. In the reaction C → D, where the value of △G is negative, the reactant has _____(more or less?) free energy than the free energy of the product.

ATP, THE ENERGY CURRENCY OF THE CELL

20. The three main parts of the ATP molecule are _____, _____, and _____.

ATP donates energy through the transfer of a phosphate group

21. Phosphate bonds in ATP are transferred by the process known as _____.

ATP links exergonic and endergonic reactions

22. Exergonic reactions are generally part of (a)_____ pathways and endergonic reactions are generally part of (b)_____ pathways.

The cell maintains a very high ratio of ATP to ADP

23. The cell cannot store _____ quantities of ATP.

ENERGY TRANSFER IN REDOX REACTIONS

Most electron carriers transfer hydrogen atoms

24. Stripped electrons and the energy they possess are transferred to an _____.

ENZYMES

All reactions have a required energy of activation

25. The activation energy of a reaction begins the reaction by using energy to _____.

An enzyme lowers a reaction's activation energy

26. Enzymes lower the activation energy to _____ the rate of the reaction.

An enzyme works by forming an enzyme-substrate complex

27. Enzymes lower activation energy by forming an unstable intermediate called the

_____.

28. When the substrate binds to the enzyme molecule, it causes a change, known as

_____.

29. Most enzyme names end in _____.

Enzymes are specific

30. Most enzymes are specific because the shape of the _____ is closely related to the shape of the substrate.

Many enzymes require cofactors

31. An organic cofactor is called an _____.

32. Some enzymes have two components, the cofactor and a protein referred to as the _____.

Enzymes are most effective at optimal conditions

33. Factors that affect enzyme activity include (list three)
_____
_____.

Enzymes are organized into teams in metabolic pathways

34. When enzymes work in teams, the (a)_____ from one enzyme-substrate reaction becomes the (b)_____ for the next enzyme-substrate reaction.

35. A series of reactions can be illustrated as A → B → C … etc. where each reaction (→) is carried out by a specific enzyme. Such a series is referred to as a _____.

The cell regulates enzymatic activity

36. Feedback inhibition is a type of enzyme regulation in which the formation of a product _____ an earlier reaction in a sequence.

Enzymes can be inhibited by certain chemical agents

37. Inhibition is _____ when the inhibitor-enzyme bond is weak.

38. _____ inhibition occurs when the inhibitor competes with the normal substrate for binding to the active site of the enzyme.

39. _____ inhibition occurs when the inhibitor binds to the enzyme at a site other than the active site.

Some drugs are enzyme inhibitors

40. Because sulfa drugs have a chemical structure similar to _____, they are able to selectively affect bacteria.

BUILDING WORDS

Use combinations of prefixes and suffixes to build words for the definitions that follow.

Prefixes	The Meaning		Suffixes	The Meaning
allo-	other, "another"		-calor(ie)	heat
ana-	up		-ergonic	work, "energy"
cata-	down		-steric	"space"
end(o)-	within			
ex(o)-	outside, outer, external			
kilo-	thousand			

Prefix	Suffix	Definition
_____	_____	1. Heat energy; the amount of heat required to raise the temperature of 1000 grams (1 kg) of water from 14.5° C to 15.5° C.
_____	-bolism	2. In living organisms, the "building up" (synthesis) of more complex substances from simpler ones.
_____	-bolism	3. In living organisms, the "breaking down" of more complex substances into simpler ones.
_____	_____	4. A spontaneous reaction that releases free energy and can therefore perform work.
_____	_____	5. A reaction that requires an input of free energy from the surroundings.
_____	_____	6. Refers to a receptor site on some region of an enzyme molecule other than the active site.

MATCHING

Terms:

a.   Catalyst
b.   Coenzyme
c.   Energy
d.   Enthalpy
e.   Entropy
f.   Enzyme
g.   Free energy
h.   Heat energy
i.   Kinetic energy
j.   Potential energy
k.   Substrate
l.   Thermodynamics

For each of these definitions, select the correct matching term from the list above.

_____ 1. A quantitative measure of the amount of randomness or disorder of a system.

_____ 2. An organic substance that is required for a particular enzymatic reaction to occur.

_____ 3. A substance on which an enzyme acts.

_____ 4. Energy in motion.

_____ 5. The study of energy and its transformations.

_____ 6. A substance that increases the speed at which a chemical reaction occurs without being used up during the reaction.

_____ 7. An organic catalyst that greatly increases the rate of a chemical reaction without being consumed by that reaction.

_____ 8. The total potential energy of a system.

_____ 9. Stored energy.

_____10. The capacity or ability to do work.

MAKING COMPARISONS

Fill in the blanks.

Chemical Reaction	Type of Metabolic Pathway	Energy Required (Endergonic) or Released (Exergonic)
Synthesis of ATP	Anabolic	Endergonic
Hydrolysis	#1	#2
Phosphorylation	#3	#4
ATP → ADP	#5	#6
Oxidation	#7	#8
Reduction	#9	#10
FAD → FADH$_2$	#11	#12

MAKING CHOICES

Place your answer(s) in the space provided. Some questions may have more than one correct answer.

_____ 1. Enzymes
    a. are lipoproteins.
    b. lower required activation energy.
    c. speed up biological chemical reactions.
    d. become products after complexing with substrates.
    e. are regulated by genes.

_____ 2. Enzyme activity may be affected by
    a. cofactors.
    b. temperature.
    c. pH.
    d. substrate concentration.
    e. genes.

_____ 3. A kilocalorie (kcal) is
    a. a measure of heat energy.
    b. an energetic electron.
    c. a way to measure energy generally.
    d. the temperature of water.
    e. an essential nutrient.

Use the following formula to answer question 4:

$$ATP + H_2O \rightarrow ADP + P \quad \Delta G = -7.3 \text{ kcal/mole}$$

_____ 4. This reaction
    a. hydrolyzes ATP.
    b. loses free energy.
    c. produces adenosine triphosphate.
    d. is endergonic.
    e. is exergonic.

_____ 5. Compliance with the second law of thermodynamics presumes that
    a. disorder is increasing.
    b. maintaining order in a system requires input of energy.
    c. entropy will decrease as order in organisms increases.
    d. all energy will eventually be useless to life.
    e. heat dissipates in all systems.

_____ 6. Kinetic energy is
    a. doing work.
    b. stored energy.
    c. in chemical bonds.
    d. energy of motion.
    e. energy of position or state.

_____ 7. In the formula $\Delta G = \Delta H - T\Delta S$
    a. free energy decreases as entropy decreases.
    b. entropy and free energy are inversely related.
    c. change in enthalpy is greater than change in free energy.
    d. temperature increase decreases free energy.
    e. increasing enthalpy increases free energy.

_____ 8. Chemical energy in molecules is
    a. kinetic energy.
    b. potential energy.
    c. released in an endergonic reaction.
    d. stored in chemical bonds.
    e. energy in motion.

_____ 9. The sum of all chemical activities in an organism is known as
    a. anabolism.
    b. catabolism.
    c. metabolism.
    d. entropy.
    e. enthalpy.

_____10. An exergonic reaction
     a. releases energy.   d. reduces total free energy.
     b. can be spontaneous.   e. produces a negative value for ΔG
     c. is a "downhill" reaction.

_____11. In the reaction **A** ⇌ **B**
     a. "**A**"is the product molecule.   d. free energy is reduced.
     b. "**B**"is the reactant molecule.   e. a tendency to accumulate "**B**" is indicated.
     c. double arrows indicate a reversible reaction.

_____12. According to the first law of thermodynamics
     a. energy conversions are never 100% efficient.   d. energy cannot be created or destroyed.
     b. during chemical reactions some energy is lost as heat increasing.   e. total energy in the universe is
     c. disorganization in the universe is decreasing

VISUAL FOUNDATIONS

Color the parts of the illustration below as indicated.

    RED   ☐   active sites
    GREEN   ☐   substrates
    YELLOW   ☐   enzyme
    BLUE   ☐   allosteric site
    ORANGE   ☐   regulator
    BROWN   ☐   cyclic AMP
    TAN   ☐   enzyme-substrate complex

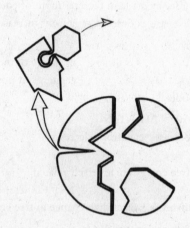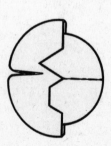

CHAPTER 8

How Cells Make ATP: Energy-Releasing Pathways

This chapter continues your study of the flow of energy through living systems. In cellular respiration, cells break down nutrients one step at a time, releasing energy from chemical bonds and transferring it to ATP where it is available for cellular work. There are three major pathways in which cells extract energy from nutrients: aerobic respiration, anaerobic respiration, and fermentation. Oxidation-reduction reactions are important in all three. Aerobic respiration, the most common pathway, involves a series of reactions in which hydrogen is transferred from a nutrient to oxygen, resulting in the formation of water. In anaerobic respiration, fuel molecules are broken down in the absence of oxygen, and an inorganic compound serves as the final hydrogen (electron) acceptor. Fermentation is a type of anaerobic respiration in which the final electron acceptor is an organic compound, and it is a much less efficient energy releasing mechanism.

REVIEWING CONCEPTS

Fill in the blanks.

INTRODUCTION

1. _____ is the exergonic aspect of metabolism involving breakdown of complex molecules.

2. _____ is the endergonic aspect of metabolism involving the synthesis of complex molecules.

3. Most anabolic reactions are endergonic and require _____ or some other energy source to drive them.

4. (a)_____ is the process where energy in (b)_____ is converted to chemical energy stored in (c)_____.

5. (a)_____ respiration requires oxygen, while (b)_____ respiration and (c)_____ do not use oxygen.

REDOX REACTIONS

6. Most cells use the catabolic process called _____ to extract free energy from nutrients, such as glucose.

7. In aerobic respiration, a fuel molecule is oxidized, yielding the by-products (a)_____ with the release of the essential (b)_____ that is required for life's activities.

THE FOUR STAGES OF AEROBIC RESPIRATION

8. The four stages of aerobic respiration of glucose are _____, _____, _____, and _____.

9. Hydrogens are removed to form reduced electron carriers in
   (a)_____ and CO_2 is produced during
   (b)_____.

In glycolysis, one molecule of glucose yields two pyruvate molecules

10. Glycolysis reactions take place in the _____ of the cell.

11. In the first phase of glycolysis, two ATP molecules are consumed and glucose is split into two _____ molecules.

12. In the second phase of glycolysis, each of the molecules resulting from splitting of glucose is oxidized and transformed into a _____ molecule.

13. Glycolysis *nets* _____ (#?) ATPs, made by
    (b)_____, and (c) _____NADH.

Pyruvate is converted to acetyl CoA

14. (a)_____(#?) NADH and (b) _____ CO_2 form during the formation of acetyl CoA from pyruvate.

The citric acid cycle oxidizes acetyl CoA

15. The eight step citric acid cycle completes the oxidation of glucose. For each acetyl group that enters the citric acid cycle, (a)_____ (#?) NAD^+ are reduced to NADH, (b)_____ (#?) molecules of CO_2 are produced, and (c)_____ (#?) $FADH_2$ is made.

16. _____(#?) acetyl CoAs are completely degraded with two turns of the citric acid cycle.

The electron transport chain is coupled to ATP synthesis

17. The hydrogens (electrons) removed during glycolysis, acetyl CoA formation, and the citric acid cycle are first transferred to the primary hydrogen acceptors (a)_____, then transferred down an (b)_____. As the electrons are passed ATP is produced by (c)_____.

18. The (a)_____ consists of a series of electron acceptors embedded in the inner membrane of mitochondria. (b)_____ is the final acceptor in the chain.

19. Electron pumping produces a (a)_____ gradient across the inner mitochondrial membrane. The concentration is high in the (b)_____ and low in the (c)_____.

20. The (a)_____ flow back across the inner membrane through (b)_____.

21. The process of _____ couples ATP synthesis with electron transport.

Aerobic respiration of one glucose yields a maximum of 36 to 38 ATPs

22. Most ATP is made by (a)_____ as opposed to a small amount made by (b)_____.

23. Mitochondrial shuttle systems harvest the electrons of _____ produced in the cytosol.

Cells regulate aerobic respiration

24. Aerobic respiration requires a steady input of fuel molecules and _____.

ENERGY YIELD OF NUTRIENTS OTHER THAN GLUCOSE

25. Human beings and many other animals usually obtain most of their energy by oxidizing _____. Amino acids are also used.

26. Amino groups are metabolized by a process called _____.

27. The _____ components of a triacylglycerol are used as fuel.

28. Fatty acids are converted into acetyl CoA by the process of _____. These molecules enter the citric acid cycle.

ANAEROBIC RESPIRATION AND FERMENTATION

29. _____ is a means of extracting energy that produces reduced inorganic substances and does not require oxygen.

30. Fermentation is an anaerobic process in which the final acceptor of electrons from NADH is an (a)_____, therefore regenerating the required (b)_____.

Alcohol fermentation and lactate fermentation are inefficient

31. When hydrogens from NADH are transferred to acetaldehyde, _____ is formed.

32. When hydrogens from NADH are transferred to pyruvate, _____ is formed.

33. Fermentation yields a net gain of only (a)_____(#?) ATPs per glucose molecule, compared with about (b)_____(#?) ATPs per glucose molecule in aerobic respiration.

BUILDING WORDS

Use combinations of prefixes and suffixes to build words for the definitions that follow.

Prefixes	The Meaning		Suffixes	The Meaning
aero-	air		-be (bios)	life
an-	without, not, lacking		-lysis	breaking down, decomposition
de-	indicates removal, separation			
glyco-	sweet, "sugar"			

Prefix	Suffix	Definition
_____	_____	1. An organism that requires air or free oxygen to live.
_____	-aerobe	2. An organism that does not require air or free oxygen to live.
_____	-hydrogenation	3. A reaction in which hydrogens are removed from the substrate.
_____	-carboxylation	4. A reaction in which a carboxyl group is removed from a substrate.
_____	-amination	5. A reaction in which an amino group is removed from a substrate.
_____	_____	6. A sequence of reactions that breaks down a molecule of glucose (a sugar) to two molecules of pyruvate.

MATCHING

Terms:

a. Aerobic respiration
b. Anaerobic respiration
c. Chemiosmosis
d. Citric acid cycle
e. Cytochromes
f. Electron transport chain
g. Ethyl alcohol
h. Facultative anaerobe
i. Fermentation
j. Lactic acid
k. Oxidation
l. Phosphorylation
m. Pyruvic acid
n. Reduction

For each of these definitions, select the correct matching term from the list above.

_____ 1. The process by which a proton gradient drives the formation of ATP.

_____ 2. Aerobic series of chemical reactions in which acetyl Co-A is completely degraded to carbon dioxide and water with the release of ATP.

_____ 3. An organism that can live in either the presence or absence of oxygen.

_____ 4. Anaerobic respiration that utilizes organic compounds both as electron donors and acceptors.

_____ 5. Oxygen-requiring pathway by which organic molecules are broken down and energy is released that can be used for biological work.

_____ 6. Produced along with carbon dioxide during fermentation.

_____ 7. The loss of electrons or hydrogen atoms from a substance.

_____ 8. The addition of a phosphate group to an organic molecule.

_____ 9. A series of chemical reactions during which hydrogens or their electrons are passed along from one receptor molecule to another, with the release of energy.

_____10. An end product of glycolysis.

MAKING COMPARISONS

Fill in the blanks.

Reactions in Glycolysis	Catalyzing Enzyme	ATPs Used (−) or Produced (+)
glucose → glucose-6-phosphate	hexokinase	− 1 ATP
glucose-6-phosphate → #1_____	Phospho-glucoisomerase	zero
#1_____ → fructose-1,6-biphosphate	#2	#3
fructose-1,6-biphosphate → #4_____ and glyceraldehydes-3-phosphate (G3P)	#5	zero
G3P → [structure: C≈O ~ Ⓟ / H—C—OH / H₂C—O—Ⓟ]	Glyceraldehydes-3-phosphate dehydrogenase	#6
[structure: C≈O ~ Ⓟ / H—C—OH / H₂C—O—Ⓟ] → #7_____	Phosphoglycero-kinase	#8
#7_____ → #9_____	Phosphoglycero-mutase	zero
#9_____ → phosphoenolpyruvate	Enolase	#10
#11_____ → pyruvate	#12	+2 ATP

MAKING CHOICES

Place your answer(s) in the space provided. Some questions may have more than one correct answer.

_____ 1. Complete aerobic metabolism yields
    a. 36 to 38 ATPs.   d. 4 ATPs from the citric acid cycle.
    b. 2 ATPs from glycolysis.   e. 2 pyruvates.
    c. 32 to 34 ATPs from electron transport and chemiosmosis.

_____ 2. Anaerobic respiration
    a. does not involve an electron transport chain.   d. involves chemiosmosis.
    b. is performed by certain prokaryotes.   e. may use nitrate as the final hydrogen acceptor.
    c. uses an inorganic substance as the final hydrogen acceptor.

_____ 3. A facultative anaerobe
    a. is capable of carrying out aerobic respiration.   d. is capable of producing CO_2.
    b. is capable of carrying out alcohol fermentation.   e. requires oxygen.
    c. is capable of producing ethanol.

_____ 4. During substrate-level phosphorylation
    a. ATP converts to ADP.   d. the electron transport chain is directly involved.
    b. ATP forms ADP.   e. chemiosmosis is directly involved.
    c some of the ATP from aerobic metabolism is produced .

_____ 5. Production of acetyl CoA from pyruvate
    a. is anabolic.   d. takes place in endoplasmic reticulum.
    b. takes place in mitochondria.   e. yields CO_2 and NADH
    c. takes place in cytoplasm.

_____ 6. The citric acid cycle begins with
    a. 2 H_2O and 4 CO_2.   d. one $FADH_2$.
    b. 6 NADH.   e. 2 acetyl CoA.
    c. oxaloacetate.

_____ 7. Glycolysis
    a. generates a net profit of two ATPs.   d. produces two pyruvates.
    b. is more efficient than aerobic respiration.   e. reduces glucose to H_2O and CO_2.
    c. takes place in the cristae of mitochondria.

_____ 8. ATP synthase
    a. is a cytochrome.   d. is a transmembrane protein.
    b. converts ATP to ADP.   e. couples protons to electrons to form water.
    c. forms channels across the inner mitochondrial membrane.

_____ 9. Chemiosmosis involves
    a. flow of protons down an electrical gradient.   d. proton channels composed of electrons.
    b. flow of protons down a concentration gradient.   e. the production of ATP.
    c. pumping of protons into the mitochondrial matrix.

_____10. In the electron transport chain of aerobic respiration
    a. oxygen is the final electron acceptor.   d. glucose is a common carrier molecule.
    b. cytochromes carry electrons.   e. electrons gain energy with each transfer.
    c. the final electron acceptor has a negative redox potential.

_____11. Cells may obtain energy from large molecules by means of

    a. catabolism.
    d. fermentation.

    b. aerobic respiration.
    e. the Kreb's cycle.

    c. anaerobic respiration.

_____12. In redox reactions _____ is/are transferred

    a. hydrogens.
    d. water.

    b. electrons.
    e. oxygen.

    c. energy.

VISUAL FOUNDATIONS

Enter the appropriate enzyme in the spaces provided.

1. _____   5. _____

2. _____   6. _____

3. _____   7. _____

4. _____   8. _____

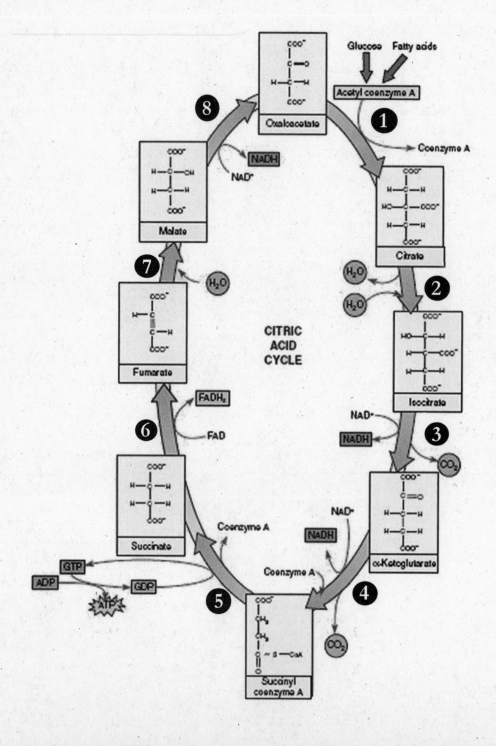

Color the parts of the illustration below as indicated.

RED ☐ electron path

GREEN ☐ complex I

YELLOW ☐ complex II

BLUE ☐ complex III

ORANGE ☐ complex IV

BROWN ☐ inner mitochondrial membrane

TAN ☐ matrix

PINK ☐ intermembrane space

VIOLET ☐ outer mitochondrial membrane

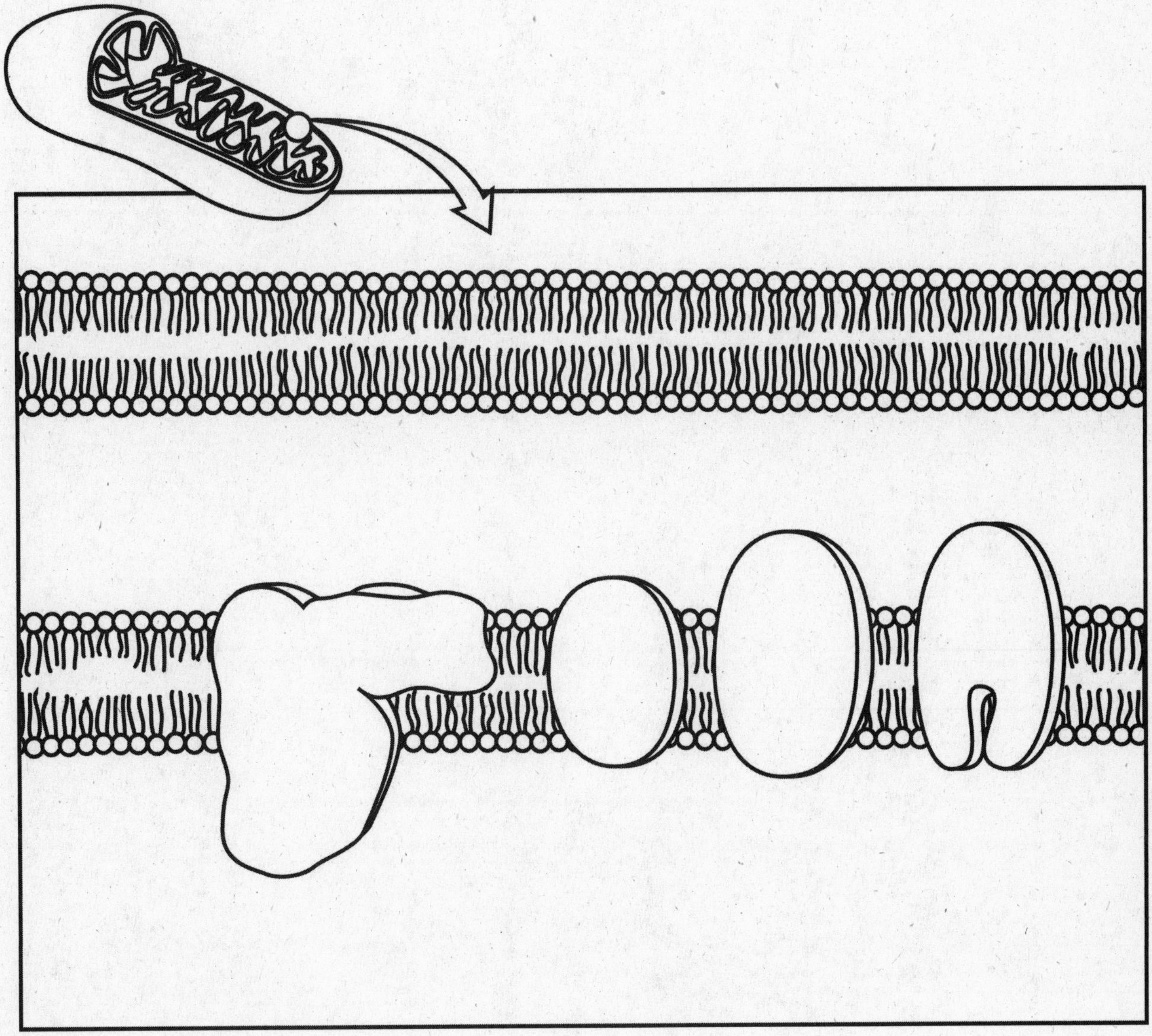

CHAPTER 9

Photosynthesis: Capturing Light Energy

This chapter concludes the overview of energy flow through living systems by examining the sequence of events in which light energy is converted into the stored chemical energy of organic molecules. This energy is what fuels the metabolic reactions discussed in the previous chapter, which sustain all life. Photosynthesis is the first step in the flow of energy through living things and involves capturing solar energy and using it to produce organic compounds from carbon dioxide and water while releasing oxygen as a by-product. Key features of light are discussed as it is key to this process. In eukaryotes, the process occurs in organized structures within chloroplasts. Photosynthesis consists of both the light-dependent and the carbon fixation reactions. There are three different pathways by which carbon dioxide is assimilated into plants. Organisms can be classified metabolically based on their carbon and energy sources. Photosynthesis is a regulated process absolutely essential for life on our planet.

REVIEWING CONCEPTS

Fill in the blanks.

INTRODUCTION

1. Using the basic raw materials (a)_____,
   photosynthetic organisms use (b) _____ energy to make
   ATP and other molecules that hold (c)_____ energy.

LIGHT AND PHOTOSYNTHESIS

2. (a)_____is the distance from one wave peak to the next. Light is
   composed of packets of energy called (b)_____. Light with
   (c)_____ has more energy associated with it.

3. When a molecule (a)_____ a photon an electron becomes energized.
   (b)_____ occurs when energized electrons return to the
   ground state, emitting their excess energy in the form of visible light. In
   photosynthesis, energized electrons leave atoms and pass to an
   (c)_____ molecule.

CHLOROPLASTS

4. Most chloroplasts are located in the (a)_____ cells of leaves.
   The (b)_____ is the fluid-filled region within the inner membrane
   of chloroplasts that contains most of the enzymes for photosynthesis.

5. _____ are flat, disk-shaped membranes in chloroplasts
   that are arranged in stacks called grana. Chlorophyll is located in these
   membranes.

Chlorophyll is found in the thylakoid membrane

6. Chlorophyll absorbs light mainly in the _____
   portions of the visible spectrum.

7. Of the several types of chlorophyll in plants,
   (a)_____ is the bright green form that initiates the light-dependent reactions, and the yellowish-green form,
   (b)_____, is an accessory pigment. Other yellow and orange accessory pigments in plant cells are (c)_____.

Chlorophyll is the main photosynthetic pigment

8. An (a)_____ is a graph that illustrates the relative absorption of different wavelengths of light by a given pigment. It is obtained with an instrument called a (b)_____.

9. An (a)_____ of photosynthesis is a measurement of the relative effectiveness of different wavelengths of light in affecting photosynthesis. It may be greater than can be accounted for by the absorption of chlorophyll alone, the difference accounted for by (b)_____ that transfer energy absorbed from the green wavelengths to chlorophyll.

OVERVIEW OF PHOTOSYNTHESIS

10. Photosynthesis involves the pigment (a)_____ capturing (b)_____ and converting (c)_____ into (d)_____ and releasing (e)_____ as a by-product.

ATP and NADPH are the products of the light-dependent reactions: an overview

11. Light-dependent reactions provide useful chemical energy for (a)_____ of photosynthetic products and occur in (b)_____.

Carbohydrates are produced during the carbon fixation reactions: an overview

12. Carbon fixation reactions transfer the energy from (a)_____ to the bonds in (b)_____ molecules and occur in the (c)_____.

THE LIGHT-DEPENDENT REACTIONS

13. These reactions use (a)_____ to generate (b)_____, which temporarily store energy.

Photosystems I and II each consist of a reaction center and multiple antenna complexes

14. The light-dependent reactions of photosynthesis begin when _____ and/or accessory pigments absorb light.

15. Chlorophylls *a* and *b* and accessory pigment molecules are organized with pigment-binding proteins in the thylakoids membrane into units called
   _____.

16. Each antenna complex absorbs light energy and transfers it to the _____, which consists of chlorophyll molecules and proteins.

17. Light energy is converted to chemical energy in the reaction centers by a series of _____.

18. The reaction center of photosystem I is made up of a pair of chlorophyll *a* molecules with an absorption peak of about 700 nm and is referred to as (a)_____. The reaction center of photosystem II is made up of a pair of chlorophyll *a* molecules with an absorption peak of about 680 nm and is referred to as (b)_____.

19. When a pigment molecule absorbs light energy, that energy is passed from one pigment molecule to another until it reaches the _____.

Noncyclic electron transport produces ATP and NADPH

20. In photosystem I, the energized electron is passed along an electron transport chain from one electron acceptor to another, until it is passed to (a)_____, an iron-containing protein, which transfers the electron to (b)_____.

21. Like photosystem I, photosystem II is activated when a pigment molecule in an _____ absorbs a photon of light energy.

22. _____ is a process that not only yields electrons, but is also the source of almost all the oxygen in the Earth's atmosphere.

23. As the electrons from photosystem II are transferred down their transfer chain, (a)_____is lost, but some is used to pump (b)_____creating a gradient used to generate (c)_____.

Cyclic electron transport produces ATP but no NADPH

24. As electrons pass from one acceptor to another in cyclic electron transport, they lose energy, some of which is used to pump protons across the (a)_____. This type of electron transport only uses (b)_____.

25. It is generally believed that ancient bacteria used _____ to produce ATP from light energy.

ATP synthesis occurs by chemiosmosis

26. Coupling ATP production to transfer of electrons energized by photons is called _____.

27. Protons accumulate in the (a)_____and flow back across the membrane through (b)_____.

28. The chemiosmotic model explains the coupling of ATP synthesis and _____.

THE CARBON FIXATION REACTIONS

29. The energy of ATP and NADPH, generated in the light-dependent reactions, is used to form organic molecules from carbon dioxide in _____.

Most plants use the Calvin cycle to fix carbon

30. This process occurs in the _____.

31. The Calvin cycle begins with the combination of one CO_2 molecule and one five-carbon sugar, (a)_____, to form a six-carbon molecule and is catalyzed by the enzyme (b)_____. The generated molecule instantly splits into two three-carbon molecules called (c)_____.

32. To produce one six-carbon carbohydrate, the light-independent reactions utilize six molecules of (a)_____, hydrogen obtained from (b)_____, and energy from (c)_____.

33. Most of the G3P generated during the Calvin cycle is used to regenerate _____.

Photorespiration reduces photosynthetic efficiency

34. Photorespiration occurs mainly during bright, hot days when plant stomata are closed. It reduces photosynthetic efficiency because CO_2 cannot enter the system, causing the enzyme (a)_____ to bind RuBP to (b)_____ instead of CO_2.

The initial carbon fixation step differs in C_4 plants and in CAM plants

35. The C_4 pathway efficiently fixes _____ at low concentrations.

36. C_4 plants initially fix CO_2 into the four-carbon molecule (a)_____ using the enzyme (b)_____.

CAM plants fix CO_2 at night

37. CAM plants open (a) _____at night and are well adapted to (b)_____ environments.

METABOLIC DIVERSITY

38. (a)_____ are organisms that cannot make their own food, unlike (b)_____ that are self nourishing.

PHOTOSYNTHESIS IN PLANTS AND IN THE ENVIRONMENT

39. Photoautotrophs are the ultimate source of almost all (a)_____ and they slow global warming by removing (b)_____from the atmosphere.

40. Molecular oxygen is constantly replenished by the _____ that releases the oxygen that all aerobic organisms require for respiration.

BUILDING WORDS

Use combinations of prefixes and suffixes to build words for the definitions that follow.

Prefixes	The Meaning		Suffixes	The Meaning
auto-	self, same		-lysis	breaking down, decomposition
hetero-	different, other		-plast	formed, molded, "body"
meso-	middle		-troph	nutrition, growth, "eat"
photo-	light			
chloro-	green			

Prefix	Suffix	Definition
_____	-phyll	1. Tissue in the middle of a leaf specialized for photosynthesis.
_____	-synthesis	2. The conversion of solar (light) energy into stored chemical energy by plants, blue-green algae, and certain bacteria.
_____	_____	3. The breakdown (splitting) of water under the influence of light energy trapped by chlorophyll.
_____	-phosphorylation	4. Phosphorylation that uses light as a source of energy.
_____	_____	5. A membranous organelle containing green photosynthetic pigments.
_____	-phyll	6. A green photosynthetic pigment.
_____	_____	7. An organism that fixes carbon, producing the organic compounds it needs.
_____	_____	8. An organism that is dependent upon the organic molecules produced by other organisms as the building blocks from which it synthesizes the carbon compounds it needs.
_____	-respirations	9. Degradation of Calvin cycle intermediates to CO_2 and water.

MATCHING

Terms:

a. C_3 pathway
b. C_4 pathway
c. Calvin cycle
d. CAM
e. Carotenoid

f. Chemoautotroph
g. Chemoheterotroph
h. Granum
i. Photoheterotroph
j. Photon

k. Photosystem
l. Stoma
m. Stroma
n. Thylakoid
o. G3P

For each of these definitions, select the correct matching term from the list above.

_____ 1. The fluid region of the chloroplast surrounding the thylakoids.

_____ 2. Interconnected system of flattened sac-like membranous structures inside the chloroplast where light energy is converted into chemical energy.

_____ 3. A stack of thylakoids within a chloroplast.

_____ 4. A yellow to orange plant pigment.

_____ 5. The usual pathway for fixing carbon dioxide in the synthesis reactions of photosynthesis.

_____ 6. A cyclic series of reactions occurring during the light-independent phase of photosynthesis.

_____ 7. A highly organized cluster of photosynthetic pigments and electron/hydrogen carriers embedded in the thylakoid membranes of chloroplasts.

_____ 8. An autotrophic organism that obtains energy from inorganic compounds.

_____ 9. A particle or packet of electromagnetic radiation.

_____10. A metabolic pathway that fixes carbon in desert plants.

_____11. Intermediate used to make carbohydrates.

_____12. Openings in plants for gas and water exchange.

MAKING COMPARISONS

Fill in the blanks.

Category	Photosynthesis	Respiration
Eukaryotic cellular site	Chloroplasts	Cytosol and mitochondria
Type of metabolic pathway (anabolic or catabolic)	#1	#2
#3	#4	Cristae of mitochondria
Source of hydrogen (electrons)	#5	#6
#7	NADP+	#8

MAKING CHOICES

Place your answer(s) in the space provided. Some questions may have more than one correct answer.

Use the following formula to answer questions 1-3:

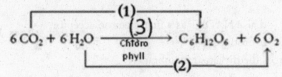

_____ 1. What kind of reaction is represented by number one (1)?

a. reduction   d. substitution

b. oxidation   e. replacement

c. neutral

_____ 2. What kind of reaction is represented by number two (2)?

a. reduction   d. substitution

b. oxidation   e. replacement

c. neutral

_____ 3. What missing component is represented by number three (3)?

a. carotene   d. palisade layer of a leaf

b. sunlight   e. chlorophyll

c. wavelengths in the range of 422-492 nm or 647-760 nm

_____ 4. Some of the requirements for the light-dependent reactions of photosynthesis include

a. water.   d. NADP.

b. photons of light.   e. oxygen.

c. carbohydrates.

_____ 5. Chloroplasts in eukaryotic cells

a. possess ATP synthase complexes.   d. contain chlorophyll.

b. have double membranes.   e. carry out photosynthesis.

c. do not exist.

_____ 6. The carbon fixation reactions of photosynthesis

a. generate PGA.   d. require ATP.

b. generate oxygen.   e. require NADPH.

c. take place in the stroma.

_____ 7. Thylakoids

a. comprise grana.   d. are located in the stroma.

b. are continuous with the plasma membrane.   e. carry out the carbon fixation reactions.

c. possess the basic fluid mosaic membrane structure.

_____ 8. Chlorophyll

a. contains magnesium in a porphyrin ring.   d. is the only pigment found in most plants.

b. dissolves in water.   e. is found mostly in the stroma.

c. is green because it absorbs green portions of the light spectrum.

_____ 9. The leaf tissue in which most of photosynthesis occurs is the

a. periderm.   d. stomata layer.

b. epidermis.   e. grana.

c. mesophyll.

_____10. C_4 fixation
    a. replaces C_3 fixation.   d. takes place in bundle sheath cells.
    b. produces PGA.   e. supplements C_3 fixation.
    c. produces oxaloacetate.

_____11. The reactions of photosystem II
    a. include splitting water with light photons.   d. use $NADP^+$ as a final electron acceptor.
    b. produce H_2O.   e. produce O_2.
    c. assist in producing an electrochemical gradient across the thylakoid membrane.

_____12. Most producers are
    a. chemosynthetic heterotrophs.   d. photosynthetic heterotrophs.
    b. chemosynthetic autotrophs.   e. photosynthetic consumers.
    c. photoautotrophs.

_____13. In plants that have chloroplasts, chlorophyll is found mainly in the
    a. thylakoid lumen.   d. stroma.
    b. thylakoid membranes.   e. intermembrane spaces.
    c. stomata.

_____14. In photosynthesis, electrons in atoms
    a. are excited.   d. are pushed to higher energy levels by photons of light.
    b. produce fluorescence.   e. release absorbed energy as another wavelength of light.
    c. are accepted by a reducing agent when they escape.

_____15. Reactions occurring during electron flow in respiration and photosynthesis are
    a. both endergonic.   d. endergonic and exergonic respectively.
    b. both exergonic.   e. neither endergonic nor exergonic.
    c. exergonic and endergonic respectively.

_____16. A/an_____spectrum measures wavelengths of light absorbed by a pigment.
    a. activity   d. fluorescence
    b. absorption   e. photosynthesis
    c. action

VISUAL FOUNDATIONS

Color the parts of the illustration below as indicated. Also label the entry site of CO_2 and the structure that transports water.

RED ☐ stoma

ORANGE ☐ palisade mesophyll

YELLOW ☐ spongy mesophyll

BLUE ☐ vein

GREEN ☐ chloroplasts

TAN ☐ upper epidermis

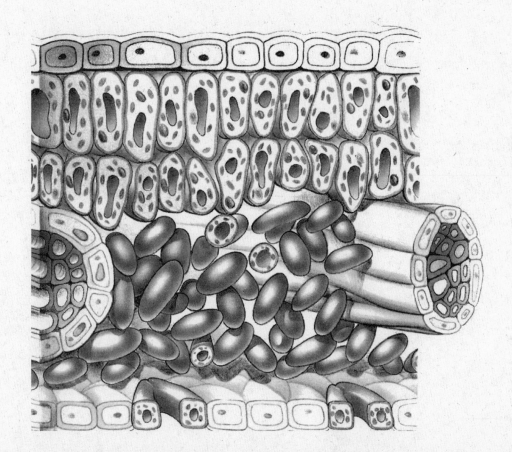

❏

CHAPTER 10

Chromosomes, Mitosis, and Meiosis

This chapter begins an examination of one of the major themes of biology – the transmission of information, specifically, the transmission of information from one generation of cells or organisms to the next. The DNA is organized into informational units called genes that control the activities of a cell. In prokaryotes, the information is contained in a single circle of DNA. In eukaryotes, it is carried in the chromosomes, made up of DNA and protein, contained within the cell nucleus. Each species is unique due to the information specified by its genes. Cell division is used for growth, repair, or reproduction. Genes are passed from parent cell to daughter cell by mitosis, a part of the cell cycle, ensuring that each new nucleus receives the same number and types of chromosomes as were present in the original nucleus. Prokaryotes use a different division process known as binary fission. The cell cycle is a highly regulated process. There are two basic types of reproduction — asexual and sexual. In asexual reproduction, a single parent cell usually splits, buds, or fragments into two or more individuals. In sexual reproduction, sex cells, or gametes, are produced by meiosis decreasing the number of chromosomes in the resulting cells by half and introducing genetic variability. When two gametes fuse, the resulting cell contains the same number of chromosomes as the parent cells. In living things, different types of sexual life cycles exist.

REVIEWING CONCEPTS

Fill in the blanks.

INTRODUCTION

1. Cell division allows for
   _____
   _____.

EUKARYOTIC CHROMOSOMES

2. DNA and associated proteins form a complex called
   _____, that makes up
   chromosomes.

DNA is organized into informational units called genes

3. Humans have about _____ genes that code for proteins.

DNA is packaged in a highly organized way in chromosomes

4. Prokaryotes usually contain one (a) _____DNA molecule, while eukaryotes have multiple (b) _____that must be (c) _____ to fit in the nucleus.

5. Positively charged histones associate with DNA forming structures called _____.

6. Nucleosomes interact with histone H1 to form the
   (a)_____. Then coiled loops form with the help of
   (c)_____proteins. (c)_____proteins help
   to form the fully condensed chromosome.

Chromosome number and informational content differ among species

7. (a)_____number is not what makes a species unique,
   instead it is the (b)_____they carry.

8. Most human body cells have exactly _____(#?)
   chromosomes.

THE CELL CYCLE AND MITOSIS

9. The period from the beginning of one cell division to the
   beginning of the next cell division is the
   _____.

Chromosomes duplicate during interphase

10. Cells spend most of their time in _____.

11. Interphase is divided into the G_1 phase, which stands for
    (a)_____ phase, the S phase, or
    (b)_____ phase, and the G_2 phase, or
    (c)_____ phase.

12. Mitosis is a (a)_____division while cytokinesis is a
    (b)_____ division.

During prophase, duplicated chromosomes become visible with the microscope

13. Sister chromatids contain (a)_____DNA sequences.
    Each chromatid includes a constricted region called the
    (b)_____.

14. Attached to each centromere is a _____, a
    multiprotein complex to which microtubules can bind.

15. Some of the microtubules radiating from each pole elongate
    toward the chromosome, forming the
    (a)_____, which separates the chromosomes
    during anaphase. These microtubules extend from a
    (b)_____, which in animal cells contains a
    pair of (c)_____.

Prometaphase begins when the nuclear envelope breaks down

16. In prometaphase, sister chromatids of each duplicated
    chromosome become attached to
    _____ at opposite poles of the cell.

Duplicated chromosomes line up on the midplane during metaphase

17. During metaphase, all the cell's chromosomes align at the cell's
    midplane, or _____.

18. (a)_____microtubules attach to the chromosomes
    while (b)_____ microtubules overlap with each other.

During anaphase, chromosomes move toward the poles

19. Once the (a)_____chromatids separate, each is again
    called a (b)_____.

20. Kinetochore microtubules shorten, or
    (a)_____, during anaphase, while polar
    microtubules (b)_____the poles apart.

During telophase, two separate nuclei form

21. Telophase is characterized by the
    _____ and a
    return to interphase-like condition.

Cytokinesis forms two separate daughter cells

22. In animal cells, actin filaments for a (a)_____
    that contracts to create the (b)_____.In plant
    cells, cytokinesis occurs by forming a (c)_____,
    a partition constructed in the equatorial region of the spindle and
    growing laterally toward the cell wall.

Mitosis produces two cells genetically identical to the parent cell

23. The regularity of the process of cell division ensures that each
    daughter nucleus receives exactly the same number and kinds of
    _____ the parent cell had.

Lacking nuclei, prokaryotes divide by binary fission

24. The asexual reproduction process in which one cell divides into
    two daughter cells is named _____.

REGULATION OF THE CELL CYCLE

25. _____can temporarily block certain key
    cell cycle events.

26. Protein kinases are enzymes that regulate by
    (a)_____ other proteins. The
    kinases that control the cell cycle, called (b)_____, are only
    active when bound to regulatory proteins called
    (c)_____.

27. (a)_____ are a group of plant hormones that
    promote mitosis both in normal growth and in wound healing. In
    animal cells, (b)_____can stimulate mitosis.

SEXUAL REPRODUCTION AND MEIOSIS

28. Offspring inherit traits that are virtually identical to those of their
    single parent when reproduction is (a)_____ and
    can be called (b)_____.

29. In (a)_____ reproduction, offspring receive genetic
    information from two parents. Haploid (n) gametes from the
    parents fuse to form a single (b)_____ (2n) cell called the
    (c)_____.

30. Homologous chromosomes are members of a pair of chromosomes that are similar in

_____.

Meiosis produces haploid cells with unique gene combinations

31. There are _____ major differences between meiosis and mitosis.

32. Meiosis typically consists of two nuclear and cytoplasmic divisions referred to as _____.

Prophase I includes synapsis and crossing-over

33. During prophase I, homologous chromosomes come to lie lengthwise side by side in a process known as

_____.

34. Homologous chromosomes exchange genetic material (crossing over) during the first meiotic (a)_____, providing more (b)_____ among gametes and offspring.

During meiosis I, homologous chromosomes separate

35. The haploid condition is established as the members of each pair of homologous chromosomes separate during the first meiotic

_____.

Chromotids separate in meiosis II

36. In metaphase I, chromatids are arranged in bundles of (a)_____, and in metaphase II, chromatids are in groups of (b)_____.

Mitosis and meiosis lead to contrasting outcomes

37. (a)_____ results in two daughter cells identical to the original cell. (b)_____ results in four genetically different, haploid daughter cells.

SEXUAL LIFE CYCLES

38. Sex cells (sperm and eggs or ova) are known as (a)_____, therefore, the formation of sex cells is referred to as (b)_____.

39. The formation of sperm is called _____.

40. The formation of eggs or ova is called

_____.

41. Plants and some algae and fungi have life cycles with (a)_____ of generations. A multicellular diploid stage is the (b)_____ generation while a multicellular haploid stage is the (c)_____ generation.

BUILDING WORDS

Use combinations of prefixes and suffixes to build words for the definitions that follow.

Prefixes	The Meaning		Suffixes	The Meaning
centro-	center		-gen(esis)	production of
chromo-	color		-mere	part
dipl-	double, in pairs		-phyte	plant
gameto-	sex cells, eggs and sperm		-some	body
hapl-	single			
inter-	between			
oo-	egg			
spermato-	seed, "sperm"			
sporo-	spore			
meta-	middle			
pro-	first			
ana-	opposite			

Prefix	Suffix	Definition
_____	_____	1. A dark staining body within the cell nucleus containing genetic information.
_____	-phase	2. The stage in the life cycle of a cell that occurs between successive cell divisions.
_____	-oid	3. An adjective pertaining to a single set of chromosomes.
_____	-oid	4. An adjective pertaining to a double set of chromosomes.
_____	_____	5. The constricted part or region of a chromosome (often near the center) to which a spindle fiber is attached.
_____	_____	6. The process by which gametes (sex cells) are produced.
_____	_____	7. The process whereby sperm are produced.
_____	_____	8. The process whereby eggs are produced.
_____	_____	9. The stage in the life cycle of a plant that produces gametes by mitosis.
_____	_____	10. The stage in the life cycle of a plant that produces spores by meiosis.
_____	-phase	11. The phase in mitosis during which the chromosomes line up along the equatorial plate.
_____	-phase	12. The phase in mitosis where sister chromatids separate.
_____	-phase	13. The phase in mitosis where chromosomes condense.

MATCHING

Terms:

a.   Anaphase
b.   Chromatin
c.   Crossing over
d.   Cytokinesis
e.   Interkinesis
f.   Kinetochore
g.   Meiosis
h.   Cyclin
i.   Mitosis
j.   Polyploid
k.   Prophase
l.   S phase
m.   Synapsis
n.   Telophase
o.   Cdk

For each of these definitions, select the correct matching term from the list above.

_____ 1. The last stage of mitosis and meiosis.

_____ 2. The phase in interphase during which DNA and other chromosomal components are synthesized.

_____ 3. Portion of the chromosome centromere to which the mitotic spindle fibers attach.

_____ 4. Process whereby genetic material is exchanged between homologous chromatids during meiosis.

_____ 5. An enzyme that regulates the cell cycle by phosphorylation.

_____ 6. DNA-protein fibers that condense to form chromosomes during prophase.

_____ 7. Process during which a diploid cell undergoes two successive nuclear divisions resulting in four haploid cells.

_____ 8. Having more than two sets of chromosomes per nucleus.

_____ 9. Division of the cell nucleus resulting in two daughter cells with the identical number of chromosomes as the parental cell.

_____10. Stage of cell division in which the cytoplasm divides into two daughter cells.

_____11. Regulatory proteins whose levels fluctuate during the cell cycle.

MAKING COMPARISONS

Fill in the blanks.

Event	Mitosis	Meiosis
Chromosome compaction (condensation)	Prophase	Prophase I, prophase II
Cytokinesis	#1	#2
Homologous chromosomes move to opposite poles	#3	#4
#5	#6	Prophase I
Cytokinesis occurs	#7	#8

Event	Mitosis	Meiosis
Chromatids separate	#9	#10
#11	Metaphase	#12
Duplication of DNA	#13	#14

MAKING CHOICES

Place your answer(s) in the space provided. Some questions may have more than one correct answer.

_____ 1. Typical somatic cells include
    a. skin cells.   d. ova.
    b. 2n cells.   e. cells resulting from oogenesis.
    c. sperm.

_____ 2. In mitosis, cells with 16 chromosomes produce daughter cells with
    a. 32 chromosomes.   d. 8 pairs of chromosomes.
    b. 8 chromosomes.   e. 4 pairs of chromosomes.
    c. 16 chromosomes.

_____ 3. In meiosis, cells with 16 chromosomes produce daughter cells with
    a. 32 chromosomes.   d. 8 pairs of chromosomes.
    b. 8 chromosomes.   e. 4 pairs of chromosomes.
    c. 16 chromosomes.

_____ 4. Gametogenesis typically involves (n=single chromosomes [haploid]; 2n=paired chromosomes [diploid])
    a. 2n to 2n.   d. mitosis.
    b. 2n to n.   e. meiosis.
    c. reduction division.

_____ 5. Eukaryotic chromosomes
    a. contain DNA and protein.   d. are uncoiled in interphase.
    b. possess asters.   e. align in the equator during prophase.
    c. are distinctly visible in a light microscope in interphase.

_____ 6. Eukaryotic DNA packaging involves
    a. nucleosomes.   d. 30nm fiber.
    b. histones.   e. polar microtubules.
    c. scaffolding proteins.

Use this list to answer questions 6-15 about meiosis:

a. interphase	d. anaphase I	g. metaphase II
b. prophase I	e. telophase I	h. anaphase II
c. metaphase I	f. prophase II	i. telophase II

_____ 7. Nuclear membranes break down.

_____ 8. Homologous chromosomes synapse.

_____ 9. Chromatids separate.

_____ 10. Members of tetrad separate.

_____11. Crossing over takes place.

_____12. Pairs of chromosomes align at equatorial plane.

_____13. Reduction of chromosome number from 2n to n.

_____14. Tetrads form.

_____15. Diploid to haploid.

_____16. Single chromosomes align at equatorial plane.

Use this list to answer questions 16-25 about mitosis:

a. interphase phase	e. telophase	h. G_2
b. prophase	f. T phase	i. S phase
c. metaphase phase	g. G_1 phase	j. M
d. anaphase		

_____17. The time between the synthesis phase and prophase.

_____18. Chromosomes begin to condense by coiling.

_____19. Condensed chromosomes uncoil.

_____20. DNA replicates.

_____21. Active synthesis and growth.

_____22. Chromosomes are lined up in a central plane.

_____23. All stages of mitosis collectively and cytokinesis.

_____24. Nuclear envelope breaks down.

_____25. The time between mitosis and start of the synthesis phase.

_____26. Chromatids divide.

VISUAL FOUNDATIONS

Color the parts of the illustration below as indicated. Also label the phase in which DNA replicates and the phase in which cells that are not dividing are arrested.

RED ☐ M phase

YELLOW ☐ G1

BLUE ☐ S

ORANGE ☐ G2

VIOLET ☐ Interphase

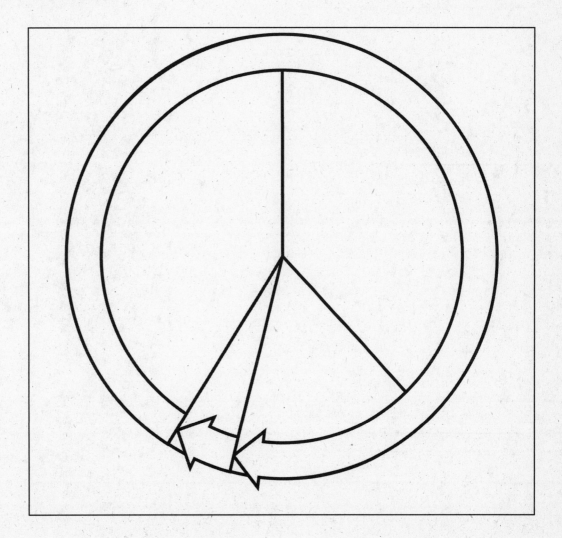

CHAPTER 11

The Basic Principles of Heredity

Now that you have a basic understanding of chromosomal behavior during sexual reproduction, we will examine the process of heredity, in which genetic information is transferred from parent to offspring. Genes that occupy corresponding loci on homologous chromosomes, called alleles, govern variations of the same characteristic. In diploid organisms, these genes exist in pairs. Scientists often study inheritance patterns by performing crosses. A monohybrid cross is a cross between two individuals carrying different alleles for a single gene locus. Similarly, a dihybrid cross is a cross between two individuals carrying different alleles at each of two gene loci. The results of monohybrid and dihybrid crosses illustrate the two basic principles of genetics known as segregation and independent assortment. The laws of probability are used to predict the results of a cross between two individuals. Special chromosomes called sex chromosomes determine the sex of most animal species and can affect inheritance of certain traits. The relationship between a single pair of alleles at a gene locus and the characteristic it controls may be simple, or it may be complex. Even the environment may influence the expression of genes.

REVIEWING CONCEPTS

Fill in the blanks.

INTRODUCTION

1. In the mid-19th century, the study of inheritance as a modern branch of science began with the work of _____.

MENDEL'S PRINCIPLES OF INHERITANCE

2. (a)_____describes the physical appearance of an organism, while (b)_____refers to its genetic makeup.

3. Offspring from the P generation cross are heterozygous; they are called the F_1 or (a)_____ generation. Offspring from the F_1 cross are called the (b)_____ generation.

4. A (a)_____ gene may mask the expression of a (b)_____ gene.

5. Mendel's hereditary factors, now called (a)_____, are a sequence of (b)_____ that contains the information to make an (c)_____.

Alleles separate before gametes are formed: the principle of segregation

6. During meiosis, members of paired genes at each allele _____ so that each gamete contains only one allele of each pair.

Alleles occupy corresponding loci on homologous chromosomes

7. The position of a gene on a chromosome is called its _____.

A monohybrid cross involves individuals with different alleles of a given locus

8. An individual is said to be (a)_____ for a particular feature when the two alleles it carries for that feature are different. They are said to be (b)_____ if the two alleles are the same.

9. A _____ predicts the ratios of genotypes and phenotypes of the offspring of a cross.

10. The phenotype of an individual does not always reveal its (a)_____. An individual expressing the dominant phenotype could be (b)_____.

11. A test cross is a cross between an individual of unknown genetic composition and a _____ individual.

A dihybrid cross involves individuals that have different alleles at two loci

12. A dihybrid cross is a cross between individuals that differ with respect to their alleles at _____.

Alleles on nonhomologous chromosomes are randomly distributed into gametes: the principle of independent assortment

13. The events of _____ are the basis for independent assortment.

Recognition of Mendel's work came during the early 20ᵗʰ century

14. The chromosome theory of inheritance can be explained by assuming that genes are _____ in specific locations along the chromosomes.

USING PROBABILITY TO PREDICT MENDELIAN INHERITANCE

15. Probability can range from (a)_____(impossible) to (b)_____(certain).

16. The _____ predicts the combined probabilities of independent events.

17. The probability of two independent events occurring together is the _____ of the probabilities of each occurring separately.

18. The _____ predicts the combined probabilities of mutually exclusive events.

The rules of probability can be applied to a variety of calculations

19. If events are truly independent, (a)_____ have no influence on the probability of the occurrence of (b)_____.

INHERITANCE AND CHROMOSOMES

Linked genes do not assort independently

20. Genes on the same chromosome are said to be _____ and do not assort independently.

21. Linked genes are recombined when chromatids exchange genetic material, a process known as _____ that occurs during meiotic prophase.

Calculating the frequency of crossing-over reveals the linear order of linked genes on a chromosome

22. A chromosome can be genetically mapped by determining the frequency of (a)_____ among genes, which can be used to generate a distance known as a (b)_____.

Sex is generally determined by sex chromosomes

23. The sex or gender of many animals is determined by the X and Y sex chromosomes. The other chromosomes in a given organism's genome are called _____.

24. The (a)_____gene causes the formation of the testes and (b)_____production.

25. When a Y-bearing sperm fertilizes an ovum, the result is a(n) (a)_____, and fertilization by an X-bearing sperm produces a(n) (b)_____.

26. X-linked recessive traits are more common in (a)_____ because they have only one (b)_____ chromosome.

27. The effect of X-linked genes is made equivalent in males and females by dose compensation, which is accomplished by a (a)_____ X-chromosome in the male or (b)_____ of one X-chromosome in the female.

28. A dense, metabolically inactive X chromosome at the edge of the nucleus in female mammalian cells is known as the _____.

EXTENSIONS OF MENDELIAN GENETICS

Dominance is not always complete

29. In genetic crosses involving (a)_____, the genotypic and phenotypic ratios are identical and the heterozygote has an (b)_____phenotype.

30. The AB blood type is an example of _____because two phenotypes are being expressed by the heterozygote.

Multiple alleles for a locus may exist in a population

31. Multiple alleles are (a)_____(#?) different alleles that can occupy the same (b)_____. However, in a single diploid individual there is a maximum of (c)____different alleles at a particular locus.

A single gene may affect multiple aspects of the phenotype

32. _____ refers to the many different effects that can often result from a given gene.

Alleles of different loci may interact to produce a phenotype

33. _____ is when one allele of a gene pair determines whether alleles of other gene pairs are expressed.

In polygenic inheritance, the offspring exhibit a continuous variation in phenotypes

34. It is called _____ when multiple independent pairs of genes have similar and additive effects on a phenotype.

Genes interact with the environment to shape phenotype

35. The range of phenotypic possibilities that can develop from a single genotype under different environmental conditions is known as the

_____.

BUILDING WORDS

Use combinations of prefixes and suffixes to build words for the definitions that follow.

Prefixes	The Meaning
di-	two, twice, double
hemi-	half
hetero-	different, other
homo-	same
mono-	alone, single, one
poly-	much, many
geno-	genetics
pheno-	to show

Prefix	Suffix	Definition
_____	-gene	1. Two or more pair of genes that affect the same trait in an additive fashion.
_____	-zygous	2. Having the same (identical) members of a gene pair.
_____	-zygous	3. Having dissimilar (different) members of a gene pair.
_____	-hybrid	4. Pertaining to the mating of individuals differing in two specific pairs of genes.
_____	-hybrid	5. Pertaining to the mating of individuals differing in one pair of genes.
_____	-zygous	6. Having only half (one) of a given pair of genes.
_____	-type	7. The genetic makeup of an individual.
_____	-type	8. Physical display of an individual's genes.

MATCHING

Terms:

a. Allele
b. Barr body
c. Dominant allele
d. Epistasis
e. Inbreeding
f. Codominance
g. Linkage
h. Locus
i. Overdominance
j. Pleiotropy
k. Recessive allele

For each of these definitions, select the correct matching term from the list above.

_____ 1. Condition in which certain alleles at one locus can alter the expression of alleles at a different locus.

_____ 2. An alternative form of a gene.

_____ 3. Condition in which a single gene produces two or more phenotypic effects.

_____ 4. A condensed and inactivated X-chromosome appearing as a distinctive dense spot in the nucleus of certain cells of female mammals.

_____ 5. The allele that is not expressed in the heterozygous state.

_____ 6. The place on a chromosome at which the gene for a given trait occurs.

_____ 7. Condition in which both alleles of a locus are expressed in a heterozygote.

_____ 8. The allele that is always expressed when it is present.

MAKING COMPARISONS

Fill in the blanks.   T = tall; t = short (complete dominance); Y = yellow; y = green (complete dominance)

	T Y	T y	t Y	t y
T Y	tall plant with yellow seeds (TT YY)	tall plant with yellow seeds (TT Yy)	tall plant with yellow seeds (Tt YY)	tall plant with yellow seeds (Tt Yy)
T y	#1	#2	#3	#4
#5	tall plant with yellow seeds (Tt YY)	#6	#7	#8
#9	#10	#11	short plant with yellow seeds (tt Yy)	#12

MAKING CHOICES

Place your answer(s) in the space provided. Some questions may have more than one correct answer.

R and r are genes for flower color. Homozygous dominant and heterozygous genotypes both have red flowers; the homozygous recessive genotype has white flowers. T and t are genes that control plant height. Homozygous dominant plants are tall, heterozygous plants are medium height, and homozygous recessive plants are short. Genes for flower color and height are on different chromosomes. Use these data and the following list to answer questions 1-9. Construct Punnett Squares as needed.

a. RRTT pink, medium	f. rrtt	k. 1:2:1:2:4:2:1:2:1	p.
b. RrTt pink, short	g. 1:1	l. red, tall	q.
c. Rrtt white, tall	h. 1:2:1	m. red, medium	r.
d. rrTT white, medium	i. 9:3:3:1	n. red, short	s.
e. rrTt white, short	j. 1:2:1:1:2:1	o. pink, tall	t.

Plants with the genotypes Rrtt and rrTT are mated. Their offspring (F$_1$) are then mated to produce another generation (F$_2$). Questions 1-5 pertain to the F$_2$ generation.

_____ 1. Which of the genotypes listed above are found among offspring?

_____ 2. What would the phenotypic ratio among offspring be if both flower color and height genes had exhibited incomplete dominance?

_____ 3. What are the genotypic and phenotypic ratios among offspring?

_____ 4. What are the genotypes found among the offspring that are not in the list?

_____ 5. What are all of the phenotypes among offspring?

Questions 6-9 pertain to the following cross: Rrtt x rrTT.

_____ 6. What is the phenotypic ratio among offspring?

_____ 7. What are the genotypes and phenotypes of the parents?

_____ 8. What is the genotypic ratio among offspring?

_____ 9. What are the genotypes and phenotypes of the offspring?

_____ 10. Which of the following is/are not consistent with Mendel's principle of dominance?

    a. All F$_1$ offspring express the dominant trait.   d. Both P generation parents are true breeding.

    b. All F$_2$ offspring express the dominant trait.   e. Only one P generation parent is true breeding.

    c. Both organisms in the P generation are homozygous.

_____ 11. Phenotypic expression

    a. always involves only one pair of genes.   d. refers specifically to the composition of genes.

    b. may involve many pairs of alleles.   e. refers to the appearance of a trait.

    c. is partly a function of environmental influences.

_____12. Pleiotropy means that a gene
   a. exhibits incomplete dominance.
   b. masks expression of other genes.
   c. has the same effect as another pair of genes.
   d. has multiple effects.
   e. is sex-influenced.

_____13. Which of the following would be false if hairy toes happened to be a recessive X-linked trait?
   a. All men would have hairy toes.
   b. No women would have hairy toes.
   c. Parents with hairy toes could have a child without hairy toes.
   d. More women than men would have hairy toes.
   e. More men than women would have hairy toes.

_____14. Linked genes are
   a. inseparable.
   b. generally on the same chromosome.
   c. in separate homologous chromosomes.
   d. in different chromatids.
   e. always separated during crossing over.

_____15. Given the following information about crossing over, determine the relative positions of three loci (X, Y, Z) on one chromosome: X and Y = 8%, Y and Z = 5%, X and Z = 3 map units.
   a. X, Y, Z
   b. X, Z, Y
   c. Z, Y, X
   d. Y, Z, X
   e. Z, Y, X

_____16. Mendel's "factors"
   a. are haploid in gametes.
   b. interact in gametes.
   c. segregate into separate gametes.
   d. are genes.
   e. affect flower color, but not seed color.

_____17. Epistasis means that alleles
   a. are sex-linked.
   b. have multiple effects.
   c. have the same effect as another pair of genes.
   d. mask the expression of another pair of genes.
   e. exhibit incomplete dominance.

_____18. Which of the following is/are false?
   a. XX is usually female.
   b. XY is usually male.
   c. birds and butterflies do not have sex chromosomes.
   d. hermaphrodites are XX or XY.
   e. XXY is usually male.

_____19. Which of the following applies if two pure-breeding P generation plants are used to ultimately produce an F_2 generation with flower colors in a ratio of 1 red: 2 pink: 1 white? (R is a gene for red; r = white)
   a. All F_1 plants were Rr.
   b. F_1 plants were red and white.
   c. At least one P generation parent was pink.
   d. F_2 genotypic ratio is 3:1.
   e. F_2 genotypic ratio is 1:2:1.

_____20. Polygenic inheritance means that a pair of genes
   a. exhibits incomplete dominance.
   b. affects expression of other genes.
   c. has a similar and additive effect as another independent pair of genes.
   d. has multiple effects.
   e. is sex-influenced.

VISUAL FOUNDATIONS

Use the cells in Row 1 as a reference. In Rows 2 and 3, draw in color the following structures: centrioles (yellow), spindle (black), properly assorted chromosomes in their appropriate colors (red and blue), and centrosomes (green).

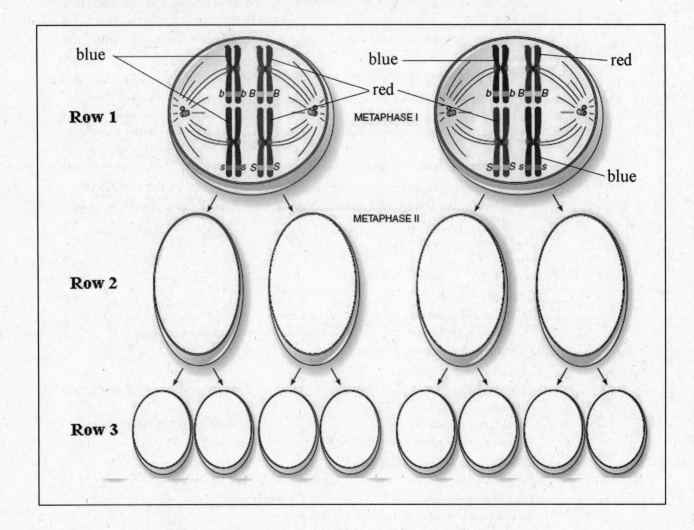

DNA: The Carrier of Genetic Information

In the previous chapter, you learned how genes are transmitted from generation to generation. This chapter addresses the chemical nature of genes and how they function. Genes are made of deoxyribonucleic acid (DNA). Each DNA molecule consists of two polynucleotide chains, held together by hydrogen bonds, arranged in a coiled double helix. The sequence of bases in DNA provides for the storage of genetic information and the uniqueness of individuals. Many proteins and enzymes are involved in the replication of DNA. When DNA replicates, the bonds between the two polynucleotide chains break, and the chains unwind and separate. Each half of the helix is then used as a template to synthesize a new strand of complementary nucleotides. The result is two DNA double helices, each identical to the original and consisting of one original strand from the parent molecule and one newly synthesized complementary strand, which is why the process is semiconservative. Since DNA is the hereditary material, repair mechanisms exist to try and correct any errors in the sequence. Replication in eukaryotic chromosomes is complicated by their linear nature.

REVIEWING CONCEPTS
Fill in the blanks.
INTRODUCTION
1. DNA is a nucleic acid which is the molecular basis of
   _____.

EVIDENCE OF DNA AS THE HEREDITARY MATERIAL
2. Scientists initially thought that (a)_____was the hereditary material because DNA was made of only (b)_____nucleotides.
DNA is the transforming principle in bacteria
3. A type of permanent genetic change in which the properties of one genetic type are conferred on another genetic type is known as
   _____.

DNA is the genetic material in certain viruses
4. _____ can reproduce by injecting only their DNA into cells, indicating that DNA is the genetic material.

THE STRUCTURE OF DNA
Nucleotides can be covalently linked in any order to form long polymers
5. Nucleotides consist of _____
   _____.
6. The bases in nucleotides include the two purines, (a)_____
   _____, and the two pyrimidines (b)_____
   _____.
7. Nucleotides are linked by a (a)_____bond called a
   (b)_____.

8. A polynucleotide chain is directional having a 5'end with a
   (a)_____ group and a 3' end with a (b)_____group.

DNA is made of two polynucleotide chains intertwined to form a double helix

9. _____ studies by Franklin and Wilkins
   showed that DNA has a helical structure with nucleotide bases stacked like
   rungs of a ladder.

10. Watson and Crick devised a DNA model based for the most part on existing
    data. Their model suggested that DNA was formed from two
    (a)_____
    arranged in a coiled double helix with the
    (b)_____forming the outside and the
    (c)_____associating as pairs along a central axis.

11. Since the two strands of the helix run in (a)_____directions, they are
    (b)_____to each other.

In double-stranded DNA, hydrogen bonds form between A and T and between G and C

12. (a)_____ hydrogen bonds form between adenine and thymine, and
    (b)_____ hydrogen bonds form between guanine and cytosine.

13. The sequence of nucleotides in one chain dictates the sequence in the other
    because of _____.

DNA REPLICATION

14. Replication of DNA is considered semiconservative because each "old"
    strand serves as a _____for the formation of a new strand.

Meselson and Stahl verified the mechanism of semiconservative replication

15. Using density gradient centrifugation, scientists can separate large molecules
    like DNA on the basis of differences in their _____.

Semiconservative replication explains the perpetuation of mutations

16. A change in the sequence of bases in a strand of DNA is known as a
    _____.

DNA replication requires protein "machinery"

17. Replication begins at (a)_____ where the DNA is
    unwound by a (b)_____. (c)_____stabilize the
    unwound strands to prevent the helix from reforming.

18. To prevent (a)_____, the enzyme
    (b)_____ generates breaks in the DNA.

19. _____ catalyze the linking together of nucleotide
    subunits.

20. Replication requires a (a)_____to occur, provides a 3' end for
    nucleotide addition, and it is generated by (b)_____.

21. DNA synthesis proceeds in a (a)____' → (b)____' direction. One strand is
    copied continuously, the other adds (c)_____ fragments
    discontinuously, which are linked together by (d)_____.

22. Replication is considered (a)_____because every origin of
    replication generates (b)____replication fork(s).

Enzymes proofread and repair errors in DNA

23. _____can proofread and correct errors in base pairing.

24. Defects in _____are linked to the development of a certain type of colon cancer.

25. The three enzymes involved in nucleotide excision repair are

_____.

Telomeres cap eukaryotic chromosome ends

26. Telomeres are (a)_____DNA sequences generated by (b)_____.

27. Programmed cell death is known as _____.

28. The shortening of telomeres may contribute to

_____.

BUILDING WORDS

Use combinations of prefixes and suffixes to build words for the definitions that follow.

Prefixes	The Meaning
a-	without, not, lacking
anti-	against, opposite of
semi-	half
tel(o)-	end

Prefix	Suffix	Definition
_____	-virulent	1. Not lethal; lacking in virulence.
_____	-parallel	2. The arrangement of the two polynucleotide chains in a DNA molecule, viz., "running" in opposite directions to each other.
_____	-conservative	3. The manner in which DNA replicates; half of the original DNA strand is conserved in each new double helix.
_____	-mere	4. The end of a eukaryotic chromosome.

MATCHING

Terms:

a. Chromatin
b. Deoxyribose
c. DNA ligase
d. DNA polymerase
e. Double helix
f. Histone
g. Mutation
h. Nucleotide
i. Purine
j. Pyrimidine
k. Origin of replication
l. Transformation

For each of these definitions, select the correct matching term from the list above.

_____ 1. The incorporation of genetic material by a cell that causes it to change its phenotype.

_____ 2. The enzyme in DNA replication responsible for base pairing.

_____ 3. The region of the DNA where the molecule "unzips" for replication to begin.

_____ 4. The bases adenine and guanine.

_____ 5. Random heritable changes in DNA.

_____ 6. A molecule composed of one or more phosphate groups, a 5-carbon sugar, and a nitrogenous base.

_____ 7. A pentose sugar lacking a hydroxyl group on carbon-2.

_____ 8. The shape of the DNA molecule.

_____ 9. Okazaki fragments would not be connected if this enzyme were missing.

_____ 10. This enzyme cannot work without an RNA primer.

MAKING COMPARISONS

Fill in the blanks.

Researcher(s)	Contribution to the Body of Knowledge About DNA
Watson and Crick	Developed a model that demonstrated how DNA can both carry information and serve as its own template for duplication
Frederick Griffith	#1
#2	Only the DNA of a bacteriophage is necessary for the reproduction of new viruses
Chargaff	#3
Franklin and Wilkins	#4
#5	Chemically identified the transforming principle described by Griffith as DNA

MAKING CHOICES

Place your answer(s) in the space provided. Some questions may have more than one correct answer.

_____ 1. Okazaki fragments are

a. joined by DNA ligase.

b. comprised of at least 3000 nucleotides.

phosphodiester linkage.

c. linked by 3' hydroxyl and 5' phosphate.

d. assembled on ribosomes.

e. joined to a growing DNA strand by a

_____ 2. The two new strands in replicating DNA "grow" as follows:

a. Leading continuously, lagging as fragments.   d. Both strands "grow" toward replication fork.

b. Leading and lagging continuously. .   e. Both strands "grow" away from replication fork.

c. Leading toward replication fork, lagging away from fork.

_____ 3. The molecules that "proofread" base pairing in replicated DNA, and correct any errors found are

a. 3',5' phosphatases.   d. DNA polymerases.

b. purine and pyrimidine polymerases.   e. DNA primases.

c. DNA ligases.

_____ 4. The scientists who demonstrated that DNA replication is semiconservative

a. Watson and Crick.   d. Avery, MacLeod, and McCarty.

b. Meselson and Stahl.   e. Franklin and Wilkins.

c. Hershey and Chase.

_____ 5. In double-stranded DNA, adenine is joined to thymine and cytosine is joined to guanine by

a. ionic bonds   d. hydrogen bonds.

b. phosphates and sugars.   e. purines and pyrimidins.

c. covalent bonds.

_____ 6. In most bacteria, such as *E. coli*, DNA is

a. a single strand.   d. absent.

b. circular.   e. associated with structural proteins.

c. a single double-stranded molecule.

_____ 7. Which of the following base pairs is/are incorrect?

a. A – T   d. T – A

b. C – A   e. T – C

c. G – C

_____ 8. The molecule(s) responsible for preventing supercoiling and preventing supercoiling & knot formation in replicating DNA is/are

a. histones.   d. topoisomerases.

b. chromatin.   e. protosomes.

c. scaffolding proteins.

_____ 9. A mutation is

a. a change in an enzyme.   d. an alteration of mRNA.

b. a change in a gene.   e. sometimes inheritable.

c. a change in DNA.

_____ 10. The 3' end of one DNA fragment is linked to the 5' end of another fragment by means of

a. DNA ligase.   d. hydrogen bonds.

b. helicase enzymes.   e. phosphodiester linkage.

c. a DNA polymerase.

_____11. The direction for synthesis of DNA is
     a. 5' → 3'.   d. 3' → 3'.
     b. 3' → 5'.   e. variable.
     c. 5' → 5'.

_____12. The genetic code is carried in the
     a. DNA backbone.   d. Okazaki fragments.
     b. sequence of bases.   e. histones.
     c. arrangement of 5', 3' phosphodiester bonds.

_____13. The double helix structure of DNA was suggested as a result of X-ray diffraction data collected by
     a. Hershey and Chase.   d. Watson and Crick.
     b. Griffith.   e. Franklin and Wilkens.
     c. Avery, MacLeod, and McCarty.

_____14. In one molecule of DNA one would expect the composition of the two strands to be
     a. both either old or new.   d. one old, one new.
     b. both all new.   e. unpredictable.
     c. both partly new fragments and partly old parental fragments.

_____15. Chargaff's rules state or infer that
     a. [A] = [T].   d. ratio of T to A = 1.
     b. [G] = [C].   e. ratio of G to C = 1.
     c. ratio of purines to pyrimidines = 1.

_____16. Deoxyribose and phosphate are joined in the DNA backbone by
     a. one of four bases.   d. phosphodiester linkages.
     b. purines.   e. the 1' carbon of the sugar.
     c. pyrimidines.

_____17. The investigators credited with elucidating the structure of DNA are
     a. Hershey and Chase.   d. Watson and Crick.
     b. Messelson and Stahl.   e. Franklin and Wilkens.
     c. Avery, MacLeod, and McCarty.

_____18. DNA polymerase requires
     a. a 3' hydroxyl group.   d. Okazaki fragments.
     b. RNA primer.   e. DNA repair.
     c. nucleotides.

VISUAL FOUNDATIONS

Color the parts of the illustration below as indicated.

RED   ☐ thymine

GREEN   ☐ adenine

YELLOW   ☐ cytosine

BLUE   ☐ guanine

ORANGE   ☐ sugar

VIOLET   ☐ phosphate

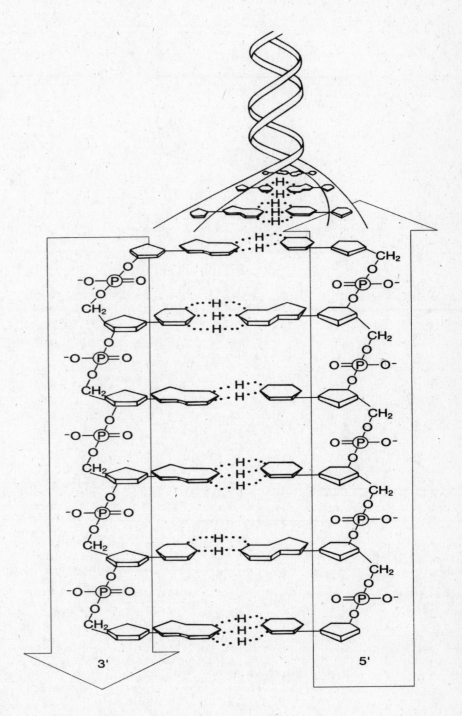

CHAPTER 13

Gene Expression

This chapter examines how cells convert the information in genes into proteins, thereby determining the phenotype of the organism. The manufacture of proteins by cells involves two major steps: the transcription of base sequences in DNA into the base sequences in RNA, and the translation of the base sequences in RNA into amino acid sequences in polypeptides. Of 64 possible codons within the genetic code, 61 code for amino acids and three serve as signals that specify the end of the coding sequence for a polypeptide chain. The genetic code is nearly universal, suggesting that it was established early in evolutionary history. Both transcription and translation involve three steps; initiation, elongation, and termination. Variations exist in gene expression between prokaryotes and eukaryotes. A gene is a DNA nucleotide sequence that carries the information to produce a specific RNA or polypeptide product. A mutation is a change in a DNA nucleotide sequence and many types and causes exist.

REVIEWING CONCEPTS

Fill in the blanks.

INTRODUCTION

1. Gene expression involves a series of steps in which the information in the sequence of _____ specifies the makeup of the cell's proteins.

DISCOVERY OF THE GENE-PROTEIN RELATIONSHIP

Beadle and Tatum proposed the one-gene, one-enzyme hypothesis

2. Neurospora is an ideal experimental organism because it grows primarily as a haploid organism, allowing researchers to immediately identify a

_____.

INFORMATION FLOW FROM DNA TO PROTEIN: AN OVERVIEW

3. RNA is different from DNA in three major ways_____

_____.

DNA is transcribed to form RNA

4. DNA is transcribed to form three main kinds of RNA. These are
_____.

5. Each tRNA is specific for only one _____.

RNA is translated to form a polypeptide

6. Each _____ in mRNA consists of three-bases that specify one amino acid in a polypeptide chain.

7. The (a)_____in a (b)_____is complementary to a codon in the (c)_____.

Biologists cracked the genetic code in the 1960s

8. The genetic code is read one triplet at a time from a fixed starting point that establishes the _____.

9. There are (a)_____(#?) possible codons. Of these, (b)_____ code for amino acids, and (c)_____(#?) function to terminate protein synthesis.

The genetic code is virtually universal

10. The genetic code being almost universal across species suggests _____.

The genetic code is redundant

11. _____ are the only amino acids that are specified by single codons.

12. The _____explains why the third base in a codon does not always matter.

TRANSCRIPTION

13. mRNA synthesis is catalyzed by DNA-dependent_____.

14. RNA is synthesized in the (a)_____direction while the DNA template is read in the (b)_____direction.

15. Any point closer to the 3' end of transcribed DNA (and therefore toward the 5' end of mRNA) is said to be (a)_____ relative to a given reference point. Conversely, areas toward the 5' end of DNA and the 3' end of mRNA are (b)_____.

The synthesis of mRNA includes initiation, elongation, and termination

16. Transcription is initiated at sites containing specific DNA sequences called the _____ regions.

17. During (a)_____RNA polymerase uses complementary base pairing to add nucleotides to the (b)_____of the RNA molecule.

18. When RNA polymerase recognizes a _____sequence in the DNA transcription ends.

Messenger RNA contains base sequences that do not directly code for protein

19. Well before the protein-coding sequences at the 5' end of mRNA, a segment called the (a)_____ contains recognition signals for ribosome binding and is followed by the (b)_____.

20. Stop codons are followed by noncoding 3' _____.

Eukaryotic mRNA is modified after transcription and before translation

21. The original transcript in eukaryotes is known as _____.

22. A cap is added to the (a)_____and a (b)_____to the (c)_____of the pre-mRNA.

23. RNA splicing is mediated by the (a)_____and removes noncoding sequences called (b)_____and links together the coding (c)_____.

TRANSLATION

24. The covalent bonds that join amino acids together are known as _____.

An amino acid is attached to tRNA before incorporation into a polypeptide

25. The specific enzymes that catalyze the formation of covalent bonds between amino acids and their respective tRNAs are the _____.

The components of the translational machinery come together at the ribosomes

26. The ribosome has (a)_____binding site for mRNA and (b)_____for tRNA. The (c)_____site contains the tRNA holding the growing polypeptide. The (d)_____site is where the tRNA enters the ribosome and the (e)_____site is where they leave.

Translation begins with the formation of an initiation complex

27. Initiation begins when (a)_____factors bind to the (b)_____ribosomal subunit.

28. The codon _____ initiates the formation of the protein-synthesizing complex.

29. The initiator tRNA coding for (a)_____then binds, followed by the (b)_____ribosomal subunit, which completes the intiation complex.

During elongation, amino acids are added to the growing polypeptide chain

30. Elongation requires (a)_____factors and energy in the form of (b)_____.

31. Peptide bond formation requires the enzyme _____.

32. The ribosome moves down the mRNA one codon by a process called (a)_____in the (b)_____direction.

One of the stop codons signals the termination of translation

33. Newly formed polypeptides are released and ribosome subunits dissociate when _____recognizes a stop codon.

34. The new polypeptide is assisted in folding by _____.

VARIATIONS IN GENE EXPRESSION

Transcription and translation are coupled in prokaryotes

35. A _____ consists of an mRNA molecule that is bound to clusters of ribosomes.

Biologists debate the evolution of eukaryotic gene structure

36. (a)_____ are regions of protein
     tertiary structure that may have specific functions and may have
     some relation to (b)_____.

Several kinds of eukaryotic RNA have a role in gene expression

37. (a)_____help to form the spliceosome while (b)_____help direct
     translation complexes to the rough ER.

38. Pre-rRNA are processed by _____.

39. RNA interference regulates (a)_____and involves
     both (b)_____.

**The definition of a gene has evolved as biologists have learned more about
genes**

40. A gene carries the information needed to produce
     _____.

The usual direction of information flow has exceptions

41. Retroviruses are viruses that require
     _____ which synthesizes DNA
     from RNA.

MUTATIONS

42. A mutation is a change in the
     _____ in DNA.

**Base substitution mutations result from the replacement of one base pair
by another**

43. (a)_____
     mutations involve a change in only one pair of nucleotides. These
     can lead to a (b)_____mutation with no effect, the
     substitution of one amino acid for another, a so called
     (c)_____ mutation or a (d)_____
     mutation involving the conversion of an amino acid-specifying
     codon to a termination codon.

Frameshift mutations result from the insertion or deletion of base pairs

44. In frameshift mutations, one of two nucleotide pairs are
     _____ from the molecule,
     altering the reading frame.

Some mutations involve mobile genetic elements

45. Mobile genetic elements can disrupt the functions of some genes
     because they are _____.

46. _____are able to increase an organism's
     ability to evolve.

Mutations have various causes

47. _____ are regions of DNA that are particularly
     susceptible to mutations.

48. Agents that cause mutations are called (a)_____
     and many are (b)_____which cause cancer.

BUILDING WORDS

Use combinations of prefixes and suffixes to build words for the definitions that follow.

Prefixes	The Meaning		Suffixes	The Meaning
anti-	against, opposite of		-gen	production of
poly-	much, many		-some	body

Prefix	Suffix	Definition
_____	-codon	1. A sequence of three nucleotides in tRNA that is complementary to ("opposite of"), and combines with, the three-nucleotide codon on mRNA.
ribo-	_____	2. An organelle (microbody) composed of RNA and proteins that functions in protein synthesis.
_____ _____		3. A submicroscopic complex consisting of many ribosomes attached to an mRNA molecule during translation.
muta-	_____	4. A substance capable of producing mutations.
carcino-	_____	5. A substance capable of producing cancer.

MATCHING

Terms:

a. Adenine
b. Codon
c. Elongation
d. Exon
e. Intron
f. mRNA
g. Promoter
h. Retrovirus
i. rRNA
j. Transcription
k. tRNA
l. Translocation
m. Uracil
n. Release factor
o. Initiation

For each of these definitions, select the correct matching term from the list above.

_____ 1. Site on DNA to which RNA polymerase attaches to begin transcription.

_____ 2. The making of mRNA from a DNA template.

_____ 3. An RNA virus that produces a DNA intermediate in its host cell.

_____ 4. A pyrimidine found in RNA.

_____ 5. A triplet of mRNA bases that specifies an amino acid or a signal to terminate the polypeptide.

_____ 6. The stage of translation during which amino acid chains are built.

_____ 7. The RNA responsible for base pairing during protein synthesis.

_____ 8. In eukaryotes, the coding region of DNA.

_____ 9. RNA that has been transcribed from DNA that specifies the amino acid sequence of a protein.

_____ 10. In eukaryotes, the region removed during splicing.

_____ 11. Recognizes stop codons.

_____ 12. The shifting of a ribosome on a mRNA.

MAKING COMPARISONS

Fill in the blanks.

mRNA Codon	DNA Base Sequence	tRNA Anticodon	Amino Acid Specified
5' — AAA — 3'	3'—TTT—5'	3'—UUU—5'	Lysine (Lys)
#1	#2	3'— UGG —5'	Threonine (Thr)
5' — GCG — 3'	#3	#4	Alanine (Ala)
5' — UGU — 3'	#5	3' — ACA — 5'	Cysteine (Cys)
#6	3' — ATT — 5', 3' — ATC — 5', 3' — ACT —5'	#7	None (stop codon)
5'—AUG—3'	#8	#9	Methionine (Met) (start codon)

MAKING CHOICES

Place your answer(s) in the space provided. Some questions may have more than one correct answer.

_____ 1. The following part of a spliceosome particle is involved in intron removal and regulation of transcription:

a. SRP RNA   d. snRNA

b. snoRNA   e. miRNA

c. siRNA

_____ 2. The _____ selectively suppresses gene expression.

a. SRP RNA   d. snRNA

b. snoRNA   e. miRNA

c. siRNA

_____ 3. Exons are

a. intervening sequences.   d. organized codes for specific amino acids.

b. special integrated coding sequences.   e. HIV resistant sequences.

c. noncoding sequences within a gene.

_____ 4. Energy for transfer of a peptide chain from the A site to the P site is provided by

    a. ATP.   d. ADP.

    b. enzymes.   e. guanosine triphosphate.

    c. GTP.

_____ 5. Transcription involves

    a. mRNA synthesis.   d. copying codes into codons.

    b. decoding of codons.   e. peptide bonding.

    c. copying DNA information.

_____ 6. Ribosomes attach to

    a. 5' end of mRNA.   d. tRNA.

    b. 3' end of mRNA.   e. RNA polymerase.

    c. an mRNA recognition sequence.

_____ 7. Translation involves

    a. hnRNA.   d. copying codes into codons.

    b. decoding of codons.   e. copying codons into codes.

    c. adapter molecules.

_____ 8. Which of the following is/are termination codons?

    a. UAG   d. GUA

    b. UUA   e. UGA

    c. UAA

_____ 9. A recognition sequence for ribosome binding to mRNA is/are

    a. aminoacyl-tRNA.   d. upstream from the coding sequence.

    b. coding sequence.   e. leader sequence.

    c. initiator.

_____ 10. An "adapter" molecule containing an anticodon and a region to which an amino acid is bonded is a/an

    a. aminoacyl-tRNA.   d. peptidyl transferase.

    b. coding sequence.   e. leader sequence.

    c. initiator.

_____ 11. Beadle and Tatum concluded that

    a. one gene affects one polypeptide.   d. mutations have no effect on enzymes.

    b. one gene affects one enzyme.   e. proteins mutate when exposed to radiations.

    c. a mutation directly affects an enzyme.

_____ 12. *Neurospora* was a good choice for the Beadle and Tatum studies because

    a. it is diploid.   d. it had no known mutant forms.

    b. its asexual spores enable genetic analysis.   e. the control strain could not use arginine.

    c. it can be cultivated on minimal medium.

_____ 13. Post-transcriptional modification and processing of mRNA takes place in

    a. cytoplasm.   d. eukaryotes only.

    b. the nucleus.   e. both prokaryotes and eukaryotes.

    c. prokaryotes only.

_____ 14. Elongation of protein synthesis involves

    a. the code AUG.   d. adding amino acids to a growing polypeptide chain.

    b. initiatior tRNA.   e. peptidyl transferase.

    c. loading initiatior tRNA on the large ribosomal subunit.

VISUAL FOUNDATIONS

Color the parts of the illustration below as indicated. Also label the 5' end of mRNA, the start codon, the leader sequence and circle the initiation complex.

RED ☐ formylated methionine tRNA
GREEN ☐ small ribosome subunit
YELLOW ☐ large ribosome subunit
PINK ☐ initiation factors
BLUE ☐ E site
ORANGE ☐ P site
VIOLET ☐ A site

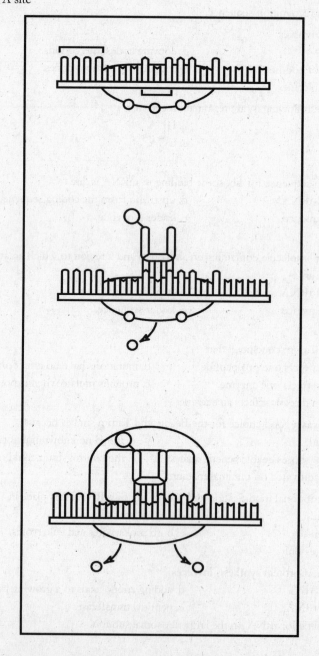

❑

CHAPTER 14

Gene Regulation

This chapter looks at the mechanisms that control gene expression, since all of the genes of a multicellular organism are not expressed all at the same time. Each type of cell in a multicellular organism has a characteristic shape, carries out very specific activities, and makes a distinct set of proteins because only part of the genetic information in a given cell is ever expressed at any given time. The control mechanisms that allow this require information in the form of signals originating either within the cell, other cells, or the environment and that interact with DNA, RNA, or protein. The mechanisms include controlling the amount of mRNA available, the rate of translation of mRNA, and the activity of the protein product. In prokaryotes, most gene regulation involves transcriptional level control of genes, commonly using operons, whose products are involved in resource utilization. In eukaryotes, gene regulation occurs at many levels, including chromatin structure, transcriptional, posttranscriptional, translational, and posttranslational.

REVIEWING CONCEPTS

Fill in the blanks.

INTRODUCTION

1. Mechanisms that regulate gene expression include:

_____
_____.

GENE REGULATION IN BACTERIA AND EUKARYOTES: AN OVERVIEW

2. _____ is the most efficient mechanism of gene regulation in bacteria.

3. Multicellular organisms focus on _____ of genes.

GENE REGULATION IN BACTERIA

4. _____ are genes that encode essential proteins that are in constant use.

Operons in bacteria facilitate the coordinated control of functionally related genes

5. A gene complex consisting of a group of structural genes with related functions and the DNA sequences responsible for controlling them is known as an _____. .

6. A group of functionally related genes may be controlled by one _____ that is located upstream from the coding sequences.

7. The _____ is a sequence of bases that switches mRNA synthesis "on" or "off."

8. An _____ inactivates a repressor in order to turn "on" a gene or operon.

9. Repressible operons are turned "off" when a repressor is activated by binding to a _____.

10. Negative control involves the binding of (a)_____while activators are important for (b)_____control.

11. Constitutive genes are transcribed at different rates based on the strength of their _____.

Some posttranscriptional regulation occurs in bacteria

12. Protein levels can be controlled by transcriptional level control. Other regulatory mechanisms that occur *after* transcription are referred to as _____.

13. _____ is when a product blocks its own production by binding to an enzyme that is required to generate that product.

GENE REGULATION IN EUKARYOTIC CELLS

14. Eukaryotic genes are typically not arranged in _____.

Eukaryotic transcription is controlled at many sites and by many different regulatory molecules

15. Highly coiled and compacted chromatin containing inactive genes is called (a)_____, whereas loosely coiled chromatin containing genes capable of transcription is referred to as (b)_____. Changing chromatin type is often done by modifying (c)_____.

16. DNA (a)_____is a way to (b)_____genes. In mammals it accounts for (c)_____, which is an example of (d)_____.

17. Genes whose products are essential for cell function can be arranged as multiple copies in (a)_____. Other genes are replicated in only certain cells by (b)_____.

18. RNA polymerase in multicellular eukaryotes binds to a portion of the promoter known as the _____.

19. (a)_____are DNA sequences that increase the rate of transcription, while (b)_____decrease it.

20. DNA binding proteins that regulate transcription are called (a)_____and are grouped together based on their type of DNA-binding (b)_____.

The mRNAs of eukaryotes have many types of posttranscriptional control

21. In eukaryotes, _____is a useful control point.

22. Through _____, the cells in each tissue produce their own version of mRNA corresponding to the particular gene.

23. The stability of _____can also affect how much protein is produced.

Posttranslational chemical modifications may alter the activity of eukaryotic proteins

24. The addition or removal of phosphate groups is an example of _____, a mechanism used by eukaryotic cells to regulate protein activity.

25. Enzymes that modify chemicals by adding phosphate groups are called _____.

26. Enzymes that modify chemicals by removing phosphate groups are called _____.

27. Protein degradation also controls activity by adding (a)_____ to proteins targeting them for destruction by the (b)_____.

BUILDING WORDS

Use combinations of prefixes and suffixes to build words for the definitions that follow.

Prefixes	The Meaning
co-	with, together, in association
eu-	good, well, "true"
hetero-	different, other
homo-	same

Prefix	Suffix	Definition
_____	-chromatin	1. Chromatin that appears loosely coiled. Because it is the only chromatin capable of transcription, it is considered to be the "good" or "true" chromatin.
_____	-chromatin	2. The inactive chromatin that appears highly coiled and compacted, and is generally not capable of transcription. (Chromatin that is "different from" or "other than" the euchromatin.)
_____	-repressor	3. A substance that, together with a repressor, represses protein synthesis in a specific gene.
_____	-dimer	4. A dimer in which the two component polypeptides are different.
_____	-dimer	5. A dimer in which the two component polypeptides are the same.

MATCHING

Terms:

a. Allosteric binding site
b. Constitutive gene
c. Enhancer
d. Gene amplification
e. Inducible system

f. Operator
g. Operon
h. Repressible system
i. Repressor
j. Transcriptional control

k. Translational control
l. Silencer

For each of these definitions, select the correct matching term from the list above.

_____ 1. One of the control regions of an operon.

_____ 2. A site located on an enzyme that enables a substance other than the normal substrate to bind to the molecule, and to change the shape of the molecule and the activity of the enzyme.

_____ 3. A regulatory protein that represses expression of a specific gene.

_____ 4. A regulatory mechanism that controls the rate at which a particular mRNA molecule is translated.

_____ 5. Process by which multiple copies of a gene are produced by selective replication, thus allowing for increased synthesis of the gene product.

_____ 6. Genes that are constantly transcribed.

_____ 7. In prokaryotes, a group of structural genes that are coordinately controlled and transcribed as a single message, plus their adjacent regulatory elements.

_____ 8. Positive regulatory elements that can be located long distances away from the actual coding regions of a gene.

_____ 9. Control in which the presence of a substrate induces the synthesis of an enzyme.

_____ 10. Negative regulatory elements that can be located long distances away from the actual coding regions of a gene.

MAKING COMPARISONS

Fill in the blanks.

Regulatory Event	Name
Protein dimers held together by leucine	Leucine zipper proteins
#1	Temporal regulation
Assist newly synthesized proteins to fold properly	#2
#3	Genomic imprinting
Degradation of a protein	#4
#5	Feedback inhibition
Help form an active transcription initiation complex thereby increasing the rate of RNA synthesis	#6
#7	Translational controls

MAKING CHOICES

Place your answer(s) in the space provided. Some questions may have more than one correct answer.

_____ 1. Genes that code for proteins that are constantly needed for survival of an organism are called
      a. promoters.   d. repressor genes.
      b. constitutive genes.   e. always "on".
      c. operons.

_____ 2. A repressor usually controls an inducible gene by
      a. keeping it "turned off".   d. inducing an antagonist.
      b. suppressing mRNA production.   e. blocking DNA replication at specific sites.
      c. reducing product resulting from an active inducer.

_____ 3. The basic way(s) that cells control their metabolic activities is/are by
      a. regulating enzyme activity.   d. controlling the number of enzyme molecules.
      b. mutations.   e. transduction.
      c. developing different genes for different purposes.

_____ 4. Feedback inhibition is an example of
      a. transcriptional control.   d. an inducible system.
      b. pretranscriptional control.   e. a repressible system.
      c. a control mechanism affecting events after translation.

_____ 5. The lactose repressor
      a. can convert to an operator.   d. becomes an activator when lactose
      is present.
      b. is several bases upstream from the operator.   e. is continuously "on".
      c. is downstream from RNA polymerase coding sequences.

_____ 6. The genes and genetic information in different cells of multicellular organisms are
      a. all slightly different.   d. identical.
      b. distinctly different.   e. almost all identical.
      c. identical in the same tissues, different in different tissues.

_____ 7. Replication of specific genes in cells that need more of them is an example of
      a. magnification.   d. a positive control system.
      b. amplification.   e. a regulon.
      c. induction.

_____ 8. Enhancers and silencers are similar because both
      a. increase transcription rate.   d. are located far from the promoter.
      b. are DNA sequences   e. decrease transcription rate
      c. are downstream from promoter.

_____ 9. The tryptophan operon in *E. coli* is
      a. usually on.   d. an inducible system.
      b. on in the absence of corepressor.   e. a repressible system.
      c. repressed only when tryptophan binds to repressor.

_____ 10. The proteasome
      a. degrades proteins.   d. recognizes ubiquitin.
      b. degrades mRNA.   e. us a posttranslational control.
      c. seem to determine the strength of a repressor.

_____11. The lactose operon in *E. coli*

    a. contains three structural genes.   d. is transcribed as part of a single RNA molecule.

    b. is turned on by a repressor.   e. is an inducible system.

    c. is a repressor.

_____12. "Zinc fingers" seem to be involved in

    a. unwinding DNA.   d. binding a regulatory protein to DNA.

    b. activating transcription.   e. blocking posttranscriptional events.

    c. insertion of a regulator domain into grooves of DNA.

_____13. A tightly coiled portion of DNA that contains active genes is

    a. called heterochromatin.   d. found only in prokaryotes.

    b. called euchromatin.   e. found only in eukaryotes.

    c. found in most eukaryotes and few prokaryotes.

VISUAL FOUNDATIONS

Color the parts of the illustration below as indicated. Also label the lactose operon, ribosomes and identify the enzymes produced in the presence of the inducer.

RED   ❑ operator region

GREEN   ❑ promoter region

YELLOW   ❑ repressor gene

BLUE   ❑ mRNA

ORANGE   ❑ inducer

BROWN   ❑ inactivated repressor protein

VIOLET   ❑ active repressor protein

TAN   ❑ structural gene

PINK   ❑ RNA polymerase

CHAPTER 15

DNA Technology and Genomics

Recombinant DNA technology involves splicing together DNA from different organisms. DNA cloning specifically allows a particular DNA sequence to be amplified to provide millions of identical copies that can be isolated in pure form, and then its functions can be studied and possibly engineered for a particular application. DNA can also be amplified in vitro using the polymerase chain reaction. DNA can be analyzed in many different ways allowing us to better understand the complex structure and control of eukaryotic genes and the roles of genes in development. The newer area of genomics also contributes to this understanding. These various techniques have led to genetic engineering — the modification of the DNA of an organism to produce new genes with new characteristics. Increasingly, DNA technology has important applications in many fields of study and business, including evolution, medicine, pharmacology, and food and forensic sciences. DNA technology has also raised important safety concerns that scientists are continually assessing.

REVIEWING CONCEPTS

Fill in the blanks.

INTRODUCTION

1. The molecular modification of an organism's DNA to produce new genes with new traits is called _____.

2. _____ is the commercial or industrial use of cells or organisms.

DNA CLONING

3. _____ cut DNA molecules at specific places, producing manageable segments that can be incorporated into vector molecules.

4. Recombinant DNA vectors are usually constructed from _____.

Restriction enzymes are "molecular scissors"

5. A _____ sequence reads the same as its complement when both are read in the 5' to 3' direction.

6. Restriction enzymes cut in a staggered fashion creating (a)_____ which can be joined to complementary sequences by (b)_____.

Recombinant DNA forms when DNA is spliced into a vector

7. Three important features of a vector
   are_____
   _____.

8. _____ are
   combination vectors with features from both bacteriophages and
   plasmids.

DNA can be cloned inside cells

9. A _____ is a collection of
   thousands of DNA fragments that represent the entire DNA in the
   genome.

10. _____ are radioactive strands of DNA
    that are complementary to specific targeted cloned fragments.
    They are used to identify a particular fragment in a large population
    of clones.

A cDNA library is complementary to mRNA and does not contain introns

11. The enzyme _____ is used to synthesize
    single-stranded complementary DNA (cDNA) on a mRNA
    template.

The polymerase chain reaction is a technique for amplifying DNA in vitro

12. The polymerase chain reaction (PCR) technique for amplifying
    DNA is advantageous because it side steps the cumbersome, time-
    consuming process of cloning DNA. PCR is an *in vitro* process
    that alternately uses (a)_____
    to replicate DNA with the use of (b)_____ to
    dissociate the replicated DNA strands for further replications.

13. The PCR technique must use a heat stable DNA polymerase
    named _____ polymerase after the organism from which it is
    obtained.

DNA ANALYSIS

Gel electrophoresis is used for separating macromolecules

14. In gel electrophoresis, nucleic acids migrate through the gel toward
    the (a)_____ of the electric field
    because they are negatively charged and will be separated based on
    (b)_____.

DNA, RNA, and protein blots detect specific fragments

15. The method of detecting DNA fragments by separating them by
    gel electrophoresis and then transferring them to a nitrocellulose or
    nylon membrane is called a(a)
    _____. When RNA is transferred
    it is a (b)_____and for a protein transfer a
    (c)_____.

Restriction fragment length polymorphisms are a measure of genetic relationships

16. In molecular biology, a _____ is the presence of inherited differences within a population.

Methods to rapidly sequence DNA have been developed

17. Automated DNA sequencing relies on the
   (a)_____ and has allowed the (b)_____ of about 800 organisms to be sequenced.

GENOMICS

18. (a)_____genomics focuses on mapping and sequencing, while (b)_____ genomics deals with the function of DNA sequencing.
   (c)_____
   genomics is used to study evolution and (d)_____studies microbial communities.

Identifying protein-coding genes is useful for research and for medical applications

19. cDNA sequences that can identify protein-coding agents are known as _____.

One way to study gene function is to silence genes one at a time

20. Short RNA molecules can be used to determine the function of a gene in _____.

21. Gene function is revealed in a procedure called _____, in which the researcher chooses and "knocks out" a single gene in an organism.

22. In _____, the gene is not completely disabled but changed enough to alter properties of the protein the gene encodes.

DNA microarrays are a powerful tool for studying how genes interact

23. _____ enable researchers to compare the activities of thousands of genes in normal and diseased cells from tissue samples.

The Human Genome Project stimulated studies of genome sequencing for other species

24. Comparing the human genome with others can identify the
(a)_____of a gene and
even the (b)_____of humans.

Several scientific fields-bioinformatics, pharmacogenetics, and proteomics-have emerged

25. (a)_____focuses on comparing nucleotide and amino acid sequence. Studying how genetic variation affects the action of drugs in individuals is (b)_____. (c)_____is the study of protein expression.

APPLICATIONS OF DNA TECHNOLOGIES

DNA technology has revolutionized medicine

26. _____, the use of specific DNA to treat a genetic disease by correcting the genetic problem, is another application of DNA technology.

27. Genetic engineering allows the production of (a)_____to treat diseases and disorders and (b)_____to help prevent infectious diseases.

DNA fingerprinting has numerous applications

28. STRs are highly _____ because they vary in length from one individual to another, making them useful in identifying individuals.

Transgenic organisms have foreign DNA incorporated into their cells

29. Transgenic organisms (plants or animals) are generally produced by injecting the DNA of a gene into a recipient's
(a)_____, but (b)_____may also be used as vectors.

30. Transgenic animals secreting foreign proteins in their milk is called_____.

31. Transgenic plants that are resistant to insect pests, viral diseases, drought, heat, cold, herbicides, and salty or acidic soil are known as
_____.

DNA TECHNOLOGY HAS RAISED SAFETY CONCERNS

32. The science of _____ uses statistical methods to quantify risks so they can be compared and contrasted.

BUILDING WORDS

Use combinations of prefixes and suffixes to build words for the definitions that follow.

Prefixes	The Meaning		Suffixes	The Meaning
bacterio-	bacteria		-phage	eat, devour
retro-	backward			
trans-	across			
kilo-	thousand			
bio-	life			

Prefix	Suffix	Definition
_____	-genic	1. Pertains to organisms that have had their genome altered by recombinant DNA technology. (Genes are taken from one organism and transferred "across" into another organism.)
_____	-virus	2. An RNA virus that makes DNA copies of itself by reverse transcription.
_____ _____		3. A virus that infects and destroys bacteria.
_____	-base	4. A unit of measurement for DNA.
_____	-technology	5. The use of cells or organisms commercially.

MATCHING

Terms:

a. Colony
b. DNA ligase
c. Genetic probe
d. Bioinformatics
e. Plasmid
f. PCR
g. Recombinant DNA
h. Restriction enzyme
i. Reverse transcriptase
j. Vector
k. Pharmacogentics
l. Metagenomics

For each of these definitions, select the correct matching term from the list above.

_____ 1. Enzyme produced by retroviruses to enable the transcription of DNA from the viral RNA in the host cell.

_____ 2. Small, circular DNA molecules that carry genes separate from the main bacterial chromosome.

_____ 3. A cluster of genetically-identical cells.

_____ 4. Bacterial enzymes used to cut DNA molecules at specific base sequences.

_____ 5. Method used to produce large amounts of DNA from tiny amounts.

_____ 6. An agent, such as a plasmid or virus, that transfers genetic information.

_____ 7. Any DNA molecule made by combining genes from different organisms.

_____ 8. A radioactively-labeled segment of RNA or single-stranded DNA that is complementary to a
          target gene.

_____ 9. Study of microbial community gene expression.

_____ 10. An enzyme used to put a DNA fragment into a BAC.

_____ 11. Comparing nucleotide sequences.

_____ 12. Studying an individual's ability to metabolize a drug.

MAKING COMPARISONS

Fill in the blanks.

Technique	Definition/Description
genetic engineering	Modifying the DNA of an organism to produce new genes with new traits
#1	Microbial proteins used to cut DNA in specific locations
probe	#2
#3	Method for replicating and isolating many copies of a specific recombinant DNA molecule.
PCR	#4
#5	Recombinant DNA vectors derived from bacterial DNA
genomic DNA library	#6
#7	The study of all proteins encoded by the human genome

MAKING CHOICES

Place your answer(s) in the space provided. Some questions may have more than
one correct answer.

_____ 1. Genes in eukaryotic organisms that do not code for proteins are
   a. introns.   d. highly polymorphic.
   b. small PCRs.   e. frequently used to diagnose genetic diseases.
   c. short tandem repeats (STRs).

_____ 2. Copies of DNA made from isolated mRNA by using reverse transcriptase are called
   a. genomic DNA.   d. complementary DNA (cDNA).
   b. recombinant DNA.   e. probes.
   c. PCRs.

_____ 3. Analysis of an individual's DNA based on his or her STRs is known as
   a. proteomics.   d. cDNA library.
   b. DNA blotting analysis.   e. genomic DNA library.
   c. DNA fingerprinting.

_____ 4. Detection of protein fragments transferred to a nitrocellulose or nylon membrane after separating them by gel electrophoresis is known as a

   a. Northern blot.   d. Western blot.

   b. Southern blot.   e. fingerprinting blot.

   c. Eastern blot.

_____ 5. The vector(s) commonly used to incorporate a DNA fragment into a carrier is/are

   a. prophages.   d. translocation microphages.

   b. *E. coli.*   e. bacteriophages.

   c. plasmids.

_____ 6. "Sticky ends" of a DNA molecules are generated by

   a. plasmozymes.   d. DNA ligase.

   b. restriction enzymes.   e. vector enzymes.

   c. DNA polymerase.

_____ 7. Restriction enzymes are normally used by bacteria to

   a. defend against viruses.   d. cut plasmids from the large chromosome.

   b. remove extraneous introns.   e. insert operators in active DNA.

   c. denature antibiotics.

_____ 8. Palindromic sequences are base sequences from one DNA strand that

   a. produce RNA polymerase.   d. cut specific fragments from DNA.

   b. form RNA complementary to DNA.   e. read in the opposite direction as its complement.

   c. can be cut by restriction enzymes.

_____ 9. cDNA libraries are compiled

   a. from introns.   d. from introns and exons.

   b. DNA copies of mRNA.   e. with the use of reverse transcriptase.

   c. from multiple copies of DNA fragments.

_____10. A base sequence that is palindromic to 3'-AGCTTAA-5' would read

   a. 3'-AGCTTAA-5'.   d. 5'-UCGAAUU-3'.

   b. 5'-TCGAATT-3'.   e. 3'-UCGAAUU-5'.

   c. 3'-TCGAATT-5'.

_____11. The Ti used to insert genes into plant cells is

   a. an RNA fragment.   d. a plasmid.

   b. not effective in dicots.   e. a vector.

   c. useful to increase yields of grain foods.

_____12. Restriction fragment length polymorphisms are

   a. complementary.   d. possibly due to mutations.

   b. RNA variants.   e. variants paired in order to obtain different clones.

   c. the result of changes in DNA.

_____13. DNA fragments for recombination are produced by cutting DNA

   a. and introducing a cleaved plasmid.   d. with bacteriophages.

   b. with restriction enzymes.   e. with DNA ligase.

   c. at specific base sequences.

_____14. Transgenic organisms are

   a. dead animals.   d. animals with foreign genes.

   b. vectors for retroviruses.   e. plants with foreign genes.

   c. used to produce recombinant proteins.

_____15. A radioactive strand of DNA that is used to locate a cloned DNA fragment is
     a. a probe.
     b. the genome.
     c. complementary to the targeted gene.
     d. a restriction enzyme.
     e. complementary to the ligase used.

_____16. RNA viruses that make DNA copies of themselves by reverse transcription are
     a. plasmids.
     b. retroviruses.
     c. double stranded.
     d. used as genetic probes.
     e. bacteriophages.

VISUAL FOUNDATIONS

Color the parts of the illustration below as indicated. Also label DNA, pre-mRNA, mature mRNA, transcription, RNA processing, and double-stranded cDNA.

RED	☐	exon
GREEN	☐	intron
YELLOW	☐	reverse transcriptase
BLUE	☐	cDNA copy of mRNA
VIOLET	☐	mRNA
PINK	☐	DNA polymerase

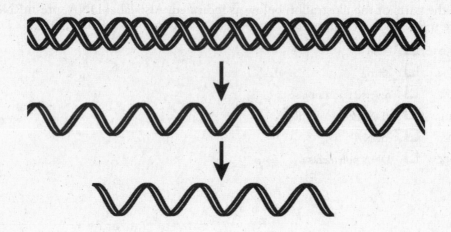

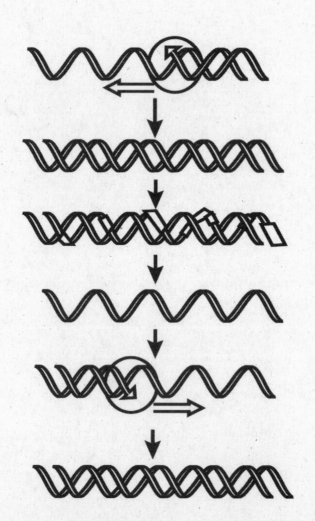

CHAPTER 16

Human Genetics and the Human Genome

The principles of genetics apply to all organisms, including humans. Human geneticists, however, cannot make specific crosses of pure human strains as geneticists do with other organisms. Instead, they must rely on studies of other organisms and population studies of large extended families. Also, knowledge of human genetics has been greatly advanced in recent years by the mapping and sequencing of the human genome and by the study of genetic diseases in humans, which can arise from abnormalities in chromosome number or structure or by single gene mutations. Gene therapy is a promising new treatment for genetic disease, but there are concerns and technical issues with the approach. New tools have been developed for genetic testing, screening, counseling, and therapy. Increasingly, work in the field of human genetics raises many ethical issues.

REVIEWING CONCEPTS

Fill in the blanks.

INTRODUCTION

1. In DNA sequencing, researchers identify the order of _____ to understand the genetic basis of human similarities and differences

STUDYING HUMAN GENETICS

Human chromosomes are studied by karyotyping

2. A _____ is an individual's chromosome composition.

3. (a)_____is another methods to analyze a karyotype using
   (b)_____to "paint" the chromosomes.

Family pedigrees help identify certain inherited conditions

4. A _____ shows the transmission of genetic traits within a family over several generations and helps determine the exact interrelationships of the DNA molecules from related individuals.

The Human Genome Project sequenced the DNA on all human chromosomes

5. The Human Genome Project sequenced _____(#?) base pairs.

6. Most of the human genome consists of
   _____
   _____.

7. _____compare genomes of a healthy individual with that of a diseased individual to look for gene variation.

Comparative genomics has revealed DNA identical in both mouse and human genomes

8. About _____ (#?) DNA segments longer than 200 base pairs are known to be 100% identical between the mouse and human genomes.

Researchers use mouse models to study human genetic diseases

9. Researchers have used _____ to produce strains of mice that are homozygous or heterozygous to genetic diseases.

ABNORMALITIES IN CHROMOSOME NUMBER AND STRUCTURE

10. _____, the presence of multiple sets of chromosomes, is usually lethal in humans, but quite common in plants.

11. The abnormal presence or absence of one chromosome in a set is called (a)_____. For the affected chromosome, the normal condition, called disomy, becomes a (b)_____ condition with the addition of an extra chromosome, and a (c)_____ condition when one member of a pair is missing. These abnormalities generally occur due to (d)_____.

Down syndrome is usually caused by trisomy 21

12. Down syndrome incidence increases with _____but no explanation for this has yet been supported.

Most sex chromosome aneuploidies are less severe than autosomal aneuploidies

13. Persons with _____ are tall, sterile males, except for an extra X chromosome.

14. Persons with _____ are sterile females without Barr bodies.

15. The _____produces tall, fertile males.

Abnormalities in chromosome structure cause certain disorders

16. _____ is the attachment of a fragment of one chromosome to a nonhomologous chromosome.

17. A _____ is loss of part of a chromosome.

18. _____ are weak points where part of a chromatid appears to be attached to the rest of the chromosome by a thin thread of DNA.

Genomic imprinting is determined by whether the inheritance is from the male or female parent

19. _____is believed to cause genomic imprinting.

GENETIC DISEASES CAUSED BY SINGLE-GENE MUTATIONS

20. PKU and akaptonuria are examples of disorders involving enzyme defects, collectively known as _____.

Most genetic diseases are inherited as autosomal recessive traits

21. Phenylketonuria results from an _____ deficiency.

22. Sickle cell anemia results from a _____ defect.

23. Cystic fibrosis results from defective _____.

24. In cystic fibrosis, a mutant protein causes the production of a very heavy _____ that eventually causes severe tissue damage.

25. Tay-Sachs disease results from abnormal (a)_____ metabolism in the (b)_____.

Some genetic diseases are inherited as autosomal dominant traits

26. Huntington's disease is caused by a rare

_____ that affects the central nervous
system.

Some genetic diseases are inherited as X-linked recessive traits

27. Hemophilia A is caused by the absence of a blood-clotting protein called

_____.

28. The X chromosome carries a large number of genes that affect (a)_____
_____, so (b)_____suffer from mental impairment
more frequently.

GENE THERAPY

29. Gene therapy is a strategy that aims to replace a (a)_____
allele with a (b)_____ allele to treat some serious genetic
diseases.

Gene therapy programs are carefully scrutinized

30. The main safety concerns with gene therapy is the toxicity of the _____.

GENETIC TESTING AND COUNSELING

Prenatal diagnosis detects chromosome abnormalities and gene defects

31. _____ may be used to diagnose prenatal genetic
diseases. It analyzes cells in amniotic fluid withdrawn from the uterus of a
pregnant woman.

32. One technique designed to detect prenatal genetic defects is
_____, which involves
inspecting fetal cells involved in forming the placenta.

33. _____is available to test for a
variety of inherited genetic conditions in individuals using IVF.

Genetic screening searches for genotypes or karyotypes

34. Newborn genetic screening is used primarily as the first step in
_____.

35. In adults, screening is used to find _____of recessive genetic
disorders.

Genetic counselors educate people about genetic diseases

36. Genetic counselors use family histories of each partner to determine the
_____ that any given offspring will inherit a particular
condition.

HUMAN GENETICS, SOCIETY, AND ETHICS

37. (a)_____individuals and (b)_____ethnic groups carry (c)_____
mutant alleles.

38. Genetic disease rates are low because the chances of reproducing with some carrying
the same harmful allele are very low, unless it is a _____,
which increases the risk.

Genetic discrimination provokes heated debate

39. Genetic discrimination is discrimination against an individual or family members because of differences from the _____ genome in that individual.

40. The GINA prohibits employers and insurance companies from
(a)_____based on information from
(b)_____.

Many ethical issues related to human genetics must be addressed

41. As human genetics becomes more important issues of _____ _____must be addressed.

_____

BUILDING WORDS

Use combinations of prefixes and suffixes to build words for the definitions that follow.

Prefixes	The Meaning
cyto-	cell
iso-	equal, "same"
poly-	much, many
trans-	across, beyond, through
tri-	three
aneu-	without

Prefix	Suffix	Definition
_____	-genetics	1. The branch of biology that uses the methods of cytology to study genetics; the study of chromosomes and their role in inheritance.
_____	-ploidy	2. The presence of multiples of complete chromosome sets.
_____	-somy	3. Condition in which a chromosome is present in triplicate instead of duplicate (i.e., the normal pair).
_____	-location	4. An abnormality in which a part of a chromosome breaks off and attaches to another chromosome.
_____	-ploidy	5. The absence or presence of an extra single chromosome.

MATCHING

Terms:

a. Amniocentesis
b. PIGD
c. Consanguineous mating
d. Disomic
e. Down Syndrome
f. Huntington's Disease
g. Karyotype
h. Klinefelter Syndrome
i. Monosomic
j. Nondisjunction
k. Nuclear sexing
l. Gene therapy

For each of these definitions, select the correct matching term from the list above.

_____ 1. A technique used to detect birth defects by sampling the fluid surrounding the fetus.

_____ 2. Abnormal separation of homologous chromosomes or sister chromatids caused by their failure to disjoin properly during cell division.

_____ 3. Genetic disease that causes mental and physical deterioration, uncontrollable muscle spasms, personality changes, and ultimately death.

_____ 4. Syndrome in which afflicted individuals have only 47 chromosomes.

_____ 5. Techniques that involve introducing normal copies of a gene into some of the cells of the body of a person afflicted with a genetic disorder.

_____ 6. Birth defect caused by an extra copy of human chromosome 21.

_____ 7. General term for the abnormal situation in which one member of a pair of chromosomes is missing.

_____ 8. Matings of close relatives.

_____ 9. Screening done to an embryo before pregnancy has actually occurred.

_____10. The number and kinds of chromosomes present in the nucleus.

MAKING COMPARISONS

Fill in the blanks.

Genetic Disease	Abnormality	Method of Transmission	Clinical Description
Cystic fibrosis	Membrane proteins that transport chloride ion malfunction	Autosomal recessive trait	Characterized by abnormal secretions in respiratory and digestive systems
Sickle cell anemia	#1	Autosomal recessive trait	#2
#3	Abnormal sex chromosome expression	#4	Small testes, sterile, abnormal breast development
Phenylketonuria (PKU)	#5	Autosomal recessive trait	Retardation caused by toxic phenylketones that damage the developing nervous system
#6	Lack of blood clotting Factor VIII	#7	Severe bleeding from even slight wounds
Tay-Sachs	#8	Autosomal recessive trait	#9
#10	A mutated nucleotide triplet (CAG) that is repeated many times	#11	Severe mental and physical deterioration, uncontrollable muscle spasms, personality changes, and ultimately death
Down syndrome	#12	Autosomal trisomy 21	#13

MAKING CHOICES

Place your answer(s) in the space provided. Some questions may have more than one correct answer.

_____ 1. The inactive X-chromosome seen as a region of darkly stained chromatin is called a

    a. translocated anomalie.   d. chromatic body.

    b. Barr body.   e. Klinefelter chromosome.

    c. artifact.

_____ 2. An individual with Turner syndrome would likely have a sex chromosome composition of

    a. XXX.   d. YO.

    b. XXY.   e. XYY.

    c. XO.

_____ 3. An individual with an XXY karyotype is

    a. male.   d. sterile.

    b. female   e. likely to be tall.

    c. fertile.

_____ 4. IVF is a popular abbreviation for

a. intravenous filtration.

b. *in vitro* fertilization.

c. a branch of stem cell research.

d. a technique for fertilization in laboratory glassware.

e. insufficient placental development leading to miscarriage.

_____ 5. A person with XYY karyotype will

a. have an enlarged head.

b. be female.

c. have 47 chromosomes.

d. have no Barr bodies.

e. have one Barr body per cell.

_____ 6. Amniocentesis involves

a. examination of maternal blood.

b. insertion of a needle into the uterus.

c. examination of fetal cells.

d. examination of karyotypes.

e. appraisal of paternal abnormalities.

_____ 7. Translocations may involve

a. loss of genes.

b. acquisition of extra genes.

c. duplication of genes.

d. loss of a chromosome.

e. acquisition of an extra chromosome.

_____ 8. A trisomic individual

a. is missing one of a pair of chromosomes.

b. may result from nondisjunction.

c. has at least three X chromosomes.

d. is a male.

e. has an extra chromosome.

_____ 9. A person with Down syndrome will likely

a. have a father with Down syndrome.

b. have trisomy 21.

c. be mentally retarded.

d. be born of a mother in her teens.

e. have 47 chromosomes.

_____ 10. The most accurate statement about the frequency of abnormal genes is that they are found in

a. certain ethnic groups.

b. certain cultural groups.

c. college professors.

d. college students.

e. everyone.

_____ 11. An ideal organism for genetic studies would be one that

a. is polygenic.

b. is heterozygous.

c. has a short generation time.

d. produces one offspring per mating.

e. can be controlled.

_____ 12. The study of human genetics relies mainly on

a. pedigrees.

b. Punnett squares.

c. fetal karyotypes.

d. distribution of a trait in a population.

e. nondisjunction.

VISUAL FOUNDATIONS

Questions 1-3 show pedigrees of the type that are used to determine patterns of inheritance.
Indicate carriers by placing a dot in the appropriate squares ([•]) and circles (⊙).
Use the following list to answer questions 1-3:

        a. x-linked recessive
        b. x-linked dominant
        c. autosomal dominant
        d. autosomal recessive
        e. cannot determine the pattern of inheritance

1. What is the pattern of inheritance for the following pedigree?

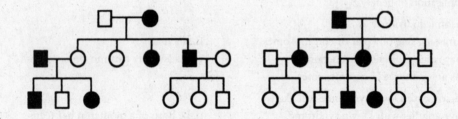

2. What is the pattern of inheritance for the following pedigree?

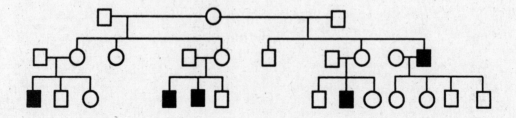

3. What is the pattern of inheritance for the following pedigree?

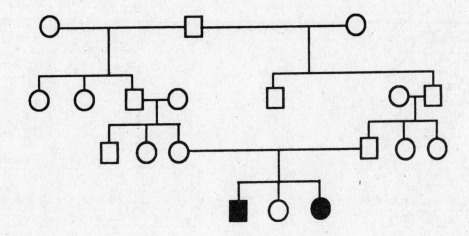

CHAPTER 17

Developmental Genetics

Development encompasses all the changes that occur in the life of an individual organism. Of particular interest is the process by which cells specialize and organize into a complex organism. Groups of cells become gradually committed to specific patterns of gene activity and progressively organized into recognizable structures. Cellular specialization is not due to the loss of genes during development, but rather to differential gene activity. Stem cells are important in this differentiation process. A variety of organisms and methodologies are used today to identify genes that control development and to determine how those genes work. Many genes important in development are quite similar in a wide range of organisms, including humans. Cancer involves altered gene expression of specific genes which leads to uncontrolled cell division.

REVIEWING CONCEPTS

Fill in the blanks.

INTRODUCTION

1. _____is the study of the genes involved in cell differentiation and development of an organism.

2. The technique to localize proteins in which a fluorescent dye is joined to an antibody that binds to a specific protein is called _____.

CELL DIFFERENTIATION AND NUCLEAR EQUIVALENCE

3. Development involves cell specialization, a developmental process involving a gradual commitment by each cell to a specific pattern of gene activity. This process is called (a)_____, and the final step in the process is referred to as (b)_____.

4. The distinctive organizational pattern, or form, that characterizes an organism is acquired through a process known as (a)_____. It proceeds through a series of steps known as (b)_____, by which groups of cells organize into identifiable structures.

5. A body of data supports the assertion that, in at least some organisms, almost all nuclei of differentiated cells are identical to each other and to the nucleus of the single cell from which they are descended, which has given rise to the concept of _____.

Most cell differences are due to differential gene expression

6. Much of the regulation that is important in development occurs at the _____ level.

A totipotent nucleus contains all the instructions for development

7. _____ cells are fully differentiated, yet, since they can still develop, they apparently still contain all of the original genetic equipment of their common single ancestral cell.

8. Genetically identical organisms, cells, or DNA molecules are called _____.

The first cloned mammal was a sheep

9. The scientists who cloned Dolly were ultimately successful because they recognized their fused cells need synchronous _____.

10. The main focus of cloning research is the production of _____, in which foreign genes have been incorporated.

Stem cells divide and give rise to differentiated cells

11. Embryonic and adult stem cells are known as _____ stem cells because they can give rise to many, but not all, of the types of cells in an organism.

12. Induced pluripotent stem cells were produced by adding _____.

13. (a)_____cloning produces a newborn human while (b)_____cloning only produces human ES or iPS cells.

THE GENETIC CONTROL OF DEVELOPMENT

A variety of model organisms provide insights into basic biological processes

14. A model organism is a species chosen for biological studies because it has

   _____

   _____

   _____.

15. A powerful research tool is to isolate _____of model organisms.

Many examples of genes that control development have been identified in *Drosophila*

16. The *Drosophila* life cycle includes (a)_____, _____, _____, stages and the (b)_____ stage generated by (c)_____.

17. The initial stages of development in *Drosophila* are under the control of _____ genes.

18. Once maternal genes have initiated pattern formation in the embryonic fruit fly, additional genes begin to act on development. Among these are the (a)_____ that begin to organize body segments and the (b)_____ that specify the developmental plan for each segment.

19. The (a)_____ is a short DNA sequence that is consistently found in association with known developmental genes and codes for a (b)_____.

Caenorhabditis elegans has a relatively rigid developmental pattern

20. These roundworms are easy to cross and study because individuals are males or _____.

21. The fates of cells in adult roundworms are predetermined by a few _____ that form in the early embryo.

22. Development in *Caenorhabditis elegans* is rigidly fixed in a predetermined pattern; that is, specific cells throughout the early embryo are already irrevocably "programmed" to develop in a particular way. This pattern of development is said to be _____.

23. Sometimes, cell differentiation is influenced by neighboring cells in a process called _____.

24. (a)_____, genetically programmed cell death, depends on proteolytic enzymes called (b)_____.

The mouse is a model for mammalian development

25. Early mice embryos can be fused to produce a _____, a term for offspring that exhibit a diversity of characteristics derived from two or more kinds of genetically dissimilar cells from different zygotes.

26. A self-regulating embryo exhibits a form of development referred to as _____. Such an embryo can still develop normally when it has extra cells or missing cells.

27. The _____receptor seems to play a role in mammalian aging, based on the knockout mouse.

Arabidopsis is a model for studying plant development, including transcription factors

28. In addition to the ABC genes, another class of genes designated _____ interacts with the B and C genes to specify the development of petals, stamens, and carpels.

CANCER AND CELL DEVELOPMENT

29. A mass of cancer cells is called a (a) _____ and when those cells spread it is referred to as (b)_____.

30. Cancer-causing genes are known as _____.

31. A gene that interacts with growth-inhibiting factors to block cell division is a _____ gene.

BUILDING WORDS

Use combinations of prefixes and suffixes to build words for the definitions that follow.

Prefixes	The Meaning		Suffixes	The Meaning
morpho-	form		-gen(esis)	production of
onco-	mass, tumor			
toti-	all			
pluri-	many			
neo-	new			

Prefix	Suffix	Definition
_____	_____	1. The development and differentiation (production) of the form and structures of the body.
_____	-gene	2. A cancer-causing gene.
_____	-potent	3. The ability of some nuclei from differentiated plant and animal cells to provide information for the development of an entire organism; such cells contain all of the genetic material that would be present in the nucleus of a zygote.
_____	-potent	4. The ability to give rise to many but not all cells in an organism
_____	-plasm	5. A mass of tissue

MATCHING

Terms:

a. Determination
b. Differentiation
c. Homeobox
d. Homeotic gene
e. Induction
f. Oncogene
g. Somatic cell
h. Stem cell
i. Totipotent
j. Transgenic
k. Tumor
l. Zygotic gene
m. Tumor suppressor gene

For each of these definitions, select the correct matching term from the list above.

_____ 1. An organism that has incorporated foreign DNA into its genome.

_____ 2. Mass of tissue that grows in an uncontrolled manner.

_____ 3. Development toward a more mature state; a process changing a young, relatively unspecialized cell to a more specialized cell.

_____ 4. The ability of a cell (or nucleus) to provide the information necessary for the development of an entire organism.

_____ 5. A phenomenon in which differentiation of a cell is influenced by interactions with particular neighboring cells.

_____ 6. A gene that controls the formation of specific structures during development.

_____ 7. Specific DNA sequence found in many genes that are involved in controlling the development of the body plan.

_____ 8. The progressive limitation of a cell line's potential fate during development.

_____ 9. Any of a number of genes that play an essential role in blocking cell division.

_____10. A body cell not involved in reproduction.

MAKING COMPARISONS

Fill in the blanks.

Germ Layer	Organ, Organ System or Tissue	Specialized Cell
Mesoderm	Skeletal muscles	Striated muscle cell
#1	#2	Skin cell
#3	Nervous system	#4
None	#5	Egg and sperm (gametes)
Endoderm	#6	Tracheal cell
#7	Urinary bladder	Epithelial cell
#8	Blood	Red blood cell

MAKING CHOICES

Place your answer(s) in the space provided. Some questions may have more than one correct answer.

_____ 1. Gap genes in *Drosophila*

a. determine segment polarity.   d. are complementary to maternal genes.

b. act on all embryonic segments.   e. are a type of segmentation gene.

c. are the first segmentation genes to act.

_____ 2. The concept of nuclear equivalence holds that

a. adult cells are identical to one another.   d. adult cells are identical to egg cells.

b. embryonic cells are identical to one another.   e. all adult somatic nuclei are identical.

c. adult somatic nuclei are identical to the nucleus of the fertilized egg.

_____ 3. If destruction of a single cell early in development results in the absence of a structure in the adult, the embryo of that organism

    a. is mosaic.

    b. is a fruit fly.

    c. contains predetermined cells.

    d. contained founder cells.

    e. is controlled to some extent by homeotic genes.

_____ 4. Embryos that act as a self-regulating whole and can accommodate missing parts

    a. are transgenic.

    b. are mosaics.

    c. have regulative development patterns.

    d. have founder cells that give rise to adult parts.

    e. contain many genomic rearrangements.

_____ 5. Morphogens are

    a. chemical agents.

    b. derived from a homeobox.

    c. determined by embryonic segmentation genes.

    d. thought to exist in _Drosophila_ eggs.

    e. unraveled portions of active DNA.

_____ 6. A group of cells grown in liquid medium from a single root cell is

    a. formed by a predetermined pattern.

    b. a mosaic.

    c. totipotent.

    d. predetermined.

    e. an embryoid body.

_____ 7. An embryo in which the fate of cells is determined early in development is said to be

    a. induced.

    b. mosaic.

    c. homeotic.

    d. transgenic.

    e. differentiated.

_____ 8. If differentiated cells can be induced to act like embryonic cells, the cells are

    a. formed by a predetermined pattern.

    b. mosaics.

    c. totipotent.

    d. predetermined.

    e. embryoids.

_____ 9. Pattern formation is a series of steps leading to

    a. determination.

    b. differentiation.

    c. predetermination.

    d. morphogenesis.

    e. the final step in cell specialization.

_____10. Differences in the molecular composition of different cells in a multicellular organism are due to

    a. loss of genes during development.

    b. transformation.

    c. regulation of the activities of different genes.

    d. nuclear equivalency.

    e. differential gene activity.

_____11. The process by which developing cells become committed to a specialization is

    a. determination.

    b. differentiation.

    c. transgenesis.

    d. morphogenesis.

    e. the first step in cell specialization.

_____12. The process by which the differentiation of an embryonic cell is influenced by neighboring embryonic cells is

    a. transformation.

    b. induction.

    c. mosaic development.

    d. an example of programmed cell death.

    e. determinate morphogenesis.

____13. Apoptosis
    a. involves caspases.
    b. occurs only in humans
    c. helps form the human hand.
    d. plays a role in development.
    e. is best studied in the mouse.

____14. Human therapeutic cloning involves
    a. creating a newborn human.
    b. supplying replacement tissues.
    c. duplicating iPS cells.
    d. world-wide opposition.
    e. duplicating ES cells.

❑

CHAPTER 18

Introduction to Darwinian Evolution

The concept of evolution is based upon past and current evidence that organisms living today evolved from earlier organisms, and it links all fields of the life sciences into a unified body of knowledge. Evolution is firmly based on the genetic principles you learned in the previous section of the book. Fieldwork conducted by Charles Darwin was instrumental in establishing that natural selection is the actual mechanism of evolution. This mechanism works because each species produces more offspring than will survive to maturity, genetic variation exists among the offspring, organisms compete for resources, and individuals with the most favorable characteristics are most likely to survive and reproduce. As a result, favorable characteristics are passed to succeeding generations. Over time, changes occur in the gene pools of geographically separated populations that may lead to the evolution of new species. A subsequent modification of Darwin's theory, the synthetic theory of evolution, explains Darwin's observation of variation among offspring in terms of mutation and recombination. An enormous body of scientific observations and experiments supports the concept of evolution. The testing of evolutionary hypotheses has demonstrated that not only is evolution a reality, but it occurs daily and hourly all around us.

REVIEWING CONCEPTS

Fill in the blanks.

INTRODUCTION

1. New species evolve from earlier species in a process called_____.
2. The concept of evolution is the cornerstone of biology, because it
   _____.

WHAT IS EVOLUTION?

3. The genetic changes that bring about evolution do not occur in individual organisms, but rather in _____.
4. A group of organisms that are capable of interbreeding are classified as a
   _____.
5. The use of bacteria to digest oil from an oil tanker spill is an example of
   _____.

PRE-DARWINIAN IDEAS ABOUT EVOLUTION

6. Aristotle was one of the first individuals to group and arrange organisms. To do this he used what he called a _____.

7. _____ was probably the first to recognize that fossils are the remains of extinct organisms.

8. _____ was the first to propose that organisms change over time as a result of some natural phenomenon.

9. _____ proposed the theory that traits *acquired* during an organism's lifetime could be passed along to their offspring.

DARWIN AND EVOLUTION

10. Charles Darwin served as the naturalist on the ship
    (a)_____ . Darwin's theory of evolution used data he collected on the similarities and differences among organisms he observed in the
    (b)_____, a chain of islands west of mainland Ecuador.

11. When horse breeders mate female and male horses to produce the strongest and fastest racing horse they are using _____ to produce these traits.

12. _____ made the important mathematical observation that populations increase in size geometrically until checked by factors in the environment.

13. An evolutionary modification that improves the chances of survival of oak trees in North
    America would be considered to be an _____.

14. Both Darwin and _____ arrived at the conclusion that evolution occurred by natural selection.

15. In 1859, Darwin published his monumental book, _____.

Darwin proposed that evolution occurs by natural selection

16. Darwin's mechanism of natural selection consists of four premises based on observations about the natural world. In summary, these are:
    (a) _____
    (b) _____
    (c) _____
    (d) _____.

The modern synthesis combines Darwin's theory with genetics

17. The modern synthesis theory incorporates the ideas and observations of
    (a) _____ and
    (b) _____.

18. The synthetic theory of evolution explains variation in terms of _____, which are inheritable and therefore potential explanations for *how* traits are passed from one generation to another.

Biologists study the effect of chance on evolution

19. Although it cannot be proven, it appears that

(a) _____ is a more important agent of evolutionary change than (b) _____.

EVIDENCE FOR EVOLUTION

The fossil record provides strong evidence for evolution

20. The fossil record demonstrates that life _____.

21. The formation of fossils is most likely to occur under conditions that favor rapid (a) _____ of the dead body and preventing body (b) _____.

22. Of the many places that organisms live, the body of those organisms living in _____ are least likely to become fossilized.

23. The earliest fossils of *H. sapiens* with anatomically modern features appeared in the fossil record approximately _____ years ago.

24. _____ are the remains of organisms that existed over a relatively short geological time period and that died and were preserved as fossils in large numbers. They are used to identify specific sedimentary layers and to arrange strata in chronological order.

The distribution of plants and animals supports evolution

25. The study of the past and present geographic distribution of plants and animals is called _____.

26. In 1915 Alfred Wegener proposed that a single land mass called Pangaea had broken apart in a process known as _____.

Comparative anatomy of related species demonstrates similarities in their structures

27. _____ features are those derived from the same structure in a common ancestor.

28. There are times when organisms living in similar environments but not in the same geographical region evolve similar adaptations. The independent evolution of similar structures is referred to as _____.

29. Non-homologous features that have similar function but evolved independently are called _____.

30. The occasional presence of "remnant organs" in organisms is to be expected as ancestral species evolve and adapt to different modes of life. Such organs or parts of organs, called _____, are degenerate and have no apparent function.

Molecular comparisons among organisms provide evidence for evolution

31. Codons code for a particular _____ in a polypeptide chain.

32. Determining the order of nucleotide bases in DNA is known as _____.

Developmental biology helps unravel evolutionary patterns

33. The evolution of new features depends on modification in
_____ that already exist.

Evolutionary hypotheses are tested experimentally

34. Rapid evolution is actually occurring on a time scale of _____ rather
than centuries.

BUILDING WORDS

Use combinations of prefixes and suffixes to build words for the definitions that follow.

Prefixes	The Meaning
bio-	life
homo-	same

Prefix	Suffix	Definition
_____	-logous	1. Pertains to having a similarity of form due to having the same evolutionary origin.
_____	-geography	2. The study of the geographical distribution of living things.
_____	-plastic	3. Pertains to features with the same or similar functions in distantly related organisms.

MATCHING

Terms:

a. Adaptation
b. Artificial selection
c. Continental drift
d. Convergent evolution
e. Evolution
f. Fossil
g. Gene pool
h. Half-life
i. Homoplastic
j. Index fossil
k. Natural selection
l. Range
m. Radioisotopes
n. Synthetic theory of evolution
o. Vestigial

For each of these definitions, select the correct matching term from the list above.

_____ 1. Used by seed companies to produce plants that are especially resistant to drought.

_____ 2. The independent evolution of structural or functional similarity in two or more organisms of widely-different, unrelated ancestry.

_____ 3. Parts or traces of an ancient organism preserved in rock.

_____ 4. Carbon-14 and uranium-235 present in different rock formations.

_____ 5. Is based upon a combination of Darwin's theory and Mendelian genetics.

_____ 6. The remains of a once functional pelvic bone.

_____ 7. The remains of an animal that can be found only in rock that is 1 million years old.

_____ 8. The genetic change in a population of organisms that occurs over time.

_____ 9. Similar in function or appearance, but not in origin or development.

_____10. The F2 generation is more reproductively fit than the F1 generation and the F1 generation is more fit than the parental generation.

_____11. Inherited variations that are favorable for survival are preserved in a population while unfavorable ones are eliminated.

MAKING COMPARISONS

Fill in the blanks.

Evidence for Evolution	How the Evidence Supports Evolution
Fossils	Direct evidence in the form of remains of ancient organisms
#1	Indicate evolutionary ties between organisms that have basic structural similarities, even though the structures may be used in different ways
#2	Indicate that organisms with different ancestries can adapt in similar ways to similar environmental demands
Vestigial organs	#3
Biogeography of plants and animals	#4
All organisms use a genetic code that is virtually identical	#5
#6	The best-adapted individuals produce the most offspring, whereas individual that are less well adapted die prematurely or produce fewer or inferior offspring
#7	Whales descended from land dwelling ancestors
#8	Some mammals found in Australia have developed similar features to mammals found in North America
DNA sequencing	#9

MAKING CHOICES

Place your answer(s) in the space provided. Some questions may have more than one correct answer.

_____ 1. Which of the following statements is/are consistent with Darwin's theory of evolution?

    a. Organisms evolve by a process of natural selection.   d. Better adapted individuals are more likely to survive.

    b. Evolution progresses toward a more perfect state.   e. New species result from a need to survive.

    c. Accumulation of adaptations in a population can lead to new species.

_____ 2. The fossil record

    a. supports contemporary theories of evolution.   d. is consistent with Darwin's theories.

    b. includes evidence of an animal's life, such as   e. includes evidence of both land and water dwelling

    how they walked.   organisms'

    c. demonstrates that life evolved through time.

_____ 3. The earliest fossils of _Homo sapiens_ with modern anatomical features appear around

    a. 1,000 years ago.   d. 5,000 BC.

    b. 10,000 years ago.   e. recorded human history.

    c. 200,000 years ago.

_____ 4. The first comprehensive theory of evolution, which proposed that organisms change as a result of natural phenomenon and not divine intervention, was proposed by

    a. Aristotle.   d. Wallace.

    b. Lamarck.   e. da Vinci.

    c. Darwin.

_____ 5. The wings of a butterfly and a bat are

    a. homoplastic but not homologous.   d. both homoplastic and homologous.

    b. homologous but not homoplastic.   e. probably derived from a common ancestor.

    c. neither homoplastic nor homologous.

_____ 6. The humerus in a bird and human are

    a. homoplastic but not homologous.   d. both homoplastic and homologous.

    b. homologous but not homoplastic.   e. probably derived from a common ancestor.

    c. neither homoplastic nor homologous.

_____ 7. The person(s) who suggested that populations increase geometrically and that their numbers must eventually be checked was

    a. Malthus.   d. Wallace.

    b. Lamarck.   e. Lyell.

    c. Darwin.

_____ 8. The person who supplied evidence that the earth was old enough to provide adequate time for evolution and the appearance of new species to have occurred is

    a. Aristotle.   d. Wallace.

    b. Lamarck.   e. Lyell.

    c. Darwin.

_____ 9. The two people who proposed that the mechanism for the evolution of organisms is by natural selection were

    a. Malthus.   d. Wallace.

    b. Lamarck.   e. Lyell.

    c. Darwin.

_____10. Structures of different organisms that have a similar form due to a common origin are

    a. homoplastic or analogous.   d. usually vestigial.

    b. artifacts.   e. found in fossils but never in contemporary organisms.

    c. homologous.

_____11. If Darwin had known and employed current biochemical techniques, it most likely would have resulted in

    a. a different theory of evolution.   d. confirmation of his theories.

    b. no theory of evolution.   e. an argument with Malthus.

    c. his conviction of immutable species.

_____12. Using microorganisms to clean up waste sites is known as

    a. biological control.   d. hazardous waste removal.

    b. reclamation.   e. bioincineration.

    c. bioremediation.

_____13. The genetic code

    a. specifies a triplet of nucleotides in DNA

    b. is virtually universal

    c. provides convincing evidence that all organisms arose from a common ancestor

    d. has been used to discredit the fossil record

    e. has provided evidence used to build cladograms

_____14. A species

    a. is a group of interbreeding organisms   d. may have multiple ancestors

    b. is a group of organisms with similar structures   e. will exhibit common behaviors

    c. is the basic group of a population

_____15. The person who proposed that all continents at one time were connected into a single land mass was

    a. Verde   d. Wegener

    b. Lyell   e. Berger

    c. Aristotle

_____16. If the half-life of the radioactive atom biology 103 is 5 years, the approximate percentage of the original amount of the isotope that will remain after thirty years is

    a. 6.0   d. 5.0

    b. 33.3   e. 1.56

    f. 6.25

VISUAL FOUNDATIONS
Color the parts of the illustration below as indicated.

RED ☐ metacarpels

GREEN ☐ ulna

YELLOW ☐ carpels

BLUE ☐ radius

BROWN ☐ phalanges

VIOLET ☐ humerus

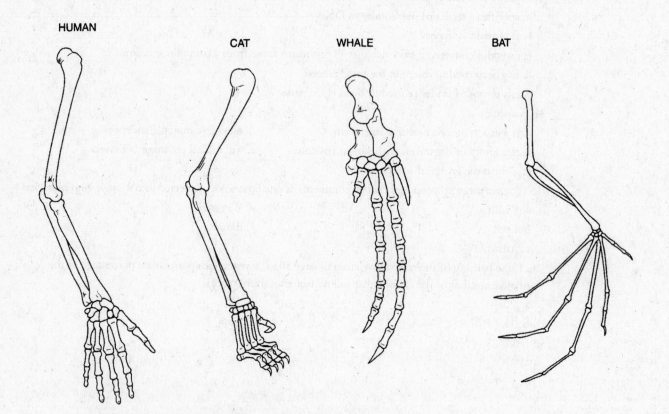

HUMAN   CAT   WHALE   BAT

CHAPTER 19

Evolutionary Change in Populations

In the previous chapter you learned that natural selection results from differential survival and reproduction of individuals, that evolutionary change is inherited from one generation to the next, that populations not individuals evolve as a result of these processes, and that the principles of inheritance you learned in Chapter 11 underlie Darwinian evolution. This chapter examines the genetic variability that occurs within a population and the evolutionary forces that act on it. The factors responsible for evolutionary change are nonrandom mating, mutation, genetic drift, gene flow, and natural selection. Genetic variation in a gene pool is the raw material for evolutionary change. Considerable variability can be observed in the gene pools of most populations.

REVIEWING CONCEPTS
Fill in the blanks.

INTRODUCTION

1. Population genetics is the study of _____
   within the population and the evolutionary forces that act on it.

GENOTYPE, PHENOTYPE, AND ALLELE FREQUENCIES

2. The gene pool of a population will include all
   _____in that population.

3. _____ is the proportion of a particular
   phenotype in a population.

THE HARDY-WEINBERG PRINCIPLE

4. In order to say that evolution is occurring within a population
   _____.

5. For any given locus the frequency of an allele can range from (a) _____ to
   (b)_____.

6. When the distribution of genotypes in a population conforms to the binomial
   equation (a)_____, the population is
   (b)_____ and is not evolving.

Genetic equilibrium occurs if certain conditions are met

7. The proportion of alleles in successive generations does not change in a
   population when certain conditions are met. Among them, briefly stated, are the
   following five:

   _____

   _____

Human MN blood groups are a valuable illustration of the Hardy-Weinberg principle

8. The alleles for the MN blood group are of special interest to geneticists because they are _____.

9. Medical evidence suggests that the MN characteristic does not produce a _____ that might affect random mating.

MICROEVOLUTION

Nonrandom mating changes genotype frequencies

10. Generation-to-generation changes in genotype frequencies within a population describe a process called _____.

11. Inbreeding refers to the mating

_____.

12. The average number of survivors among offspring from a given genotype compared to other genotypes is a measure of

_____.

13. Selection of mates based on a phenotype is known as _____.

14. Assortative mating changes allele frequencies only (a) _____ while inbreeding affects genotype frequencies (b) _____.

MUTATION INCREASES VARIATION WITHIN A POPULATION

15. A _____ is an unpredictable change in DNA.

IN GENETIC DRIFT, RANDOM EVENTS CHANGE ALLELE FREQUENCIES

16. Production of random evolutionary changes in small breeding populations is known as _____.

17. Genetic drift tends to (a)_____ genetic variations within a population, although it tends to (b) _____ genetic differences among different populations.

18. There are times when a population may suddenly and randomly decrease in size. Such a dramatic change in total population number can lead to a (a)_____ which will have an impact on (b)_____ as the population increases in size again.

19. The founder effect occurs when _____.

GENE FLOW GENERALLY INCREASES VARIATION WITHIN A POPULATION

20. When individuals of breeding age migrate from one country to another there a corresponding movement of alleles. This movement of alleles referred to as _____ has significant evolutionary consequences.

NATURAL SELECTION CHANGES ALLELE FREQUENCIES IN A WAY THAT INCREASES ADAPTATION

21. Selection against phenotypic extremes, thereby favoring intermediate phenotypes, is called _____.

22. Selection that favors one particular phenotype over another is known as

_____.

23. Selection that favors phenotypic extremes is called

_____.

GENETIC VARIATION IN POPULATIONS

Genetic polymorphism can be studied in several ways

24. Much of genetic polymorphism is not evident, because it doesn't produce distinct

_____.

Balanced polymorphism exists for long periods

25. _____ occurs when two or more alleles persist in a population over many generations as a result of natural selection.

26. _____ occurs when a genotype such as Aa has a higher degree of fitness than either AA or aa.

27. Selection that acts to decrease the frequency of the more common phenotypes and increase the frequency of the less common phenotypes is called

_____.

Neutral variation may give no selective advantage or disadvantage

28. Variation that does not alter the ability of an individual to survive and reproduce, and is therefore not adaptive, is called _____.

Populations in different geographical areas often exhibit genetic adaptations to local environments

29. Along with population variation, genetic differences often exist among different populations within the same species, a phenomenon known as

_____.

BUILDING WORDS

Use combinations of prefixes and suffixes to build words for the definitions that follow.

Prefixes	The Meaning		Suffixes	The Meaning
micro-	small		-typ(e)	form
pheno-	visible			

Prefix	Suffix	Definition
_____	-evolution	1. Changes in allele frequencies over successive generations; involves small changes within a population.
_____	_____	2.. An organism's visible form or characteristics.

MATCHING

Terms:

a.   Assortative mating
b.   Balanced polymorphism
c.   Directional selection
d.   Disruptive selection
e.   Founder effect
f.   Gene flow
g.   Gene pool
h.   Genetic drift
i.   Genetic polymorphism
j.   Hardy-Weinberg principle
k.   Heterozygote advantage
l.   Natural selection
m.   Population
n.   Stabilizing selection

For each of these definitions, select the correct matching term from the list above.

_____ 1. Natural selection that favors intermediate variants and acts against extreme phenotypes.

_____ 2. A phenomenon in which organisms that are heterozygotes are more likely to demonstrate reproductive success than individuals that are homozygotes.

_____ 3. A phenomenon that results in changes in allele frequency in a population from one generation to the next.

_____ 4. A phenomenon that occurs as a result of individuals from Canada migrating and interbreeding with individuals from Finland.

_____ 5. The gradual replacement of one phenotype with another due to environmental change.

_____ 6. A phenomenon in which natural selection results in multiple alleles being maintained in a population over several generations.

_____ 7. Genetic drift that results from a small number of individuals colonizing a new area.

_____ 8. A phenomenon in which individuals who have black hair and are six feet tall interbreed with other individuals having the same phenotype.

_____ 9. The principle that in a randomly-mating large population the relative frequencies of allelic genes do not change from generation to generation.

_____10. A group of organisms of the same species that live in the same geographical area at the same time.

_____ 11. A phenomenon in which organisms in a population with certain characteristics are more likely to show reproductive success than other organisms in the same population.

MAKING COMPARISONS

Fill in the blanks.

Category	Contributes to Evolutionary Change (yes or no)
Natural selection	#1
Gene flow	#2
Random mating	#3
Genetic drift	#4
Mutation	#5
Small population size	#6
Geographic stability	#7
Genetic polymorphism	#8
Neutral variation	#9

MAKING CHOICES

Place your answer(s) in the space provided. Some questions may have more than one correct answer.

_____ 1. The proportion of alleles in a population does not change if there is/are

a. only 10% of the alleles mutating in each generation.   d. mating with individuals in a similar population.

b. random mating.

c. a large number of individuals in the population.   e. natural selection of advantageous alleles.

_____ 2. When gradual environmental changes favor intermediate phenotypes in a normal distribution curve, the result in time is likely to be

a. stabilizing selection.   d. disruptive selection.

b. no selection.   e. directional selection.

c. a shift in allele frequencies.

_____ 3. Changes in allele frequencies within a population occur during

a. genetic drift.   d. p^2 shifts.

b. macroevolution.   e. q^2 shifts.

c. microevolution.

_____ 4. A relatively quick, extreme environmental change that favors several phenotypes at the expense of the mean phenotype will likely result in

a. stabilizing selection.   d. disruptive selection.

b. no selection.   e. directional selection.

c. a shift in allele frequencies.

_____5. Positive reproductive success of individuals who are DD compared to those who are Dd best describes
  a. positive assortative mating.   d. genetic polymorphism.
  b. natural selection.   e. a heterozygote advantage.
  c. disruptive selection.

_____6. In populations, all new alleles occur as a result of
  a. gene flow.   d. genetic drift.
  b. natural selection.   e. the heterozygote advantage.
  c. mutations.

_____7. Random sexual reproduction among members of a large population that occurs in the absence of any selective pressure on the gene pool ultimately leads to
  a. a Hardy-Weinberg value of 1.   d. non assortative mating.
  b. changes in gene frequencies.   e. a change in the frequency of 2pq, but not p^2 or q^2.
  c. generations of unchanged allele frequencies.

_____8. The ultimate result of a beekeeper having 75% of her bees freeze to death in a severe snow storm will be

  a. a bottleneck   d. stabilizing selection

  b. frequency-dependent selection   e. disruptive selection

  c. genetic drift

_____9. The gradual change in the size and shape of a species of berry bush that occurs as you go from the top of the Grand Canyon into the valley is an example of

  a. a cline   d. genetic drift

  b. a neutral variation   e. disruptive selection

  c. frequency-dependent selection

_____10. Genetic equilibrium occurs as a result of

  a. migration   d. natural selection

  b. nonrandom mating   e. no net mutations occurring

  c. a large population size

Use the following information, the Hardy-Weinberg equation, and the list below to answer questions 11-16. The phenotype coded for by the genotype "bb" is found among 900 individuals in a population of 10,000 randomly mating individuals.

a. 0.01	f. 0.12	k. 0.39	p. 0.60	u. 4.0
b. 0.02	g. 0.18	l. 0.40	q. 0.65	v. 9.0
c. 0.04	h. 0.22	m. 0.42	r. 0.70	w. 35.0
d. 0.08	i. 0.30	n. 0.49	s. 0.81	x. 49.0
e. 0.09	j. 0.33	o. 0.50	t. 2.0	y. 91.0

_____11. Frequency of homozygously recessive individuals.

_____12. Frequency of homozygously dominant individuals.

_____13. Frequency of q.

_____14. Percent of individuals containing one or more of the dominant alleles.

_____12. Frequency of p.

_____13. Frequency of heterozygous individuals.

VISUAL FOUNDATIONS

Color the parts of the illustration below as indicated. Also label directional selection, disruptive selection, and stabilizing selection.

RED ☐ axis indicating number of individuals

GREEN ☐ axis indicating phenotype variation

YELLOW ☐ normal distribution

BLUE ☐ less favored phenotypes

ORANGE ☐ favored phenotypes

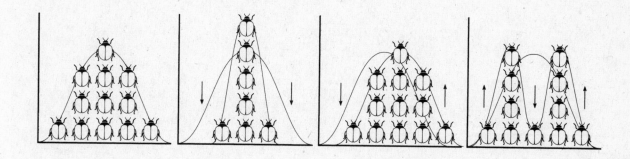

Speciation and Macroevolution

In the previous chapter, you learned how populations evolve. This chapter examines the evolution of species and macroevolution – the dramatic evolutionary changes that occur over long time spans. A biological species is generally viewed as a group of sexually reproducing organisms that share a common gene pool and that are reproductively isolated from other organisms. Most species have two or more mechanisms that serve to isolate and preserve the integrity of their gene pool. If two different species should attempt to mate, prezygotic mechanisms prevent fertilization from taking place. Other, postzygotic, mechanisms, ensure reproductive failure should fertilization occur. Speciation, or the development of a new species, most commonly occurs when a group of reproducing individuals of a population becomes geographically separated from the rest of the species and subsequently evolves. A new species can also evolve, however, within the same geographical region as its parent species. There are two theories related to the pace of evolution. One is gradualism, which says that there is a slow, steady change in species over time, and the other is the punctuated equilibrium, which says there are long periods of little evolutionary change followed by short periods of rapid speciation. Macroevolution includes the origin of large-scale phenotypic changes and evolutionary novelties (e.g., wings with feathers), the evolution of many species from a single species, and the extinction of species.

REVIEWING CONCEPTS

Fill in the blanks.

INTRODUCTION

1. Although there is an estimated (a) _____ species living today, more than (b) _____% of all species that ever existed are extinct.

WHAT IS A SPECIES?

2. The 18th-century biologist, (a) _____, is generally considered the founder of modern taxonomy. He used structural features to separate plants and animals into species. This method of classification is known as the (b) _____.

3. Although Linnaeus' method is still very important in describing species, it alone is not enough to distinguish individual species. Biologists now define a species as a group of (a)_____ isolated organisms with a common (b)_____.

4. Even the definition of a species used by modern biologists is not perfect because it applies only to (a) _____.

5. A taxonomic subdivision of a species is called a _____.

REPRODUCTIVE ISOLATION

Prezygotic barriers interfere with fertilization

6. Reproductive isolating mechanisms help to preserve the _____ of an individual species.

7. Prezygotic isolating mechanisms prevent the formation of an _____ zygote.

8. Prezygotic isolating mechanisms include (a) _____, which occurs because two groups reproduce at different times; (b)_____, which occurs when individuals display different, incompatible courtship patterns; (c)_____, which occurs when anatomical differences thwart successful mating; and (d)_____, which occurs when chemical differences between gametes prevents interspecific fertilization.

Postzygotic barriers prevent gene flow when fertilization occurs

9. Embryos are aborted due to _____.

10. _____ occurs when the gametes produced by interspecific hybrids are abnormal and nonfunctional.

11. On occasion an interspecific hybrid is formed and the F1 generation is able to reproduce. However, the continued reproductive success of the hybrid is prevented as a result of _____.

Biologists are discovering the genetic basis of isolating mechanisms

12. _____ is a sperm protein in abalone that attaches to a specific receptor protein located on the egg envelope and then produces a hole in the envelope that permits the sperm to penetrate the egg.

SPECIATION

13. Two new species are formed from a single, original species when two (a) _____ of that species become (b) _____ from one another.

Long physical isolation and different selective pressures result in allopatric speciation

14. Allopatric speciation occurs when one population becomes _____from the rest of the species and subsequently evolves.

15. Only in natural selection is the change in _____ adaptive.

Two populations diverge in the same physical location by sympatric speciation

16. Sympatric speciation usually occurs *within* a geographical region as a result of two things (a) _____ and (b) _____.

17. Of all the organisms, sympatric speciation is most common in _____.

18. Possession of more than two sets of chromosomes is called (a) _____ which is common in plants. However, when it results from the pairing of chromosomes from different species, it is known a (b)_____.

The study of hybrid zones has made important contributions to what is known about speciation

19. The _____ is an area of overlap in which populations, subspecies, or species come into contact and can interbreed.

20. _____ describes an increase in reproductive isolation that can occur over time in a hybrid zone.

21. _____ is said to occur when the hybrids in a hybrid zone have greater reproductive fitness than either of the two parental species.

THE RATE OF EVOLUTIONARY CHANGE

22. The theory of _____ supports the concept that evolution proceeds in rapid bursts of changes which are followed by long periods of inactivity or stasis.

23. The theory of _____ supports the concept that populations slowly and steadily diverge from one another by the accumulation of adaptive characteristics within a population.

MACROEVOLUTION

Evolutionary novelties originate through modifications of pre-existing structures

24. Evolutionary novelties are variations of pre-existing structures called _____.

25. Evolutionary "novelties" originate from mutations that alter developmental pathways. For example, (a) _____ occurs when developing body parts grow at different rates, and (b)_____ results from differences in the *timing* of development.

Adaptive radiation is the diversification of an ancestral species into many species

26. (a)_____ are new ecological roles made possible by an adaptive advancement. When an organism with a newly acquired evolutionary advancement assumes a new ecological role made possible by its advancement(s), its diversification is known as (b) _____.

Extinction is an important aspect of evolution

27. Continuous, ongoing, and relatively low frequency extinction is called _____.

28. _____ extinction is a relatively rapid and widespread loss of numerous species.

Is microevolution related to speciation and macroevolution?

29. Biologists hypothesize that (a)_____ processes explain (b)_____ patterns.

BUILDING WORDS

Use combinations of prefixes and suffixes to build words for the definitions that follow.

Prefixes	The Meaning	Suffixes	The Meaning
allo-	other, different	-metric	unit of length
macro-	large, long, great, excessive	-morph(ic)	form
paedo-	child	-patri(c)	native land
poly-	much, many	-ploidy	number of chromosome sets in a
genome			
sym-	together		

Prefix	Suffix	Definition
_____	_____	1. Retaining juvenile characteristics as an adult
_____	_____	2. The presence of multiples of complete chromosome sets.
_____	_____	3. Originating in or occupying different (other) geographical areas.
_____	_____	4. Having multiple sets of chromosomes from two different species
_____	-evolution	5. Large-scale evolutionary change.
_____	_____	6. Varied rates of growth for different parts of the body during development (some parts grow at rates different from other parts).

MATCHING

Terms:

a. Adaptive radiation
b. Allopatric speciation
c. Behavioral isolation
d. Biological species
e. Extinction
f. Fusion
g. Gametic isolation

h. Gradualism
i. Habitat isolation
j. Hybrid breakdown
k. Hybrid inviability
l. Hybrid sterility
m. Hybridization
n. Mechanical isolation

o. Phylogenetic species
p. Punctuated equilibrium
q. Reinforcement
r. Reproductive isolation
s. Speciation
t. Sympatric speciation
u. Temporal isolation

For each of these definitions, select the correct matching term from the list above.

_____ 1. A postzygotic isolating mechanism in which the F1 hybrids reproduce but the F2 generation cannot.

_____ 2. The rapid appearance of new species in the fossil record.

_____ 3. The evolution of a new species within the same geographical region as the parent species.

_____ 4. The evolution of many related bird species from a single ancestral bird species in a relatively short period of time.

_____ 5. A prezygotic isolating mechanism in which sexual reproduction between two individuals cannot occur because of differences in compatibility of the reproductive organs.

_____ 6. A model of evolution in which the hybrids have less reproductive success than either parental group.

_____ 7. Sexual reproduction between individuals from closely related species.

_____ 8. Small chickadee birds nesting in a tree do not mate with small chickadee birds nesting on the ground.

_____ 9. A group of individuals determined by the number of appendages they have.

_____10. The end of a lineage, occurring when the last individual of a species dies.

_____11. The zygote produced as the result of a seal mating with a walrus dies before it can fully develop.

MAKING COMPARISONS

Fill in the blanks.

Reproductive Isolating Mechanism	How It Works
Behavioral isolation	Similar species have distinctive courtship behaviors
#1	Prevents the offspring of hybrids that are able to reproduce successfully from reproducing past one or a few generations
Gametic isolation	#2
#3	Interspecific hybrid survives to adulthood but is unable to reproduce successfully
Mechanical isolation	#4
#5	Interspecific hybrid dies at early stage of embryonic development
#6	Similar species reproduce at different times
Habitat isolation	#7
#8	Barrier to reproduction that prevents the formation of a zygote after two different species have mated

MAKING CHOICES

Place your answer(s) in the space provided. Some questions may have more than one correct answer.

_____ 1. When large-scale phenotypic changes in populations justify placing them in taxonomic groups at the species level or higher it is known as
   a. macroevolution.
   b. paedomorphosis.
   c. polymorphic speciation.
   d. macromorphism.
   e. adaptive radiation.

_____ 2. The process by which an ancestral species evolves into many new species is known as
   a. macroevolution.
   b. paedomorphosis.
   c. polymorphic speciation.
   d. macromorphism.
   e. adaptive radiation.

_____ 3. A species typically has
   a. a common gene pool.
   b. an isolated gene pool.
   c. the capacity to reproduce with other species in the laboratory.
   d. members with different morphological characteristics.
   e. become reproductively isolated.

_____ 4. The fact that dogs come in many sizes, shapes, and colors supports the contention that
   a. dogs have one gene pool.
   b. dogs can produce hybrids.
   c. physical appearance alone is not enough to define a species.
   d. reproduction within a species produces sterile offspring.
   e. all dogs belong to one species.

_____ 5. A population that evolves within the same geographical region as its parent species is an example of
    a. cladogenesis.
    b. character displacement.
    c. anagenesis.
    d. sympatric speciation.
    e. allopatric speciation.

_____ 6. If two closely related species produce a fertile interspecific hybrid, the hybrid is the result of
    a. anagenesis.
    b. phyletic evolution.
    c. cladogenesis.
    d. diversifying evolution.
    e. allopolyploidy.

_____ 7. If two distinct populations of organisms occasionally interbreed in the wild, they are considered to be
    a. separate species.
    b. one species.
    c. reproductively isolated.
    d. one gene pool.
    e. evolving.

_____ 8. If flower-loving female botany majors were required to smell a flower before male botany majors would mate with them, and if male zoology majors would only mate with females that did not sniff flowers, zoology majors and botany majors would likely become
    a. separate species.
    b. one species.
    c. reproductively isolated.
    d. one gene pool.
    e. behaviorally isolated.

_____ 9. Evolution of organisms occurs by means of
    a. changes in individual organisms.
    b. uniformitarianism.
    c. changes in gene frequencies in the gene pool.
    d. extinctions.
    e. changes in populations.

_____10. A population that evolves as the result of its geographic separation from the rest of the species is an example of
    a. cladogenesis.
    b. character displacement.
    c. anagenesis.
    d. sympatric speciation.
    e. allopatric speciation.

_____11. If college-educated persons mated only at night and noncollege-educated persons mated only at the lunch hour, and no amount of laboratory experimentation could change these habits, these two populations would be considered
    a. separate species.
    b. one species.
    c. reproductively isolated.
    d. members of one gene pool.
    e. temporally isolated.

_____12. Hybrid breakdown affects the
    a. P_1 generation.
    b. F_1 generation.
    c. F_2 generation.
    d. F_3 generation.
    e. course of evolution.

_____13. The death of interspecific embryos during development is
    a. called hybrid sterility.
    b. one type of postzygotic barrier.
    c. a form of anagenesis.
    d. called hybrid inviability.
    e. allopatric dissociation.

_____ 14. Containing multiple sets of chromosomes from two or more species would make an organism a/an
    a. hybrid.
    b. autopolyploid.
    c. polyploidy.
    d. allopolyploid.
    e. a species formed by hybridization.

_____ 15. Ectones are

    a. areas of transition between two different environments.   d. endemic to North America.

    b. stable hybrid zones.   e. an example of polyploidy in action.

    c. areas of parental reproductive success.

VISUAL FOUNDATIONS

Color the parts of the illustration below as indicated. Also label gradualism and punctuated equilibrium.

RED   ☐   original species

GREEN   ☐   last species to evolve

YELLOW   ☐   time axis

BLUE   ☐   first species to evolve

ORANGE   ☐   structural change axis

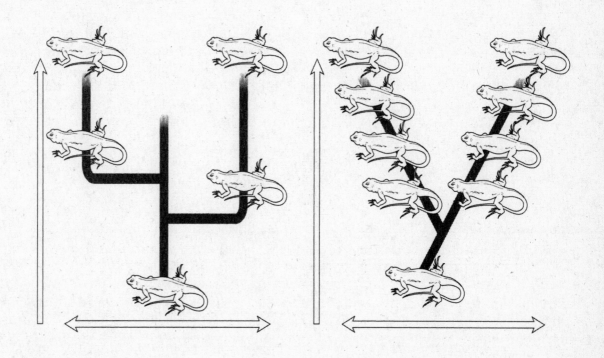

CHAPTER 21

The Origin and Evolutionary History of Life

The previous three chapters dealt with the evolution of organisms, but what about the evolution of life in the first place? This chapter addresses how life began and traces its long evolutionary history. Life on earth developed from nonliving matter. Exactly how this process, called chemical evolution, occurred is not certain. Current models suggest that small organic molecules formed spontaneously, accumulated, and became organized into complicated assemblages. Cells then evolved from these macromolecular assemblages. This process occurred in the presence of little or no oxygen and in the presence of energy, the chemical building blocks of organic molecules, and sufficient time for the molecules to accumulate and react. The first cells to evolve were anaerobic and prokaryotic. They are believed to have obtained their energy from organic compounds in the environment. Later, cells evolved that could obtain energy from sunlight. Photosynthesis produced enough oxygen to change the atmosphere significantly and thereby alter the evolution of early life. Organisms evolved that had the ability to use oxygen in cell respiration. Eukaryotes are believed to have evolved from prokaryotes. Mitochondria, chloroplasts, and certain other organelles may have originated from symbiotic relationships between two prokaryotic organisms.

REVIEWING CONCEPTS

Fill in the blanks.

INTRODUCTION

1. _____ is the process by which life developed from nonliving matter.

CHEMICAL EVOLUTION ON EARLY EARTH

2. Earth is estimated to be about _____ billion years old.

3. The four basic requirements for the chemical evolution of life are:
   _____
   _____.

Organic molecules formed on primitive Earth

4. Two hypotheses have been proposed to explain the have organic molecules may have originated. The (a) _____ hypothesis proposes that these molecules formed near the surface of the earth and the (b) _____ hypothesis proposes that the organic precursors formed near cracks on the ocean floor.

5. The idea that organic molecules could form from nonliving components on primitive Earth originated in the early 20th century with the Russian biochemist (a)_____ and the Scottish biologist (b)_____.

6. In the 1950s, (a)_____ and (b)_____ constructed conditions in the laboratory that were thought to prevail on primitive Earth. Their experiment investigated and atmosphere of water and the three gases (c)_____. These and other experiments have produced a variety of organic molecules.

THE FIRST CELLS

6. _____ are assemblages of organic polymers that form spontaneously, are organized, and to some extent resemble living cells.

7. _____ are formed by adding water to abiotically produced peptides.

Molecular reproduction was a crucial step in the origin of cells

8. (a)_____ may have been the first polynucleotide to carry "hereditary" information. Interestingly, some forms of this molecule, called (b)_____, can catalyze the formation of more forms.

Biological evolution began with the first cells

9. (a)_____, or ancient remains of microscopic life, suggest that cells may have been thriving as long as (b)_____ billion years ago.

The first cells were probably heterotrophic

10. The first cells were most likely (aerobic or anaerobic?) (a)_____ and (prokaryotic or eukaryotic?) (b)_____.

11. Organisms that use photosynthesis to produce their own raw materials such as glucose are called _____.

12. The first photosynthetic organisms to obtain hydrogen electrons by splitting water were the _____.

Aerobes appeared after oxygen increased in the atmosphere

13. _____ are organisms that can only survive in the absence of oxygen.

14. The evolution of photosynthesis ultimately changed early life and afforded vastly

new opportunities for diversity because it generated _____.

Eukaryotic cells descended from prokaryotic cells

15. Eukaryotes appeared in the fossil record about _____billion years ago.

16. The _____ hypothesis proposes that organelles such as mitochondria and chloroplasts may have evolved from a symbiotic relationship between two prokaryotes.

THE HISTORY OF LIFE

17. _____are used to date the same layer of rock that were deposited around the world.

18. The largest division of geologic time is called an _____.

19. Geologic periods are divided into _____.

Rocks from the Ediacaran period contain fossils of cells and simple animals

20. The _____eon came after the Archean eon and in this eon the rocks are less altered by heat and pressure.

21. The oldest known fossils of multicellular animals are called _____ fossils.

A diversity of organisms evolved during the Paleozoic era

22. The Paleozoic era began about _____ years ago with a profound burst of evolution.

23. Later, in the (a)_____ period, the jawless, bony-armored fish called (b)_____ appeared.

24. Jawed fishes first appeared in the _____ period.

25. The age of the fishes occurred during the _____ period.

Dinosaurs and other reptiles dominated the Mesozoic era

26. The Mesozoic era began about _____ years ago.

27. The Mesozoic era consisted of the three periods: (a)_____ when reptiles underwent adaptive radiation and mammals appeared, (b)_____ when large dinosaurs and the first toothed birds appeared, and (c)_____ during which large dinosaurs and toothed birds became extinct.

The Cenozoic era is the Age of Mammals

28. The Cenozoic era can be divided into three periods:

(a)_____

(b) _____

(c) _____

BUILDING WORDS

Use combinations of prefixes and suffixes to build words for the definitions that follow.

Prefixes	The Meaning	Suffixes	The Meaning
aero-	air	-be (bios)	life
an-	without, not, lacking	-bio(nt)	life
auto-	self, same	-troph	nutrition, growth, "eat"
Pre-	before, prior to		
hetero-	different, other		
proto-	first, earliest form of		

Prefix	Suffix	Definition
_____	_____	1. An organism that produces its own food.
_____	_____	2. Spontaneous assemblages of organic polymers that may have been involved in the evolution of the earliest forms of life.
_____	_____	3. An organism that is dependent upon other organisms for food, energy, and oxygen.
_____	-cambrian	4. The geologic time period prior to the Cambrian.
_____	_____	5. An organism that requires air or free oxygen to live.
_____	-aerobe	6. An organism that does not require air or free oxygen to live.

MATCHING

Terms:

a. Autotroph
b. Cenozoic era
c. Chemical evolution
d. Cyanobacteria
e. Endosymbiont
f. Epoch
g. Era
h. Heterotroph

i. Iron-sulfur world hypothesis
j. Mesozoic era
k. Microfossil
l. Microsphere
m. Paleozoic era
n. Period
o. Prebiotic soup hypothesis

p. Ribozyme
q. Stromlaolite

For each of these definitions, select the correct matching term from the list above.

_____ 1. A geological time period that is a subdivision of an era, and is divided into epochs.

_____ 2. Term applied to divisions of geological time that are divided into periods.

_____ 3. The "Age of Mammals."

_____ 4. A rocklike column composed of many minute layers of prokaryotic cells; a type of fossil evidence.

_____ 5. A type of protobiont formed by adding water to abiotically-formed polypeptides.

_____ 6. A major interval of geological time; a subdivision of a period.

_____ 7. An organism that lives symbiotically inside a host cell.

_____ 8. The remains of a microscopic organism.

_____ 9. A molecule of RNA that has catalytic ability.

_____10. Organism that consume preformed molecules.

_____11. Hypothesis that proposes that organic precursors formed on the ocean floor.

_____12. Era that began with the appearance of bacteria and cyanobacteria and ended with the diversification of conifers.

MAKING COMPARISONS

Fill in the blanks.

Biological Event	Period	Era
First fishes appear	#1	#2
The first mammals and first dinosaurs	#3	#4
Dinosaurs peak and become extinct	#5	#6
Age of Marine Invertebrates	#7	#8
Age of *Homo sapiens*	#9	#10
Amphibians and wingless insects appear	#11	#12
Apes appear	#13	#14
Reptiles appear	#15	#16

MAKING CHOICES

Place your answer(s) in the space provided. Some questions may have more than one correct answer.

_____ 1. The earth's surface is protected from the mutagenic effects of ultraviolet radiation by a layer of

    a. CO_2.
    b. N_2.
    c. H_2.
    d. O_3.
    e. H_2O.

_____ 2. It is thought that the mitochondra in eukaryotic cells evolved from

    a. anaerobic bacteria.
    b. aerobic bacteria.
    c. endosymbionts.
    d. primitive stem cells.
    e. intracytoplasmic protozoans.

_____ 3. The Earth is thought to be about how many years old?

    a. 10-20 billion.
    b. 4,000.
    c. 3.5 billion.
    d. 4.6 billion.
    e. 10-20 million.

_____ 4. About how many years passed between the Earth's formation and the appearance of the earliest life?

    a. 10-20 million
    b. 100-200 million
    c. 800-900 million
    d. five billion
    e. 10 billion

_____ 5. Earth's early atmosphere became oxygenated through the photosynthetic activity of

    a.. purple sulfur bacteria
    b. cyanobacteria.
    c. water-splitting autotrophs.
    d. green sulfur bacteria.
    e. hydrogen-sulfide bacteria.

_____ 6. Aerobes are generally better competitors than anaerobes because aerobic respiration

    a. is more efficient.
    b. splits water.
    c. is confined to symbiotic mitochondria.
    d. extracts more energy from a molecule.
    e. has had a longer period of evolution.

_____ 7. The early Earth's atmosphere included
   a. CO$_2$.
   b. N$_2$.
   c. H$_2$.
   d. CO.
   e. H$_2$O vapor.

_____ 8. The basic requirements for chemical evolution include
   a. DNA.
   b. oxygen.
   c. energy.
   d. water.
   e. time.

_____ 9. Experimenters that simulated conditions thought to have existed on early Earth found that these conditions could produce
   a. DNA.
   b. RNA.
   c. nucleotide bases.
   d. proteins.
   e. amino acids.

_____ 10. Strong support for the endosymbiotic theory of eukaryotic cell origins is derived from the fact that mitochondria and chloroplasts have
   a. unique DNA.
   b. two membranes.
   c. ribosomes.
   d. tRNA.
   e. endosymbionts.

_____ 11. The consensus among scientists is that the first cells were
   a. aerobic autotrophic prokaryotes.
   b. anaerobic autotrophic prokaryotes.
   c. aerobic heterotrophic prokaryotes.
   d. anaerobic heterotrophic prokaryotes.
   e. anaerobic heterotrophic eukaryotes.

_____ 12. Which of the following era/period combinations is/are mismatched?
   a. Cenozoic/Paleogene
   b. Paleozoic/Miocene
   c. Mesozoic/Jurassic
   d. Cenozoic/Cambrian
   e. Paleozoic/Devonian

_____ 13. The Paleozoic era
   a. began about 542 mya.
   b. lasted about 291 million years.
   c. consists of six periods.
   d. includes the Cambrian period.
   e. gave rise to many new species.

_____ 14. Life evolving from nonliving matter describes
   a. protobionts.
   b. how the first cells evolved.
   c. microspheres.
   d. the appearance of microfossils.
   e. chemical evolution.

_____ 15. Stromatolites
   a. are found in the U.S. and Canada
   b. may be 3 billion years old
   c. are evidence of eukaryotes
   d. are fossils.
   e. are evidence of early prokaryotes

_____ 16. The first photosynthetic organisms appeared about
   a. 2. Bya
   b. 3 bya
   c. 4 bya
   d. 2.5 bya
   e. 3.5 bya

_____ 17. Experiments investigating the hypothesis that molecules could form in a reducing atmosphere like that present on the primitive Earth were conducted by

    a. Haldane   d. Miller

    b. Urey   e. Oparin

    e. Clay

_____ 18. The oldest eon is

    a. Ordovician   d. Proterozoic

    b. Phanerozoic   e. Cenozoic

    c. Archaean

Color the parts of the illustration below as indicated.

RED   ❑   DNA
GREEN   ❑   chloroplast
YELLOW   ❑   aerobic bacterium
BLUE   ❑   nuclear envelope
ORANGE   ❑   mitochondrion
BROWN   ❑   endoplasmic reticulum
TAN   ❑   photosynthetic bacterium

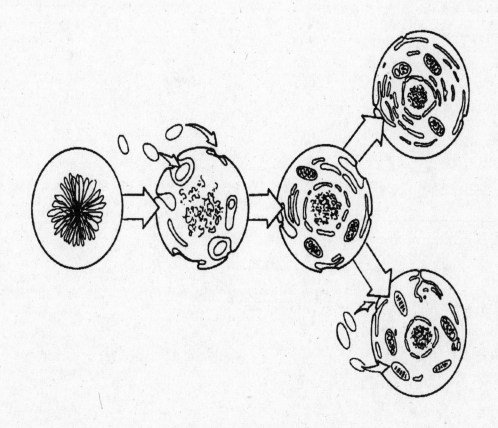

CHAPTER 22

❑

The Evolution of Primates

This is the last of five chapters that address the concept of evolution – the concept that links all fields of the life sciences into a unified body of knowledge. This chapter addresses the evolution of primates, with particular focus on humans and their ancestors. Primates evolved from small, tree-dwelling, shrew-like placental mammals. The first primates to evolve were the lemurs, galagos, and lorises. These were followed by the tarsiers and anthropoids. The early anthropoids branched into two groups, the New World Monkeys and the Old World Monkeys. The latter gave rise to apes that, in turn, gave rise to the hominins (humans and their ancestors). Hominin evolution began in Africa. The first hominin to have enough human features to be placed in the same genus as modern humans is *Homo habilis*. Anatomically modern *Homo sapiens* appeared approximately 195,000 years ago. The evolutionary increase in human brain size made cultural evolution possible.

INTRODUCTION

1. Paleontologists hypothesize that the first primates descended from small, shrew-like _____.

2. Many traits of the living primate species are a reflection of their _____.

PRIMATE ADAPTATIONS

3. Humans and other primates have hands that (a) _____ and feet that (b) _____.

4. The first primates appeared on earth about _____ years ago.

5. Eyes located on the front of the head allows primates to _____.

PRIMATE CLASSIFICATION

6. The three suborders in the order Primates are

   (a)_____ which includes lemurs, lorises, and galagos,

   (b)_____ which includes tarsiers, and

   (c)_____ which includes monkeys, apes, and humans.

Suborder Anthropoidea includes monkeys, apes, and humans

7. Fossil evidence indicates that anthropoids originated in either

   (a) _____ or (b) _____.

8. A significant difference between anthropoids and other primates is the size _____.

9. A few monkeys have tails capable of wrapping around branches and serving as fifth limbs known as _____.

REVIEWING CONCEPTS

Fill in the blanks.

Apes are our closest living relatives

10. Apes and humans comprise a group called the _____.

11. The five living genera of hominoids are

    _____

    _____.

12. Gibbons can arm-swing from limb to limb. Another name for this activity is
    _____.

13. Gorillas use quadrupedal locomotion also known as
    _____.

HOMININ EVOLUTION

14. In apes the large hole in the base of the skull through which the spinal cord connects to the brain is called the _____.

15. Apes have prominent _____ above the eye sockets which are absent in humans.

The earliest hominins may have lived 6 mya to 7 mya

16. Earliest hominids most likely appeared about _____ years ago.

17. Hominid evolution probably began on the continent of _____.

Australopithecines, Austalopithecus, and Paranthropus are auatralopithecines, or "southern apes"

18. _____ is considered to be close to the "root" of the human family tree.

19. The hominin species *Australopithecus anamensis* exhibits distinct phenotypic differences between the two sexes, a condition known as
    _____.

20. "Lucy" was one of the most ancient hominins, a member of the species
    _____.

Homo habilis is the oldest member of genus *Homo*

21. *Homo habilis* was discovered in the country of _____ in Africa.

22. *Homo habilis* appeared about _____ years ago. *H. habilis* was the first hominin to consciously design useful tools.

Homo ergaster may have arisen from H. habilis

23. New discoveries indicate that the early fossils of *H. erectus* really represent two different species. (a) _____ an earlier African species and (b) _____ a later eastern Asian offshoot.

Homo erectus **probably evolved from** *H. ergaster*

24. Fossils of *Homo erectus* were originally found in _____.

25. *Homo erectus* is about _____ years old. It had a larger brain than *H. habilis* and was the first hominin to have fewer differences between the sexes. *H. erectus* made more advanced tools and used fire.

Archaic *Humans date from about 1.2 mya to* 200,000 years ago

26. Archaic humans are descendants of _____.

27. The oldest archaic human fossils, *H. antecessor,* were discovered in
_____.

Neanderthals appeared approximately 250,000 years ago

28. Neanderthal fossils were first discovered in _____.

29. Neanderthals are considered to be an evolutionary _____.

Scientists have reached a near consensus on the origin of modern *Homo sapiens*

30. The "_____" model is the main explanation for the
origin of modern humans living around the world.

CULTURAL CHANGE

31. At the DNA level, humans are approximately (a) _____% identical to gorillas
and (b) _____% identical to chimpanzees.

32. Human culture is generally divided into three stages which are

(a)

_____

(b)

_____

(c)

_____.

Development of agriculture resulted in a more dependable food supply

33. Evidence has shown that humans had begun to cultivate crops approximately
_____ years ago.

Human culture has had a profound impact on the biosphere

34. The UN projects that an additional _____ people will
be added to the world population by the year 2050.

35. Cultural evolution has resulted in large-scale disruption and degradation of the
_____.

BUILDING WORDS

Use combinations of prefixes and suffixes to build words for the definitions that follow.

Prefixes	**The Meaning**		**Suffixes**	**The Meaning**
arbor-	tree-like		-eal	pertaining to
anthrop(o)-	human		-oid	resembling, like
bi-	twice, two		-ped(al)	foot
homin-	man			
pro-	"before"			
quadr(u)-	four			
supra-	above			

Prefix	**Suffix**	**Definition**
_____	_____	1. Member of the suborder Anthropoidea; animals that "resemble" humans.
_____	_____	2. Pertains to an animal that walks on four feet.
_____	_____	3. Pertains to an animal that walks on two feet.
_____	_____	4. Pertains to animals that live in trees.
_____	-orbital	5. Situated above the eye socket.

MATCHING

Terms:

a. Arboreal
b. *Australopithecus*
c. Brachiate
d. Foramen magnum
e. Hominin
f. *Homo erectus*
g. *Homo habilis*
h. *Homo sapiens*
i. Neandertals
j. Orroin
k. Paleoanthropologist
l. Prehensile

For each of these definitions, select the correct matching term from the list above.

_____ 1. Any of a group of extinct and living humans.

_____ 2. The first hominin in the line leading up to *H. sapiens*.

_____ 3. Tree-dwelling.

_____ 4. One of the earliest know hominins.

_____ 5. Adapted for grasping by wrapping around an object.

_____ 6. A scientist that studies human evolution.

_____ 7. The first hominid to have enough human features to be placed in the same genus as modern humans.

_____ 8. To swing arm to arm, from one branch to another.

_____ 9. The opening in the base of the vertebrate skull through which the spinal cord passes.

_____ 10. This species could survive in cold areas, obtained food by hunting, and lived in shelters.

MAKING COMPARISONS

Fill in the blanks.

Hominid	Time of Origin	Characteristic(s)
Orrorin	About 6 million years ago (mya)	One of the earliest known hominins
#1	About 3.6 mya	Small brain, walked upright, large canine teeth, jutting jaw, no evidence of tool use (famous 3.2 mya example was nicknamed "Lucy")
#2	250,000 years ago	Disappeared mysteriously approximately 28,000 years ago
#3	About 4.8 mya	Close to the "root" of the human family tree
#4	3 mya	Walked erect, humanlike hands and teeth, likely omnivorous, likely derived from *A. afarensis*
Homo habilis	#5	#6
Sahelanthropus sp.	#7	The earliest known hominid, close to the last common ancestor of hominids and chimpanzees
Homo erectus	#8	#9
Homo antecessor	#10	Oldest archaic human fossils found in Europe, practiced cannibalism
#11	4.2 mya	Apelike and human like features, marked phenotypic differences show sexual dimorphism

MAKING CHOICES

Place your answer(s) in the space provided. Some questions may have more than one correct answer.

_____ 1. The suborders of primates are

    a. Hominoids.   d. Tarsiiformes.

    b. Hominins.   e. Anthropoidea.

    c. Prosimii.

_____ 2. The group(s) to which humans and their ancestors belong is/are

    a. Hominoids.   d. Tarsiiformes.

    b. Hominins.   e. Anthropoidea.

    c. Prosimii.

_____ 3. The group(s) to which apes belong is/are

    a. Hominoids.   d. Tarsiiformes.

    b. Hominins.   e. Anthropoidea.

    c. Prosimii.

_____ 4. Both apes and humans
   a. are hominoids.   d. lack tails.
   b. are hominids.   e. have opposable thumbs.
   c. can brachiate.

_____ 5. The first hominin that migrated to Europe and Asia is
   a. _H. habilis._   d. australopithecines.
   b. _H. erectus._   e. "Lucy" and her descendants.
   c. _H. sapiens._

_____ 6. Archaic humans appeared about how many years ago?
   a. 20,000-40,000   d. 800,000-900,000
   b. 100,000-150,000   e. one million
   c. 200,000-400,000

_____ 7. The Peking man and Java man are classified as
   a. _H. habilis._   d. apes.
   b. _H. erectus._   e. hominids.
   c. _H. sapiens._

_____ 8. The "Neandertal man" appeared about _____ years ago.
   a. 800,000   d. 1 million
   b. 250,000   e. 2 million
   c. 30,000

_____ 9. The earliest hominin species to be placed in the genus _Homo_ is
   a. _habilis._   d. australopithecines.
   b. _erectus._   e. "Lucy" and her descendants.
   c. _sapiens._

_____ 10. The earliest hominins
   a. had short canines.   d. appeared about 6-7 mya.
   b. are in the genus _Homo._   e. evolved in Africa.
   c. are in the species _sapiens._

_____ 11. Based on molecular similarities and other characteristics, it is thought that the nearest living relative of
   humans is the
   a. gorilla.   d. chimpanzee.
   b. monkey.   e. ape.
   c. gibbon.

_____ 12. Which of the following support(s) the "out-of-Africa hypothesis?"
   a. modern _H. sapiens_ evolved from _Homo neanderthalensis_   d. racial differences.
   b. newly evolved _H. sapiens_ migrated to Europe   e. paternal mitochondrial DNA.
   c. results from recent molecular analyses

_____ 13. The immediate ancestor to the genus _Homo_ is
   a. Prosimii.   d. primitive apes.
   b. Tarsiiformes.   e. Therapsids.
   c. Australopithecines.

_____ 14. The earliest Hominins may have lived as long as
   a. 3-4 mya.   d. 6-7 mya.
   b. 4-5 mya.   e. 7-8 mya.
   c. 5-6 mya.

____ 15. Old world monkeys

    a. have a prehensile tail

    b. have a have a fully opposable thumb

    c. are found in tropical environments

    d. have nostrils that are close together

    e. are very social animals

____ 16. In a cladogram demonstrating primate evolution, the first outgroup is the

    a. tarsiers

    b. chimpanzees

    c. New World monkeys

    d. gibbons

    e. lemurs

VISUAL FOUNDATIONS

Color the parts of the illustration below as indicated.

VIOLET ☐ cerebrum

GREEN ☐ jaw (mandible bone)

BLUE ☐ supraorbital ridge

YELLOW ☐ canine teeth

ORANGE ☐ incisor teeth

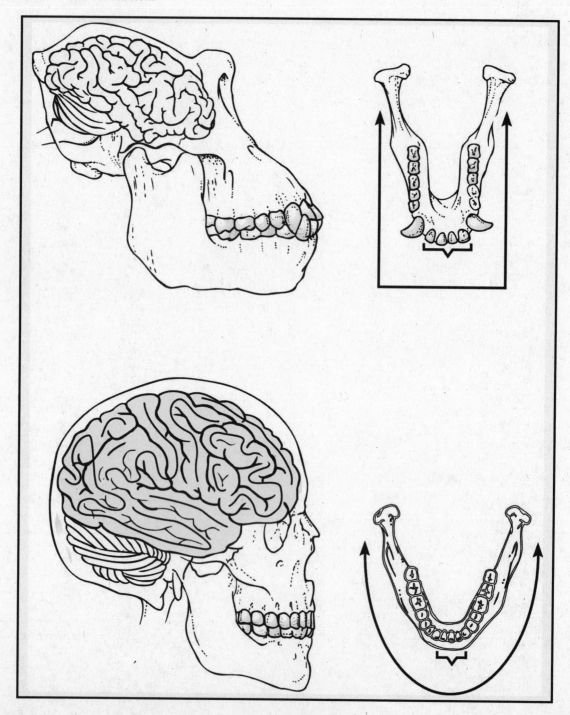

Color the parts of the illustration below as indicated. Label the anatomical difference between human and gorilla on each skeletal element.

RED ☐ pelvis
GREEN ☐ skull
YELLOW ☐ first toe
BLUE ☐ spine
ORANGE ☐ arm
BROWN ☐ leg

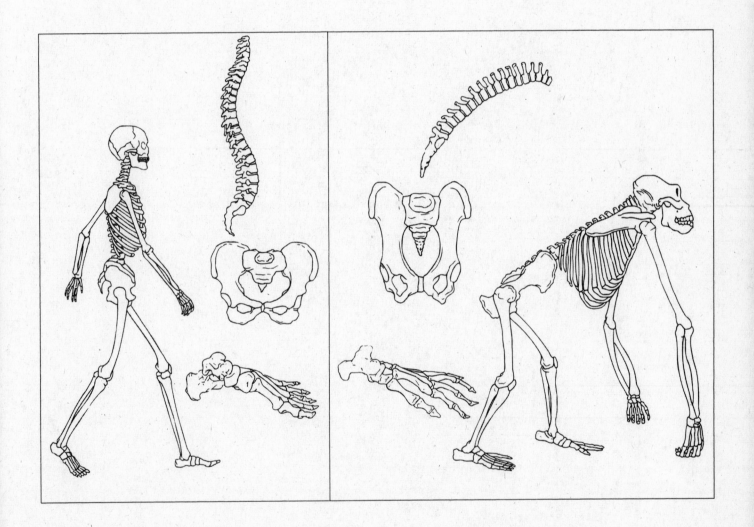

CHAPTER 23

Understanding Diversity: Systematics

Evolution, the cumulative changes that occur in the gene pool of a population over time, has been the central focus of diversity. This section looks at The Diversity of Life which occurs as a result of differences in the gene pool and gene expression between species. This chapter provides an overview of some of the methods and approaches scientists can use to classify organisms. Biologists rely on a combination of molecular data and structural comparisons to infer relationships between organisms and reconstruct evolutionary history. Systematics is the scientific study of the diversity of organisms and their evolutionary relationships. Taxonomy, the science of naming, describing, and classifying organisms, is an important part of systematics. Since the mid-18th century biologists have classified organisms using the binomial system developed by Carolus Linnaeus. This system uses a two-part name to describe an individual species. The first name, a noun, is the genus and the second, an adjective, is the specific epithet. To accurately identify a species both names must be used together and never alone. Classification is hierarchical beginning with a single organism, the species, and ending with the largest, most inclusive group, the domain. The intermediate groups are made up of related individuals. For example a group of genera constitutes a family and a group of families constitutes an order. In establishing a "tree of life" systematists rely heavily on evolutionary change and relationships and they look for evidence that suggests groups of organisms are related to a common ancestor. It is important to remember that taxonomists continue to make changes and adjustments to phylogenetic relationships as new information becomes available.

REVIEWING CONCEPTS

Fill in the blanks.

INTRODUCTION

1. The variety of living organisms and ecosystems is referred to as _____.

CLASSIFYING ORGANISMS

2. _____ is the study of diversity and of organisms and heir evolutionary relationships.

3. _____ is the science of naming, describing, and classifying organisms.

Organisms are named using a binomial system

4. Carolus Linnaeus designed a classification system, the _____ of nomenclature, in which each species or organism is identified using a unique two-part name.

5. For each species the two-part name consists first of the (a)_____, which is capitalized, followed by the (b)_____ which is not capitalized.

Each taxonomic level is more general than the one below it

6. The classification system consists of several levels of organization. A group of closely related species is placed in the same genus and a group of closely related genera are grouped together in the same (a)_____ which in turn are grouped and placed in orders. Orders would be grouped in the next level to form a (b)_____, which would be grouped to form a (c)_____.

7. The broadest, most inclusive group is called a _____.

8. In the classification system, each grouping, or level, such as the species, genus, order, is called a _____.

DETERMINING THE MAJOR BRANCHES IN THE TREE OF LIFE

SYSTEMACTICS IS AN EVOLVING SCIENCE

9. Before the development of the microscope, organisms were only placed into kingdom: (a)_____ or (b) _____.

10. In the late 1800s a third kingdom, _____, was proposed that was to include bacteria and microorganisms.

11. _____ are non-photosynthetic organisms that absorb nutrients and are placed in a kingdom of the same name.

12. Kingdom (a)_____ includes organisms that lack a distinct, membrane bound nucleus. This kingdom has now been divided into two domains: (b)_____ and (b)_____.

Some biologists are moving away from Linnaean categories

13. In the PhyloCode approach to classification, a group of organisms that share characters inherited from a common ancestor is called a _____.

Phylogenetic trees show hypothesized evolutionary relationships

14. A phylogenetic tree that uses clades to show evolutionary relationships is called a _____.

Systematists continue to consider other hypotheses

15. Genes are transmitted from parents to offspring in a process called _____.

16. Gene swapping between organisms in one taxon and related organisms in another taxon is called _____.

RECONSTRUCTING EVOLUTIONAY HISTORY

17. Modern classification is based on the reconstruction of _____, or phylogeny, as it is called.

18. A _____ describes a group of individuals of the same species living in the same area.

Homologous structures are important in determining evolutionary relationships

19. _____ describes the independent evolution of similar structures in distantly related organisms.

20. A characteristic that superficially appears homologous but is actually independently acquired by convergent evolution or reversal is described as exhibiting _____.

Shared derived characters provide clues about phylogeny

21. Features that appeared a long time ago and remain in the descendents of that organism are called _____.

22. Features that occurred in a recent common ancestor and remain in the descendents of that organism are called_____.

Biologists carefully choose taxonomic criteria

23. Organisms are typically classified on the basis of a _____ of traits rather than on any single trait.

Molecular homologies help clarify phylogeny

24. An organism's unique DNA and RNA sequences can be used as a _____ to identify that organism.

25. The science of _____ focuses on molecular structure to identify evolutionary relationships.

26. Macromolecules that have a similar subunit structure and that are functionally similar in two different groups are considered to be _____.

27. Comparisons of the similarities in the
(a)_____ of proteins, and the
(b)_____of nucleotides can be used to confirm evolutionary relationships.

Taxa are grouped based on their evolutionary relationships

28. When the organisms in a given taxon include the ancestral species and all of its descendents, the group is said to be _____.

29. When the organisms in a given taxon do not include all of the descendents of an ancestral species, the group is said to be _____.

30. When the organisms in a given taxon do not share a common ancestor, the group is said to be _____.

CONSTRUCTING PHYLOGENETIC TREES

31. The phenetic system is a numerical taxonomy in which organisms are grouped according to the number of _____ they share.

32. The two major approaches to systematics are (a) _____ and (b) _____.

Outgroup analysis is used in constructing and interpreting cladograms

33. An (a) _____ is a taxon that is considered to have diverged earlier than the taxa under investigation which is called the (b) _____.

A cladogram is constructed by considering shared derived characters

34. To be a valid monophyletic clade, all members must share at least one _____.

35. Membership in a clade cannot be established by shared _____.

In a cladogram each branch point represents a major evolutionary step

36. Relationships between taxa are determined only by tracing along the branches to the

_____.

37. A clodogram does not establish direct

_____ relationships among taxa.

Systematists use the principles of parsimony and maximum likelihood to make decisions

38. The principle of _____ is based on the experience that the simplest explanation is probably the correct one.

39. _____ is a statistical method of systematics that depends on probability.

BUILDING WORDS

Use combinations of prefixes and suffixes to build words for the definitions that follow.

Prefixes	The Meaning
mono-	alone, single, one
para-	alongside, beside, near
poly-	much, many
sub-	under, below

Prefix	Suffix	Definition
_____	-phylum	1. A distinctive subunit of a phylum.
_____	-phyletic	2. Refers to a taxon in which the group includes some but not all the descendents of a common ancestor.
_____	-phyletic	3. Refers to a taxon consisting of several evolutionary lines and not including a common ancestor.

MATCHING

Terms:

a. Ancestral character
b. Clade
c. Classification
d. Derived character
e. Genus
   Horizontal gene transfer

   Kingdom
f. Order
   Parsimony
g. Phenetics
h. Phylum
i. Species

   Specific epithet
j. Systematics
k. Taxon
l. Taxonomy
   Vertical gene transfer

For each of these definitions, select the correct matching term from the list above.

_____ 1. The arranging of organisms into groups using similarities and evolutionary relationships among lineages.

_____ 2. The science of naming, describing, and classifying organisms.

_____ 3. The noun part of the binomial system used to describe organisms.

_____ 4. A taxon that comprises related classes.

_____ 5. A formal grouping of organisms such as a class or a family.

_____ 6. A monophyletic group of organisms sharing a common ancestor.

_____ 7. The systematic study of organisms based on similarities of many characters.

_____ 8. The transfer of genes between different species.

_____ 9. A recently evolved characteristic found in a clade.

_____ 10. Using the simplest explanation of the available data to classify organisms.

MAKING COMPARISONS

Fill in the blanks.

Domain	Kingdom	Characteristics
Bacteria	Bacteria	Unicellular prokaryotes, generally with cell walls composed of peptidoglycan
#1	#2	Unicellular prokaryotes lacking peptidoglycan in cell walls
#3	#4	Eukaryotic, generally unicellular protozoa, algae, slime molds, water molds
#5	#6	Heterotrophic eukaryotes with cell walls
#7	#8	Heterotrophic eukaryotes without cell walls
#9	#10	Autotrophic eukaryotes with cell walls

MAKING CHOICES

Place your answer(s) in the space provided. Some questions may have more than one correct answer.

_____ 1. The study of the diversity of organisms and their evolutionary relationships is

    a. taxonomy.
            d. determinism.

    b. nomenclature.
            e. classification.

    c. systematics.

_____ 2. _____ is the classification of organisms based on the number of characters they share

    a. Taxonomy
            d. Phenetics

    b. Phylogeny
            e. Evolutionary systematics

    c. Systematics

_____ 3. Which of the following groups includes the largest number of organisms?

    a. class
            d. phylum

    b. phylum
            e. family

    c. order

_____ 4. The systematist would be most interested in

    a. lumping taxa.
            d. ontogeny.

    b. splitting taxa.
            e. ancestries.

    c. evolutionary relationships.

_____ 5. Which organisms are placed in kingdom Prokaryotae?

    a. bacteria
            d. archea

    b. plants
            e. protists

    c. fungi

_____ 6. Which classification scheme groups organisms into clades based on evolutionary relationships?
a. barcode
b. phenetics
c. cladism
d. parsimony
e. PhyloCode

_____ 7. Shared ancestral characters
a. can be used to determine when two groups diverged.
b. can be used to distinguish various classes of vertebrates.
c. are evidence of convergent evolution.
d. will be homoplastic.
e. are used in establishing barcodes.

_____ 8. Which of the following is/are used extensively by the molecular biologist to determine relationships?
a. carbohydrates.
b. nucleic acids.
c. lipids.
d. proteins.
e. hormones.

_____ 9. The system of binomial nomenclature was developed by
a. Charles Darwin.
b. Carolus Linnaeus.
c. Paul Hebert.
d. Carl Woese.
e. Watson and Crick.

_____10. Which statement is incorrect?
a. A cladogram tells us how recently two groups shared a common ancestor.
b. Systematists use statistical probability, maximum likelihood, in establishing taxonomic relationships.
c. A specific epithet can be used in more than one binomial classification.
d. Protists are no longer placed in a single kingdom.
e. A taxon is a formal grouping of organisms that cannot be separated into subgroups.

VISUAL FOUNDATIONS

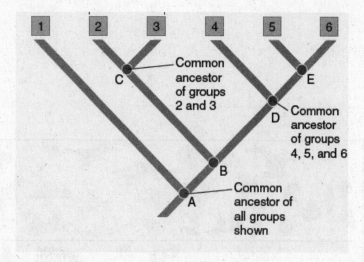

Use the above figure to answer the following questions.

1. B would be the common ancestor for which monophyletic taxon?

2. D would be the common ancestor for which monophyletic taxon?

3. Why does placing groups 5 and 6 in to a single taxon make it a monophyletic taxon?

4. Why does placing groups 1 and 2 into a single taxon make it a paraphyletic taxon?

5. Why does placing groups 2 and 4 into a single taxon make it a polyphyletic taxon?

Color the parts of the illustration below as indicated. Inside each large box, write the specific name of each category used in classifying the depicted organisms, and also indicate the specific epithet of the organism used to illustrate this hierachical organization in the species box.

RED ☐ species
TAN ☐ domain
GREEN ☐ phylum
YELLOW ☐ class
BLUE ☐ kingdom
ORANGE ☐ order
VIOLET ☐ family
PINK ☐ genus

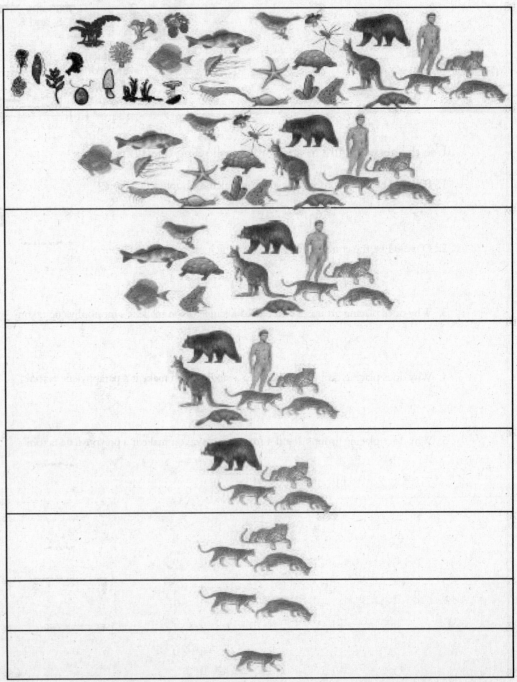

CHAPTER 24

Viruses and Subviral Agents

This chapter examines the diversity and characteristics of viruses and the still smaller infective agents, satellites, viroids and prions. Viruses and subviral agents impact living organisms. Viruses themselves have some characteristics of living things, for example they are able to replicate, but they cannot survive and exist independently in the way cells can. Viruses consist of a core of DNA or RNA surrounded by a protein coat. Some are surrounded by an outer envelope too. Whereas some viruses may or may not kill their hosts, others invariably do. Some integrate their DNA into the host DNA, conferring new properties on the host. Viruses cause serious diseases in plants and animals. Satellites are subviral agents that depend upon co-infection of a host cell with a helper virus to facilitate their own replication. Viroids are the smallest known pathogens and they consist of RNA with no protein coat. Viroids are generally found within the host cell nucleus and appear to interfere with gene replication. Prions appear to consist of only protein. When present, prions can cause cells to malfunction.

REVIEWING CONCEPTS
Fill in the blanks.

INTRODUCTION

1. Pathogens are infective agents such as viruses or bacteria that cause _____.

2. Viruses range in size from _____ to _____ nm.

THE STATAUS AND STRUCTURE OF VIRUSES

A virus consists of nucleic acid surrounded by a protein coat

3. The core of a virus is a _____.

4. Viruses are surrounded by a protein coat called a _____.

5. A complete virus particle that is outside of the host cell is called a _____.

6. The protective coat of a virus is called a _____.

7. The shape of a virus is determined by the _____.

8. Enveloped viruses have an outer membranous envelope surrounding the capsid. The source of the envelope is actually the _____.

CLASSIFICATION OF VIRUSES

9. Viruses can be classified based on the types of organisms they infect or on their _____.

VIRAL REPLICATION

10. Viruses replicate inside host cells by taking over the

    (a) _____ and (b)

    _____ mechanisms of the host cell.

Bacteriophages infect bacteria

11. The most common structure of a bacteriophage consists of a long nucleic acid molecule coiled within a _____.

Viruses replicate inside host cells

12. The two types of reproductive cycles observed among viruses are

    (a)_____ and

    (b)_____.

13. Viruses that have only a lytic cycle are said to be

    _____.

14. The five steps that are typical of lytic viral reproduction are

    _____

    _____

    _____.

15. Bacteria protect themselves from viral infections by producing

    _____ that prevents the phage DNA

    from replicating.

16. In lysogenic cycles, the viral genome usually becomes incorporated into the host bacterial DNA. Viruses that do not always destroy the host cell are referred to as _____ viruses.

17. Viral DNA that is integrated into host DNA is called a

    (a)_____, and bacterial cells carrying integrated

    viral DNA are called (b)_____.

18. _____ occurs when bacteria exhibit new properties as a result of integration of temperate viral DNA.

VIRAL DISEASES

19. Viruses that cause a disease are called

    _____.

Viruses cause serious plant diseases

20. Most plant viruses have (a) _____ but do not have

    (b) _____.

21. The genome of most plant viruses consists of

    _____.

22. Viruses can only penetrate plant cells if

    _____.

Viruses cause serious diseases in animals

23. The intentional use of microorganisms or toxins to cause death or disease in humans or in the organisms upon which humans depend is called

    _____.

24. RNA viruses called _____ transcribe their RNA genome into a DNA strand that is then used as a template to produce more viral RNA. HIV is one such virus.

25. The DNA polymerase used by retroviruses in transcription is called

    _____.

EVOLUTION OF VIRUSES

26. According to the _____ hypothesis, viruses may have originated as mobile genetic elements such as transposoms or plasmids.

27. According to the _____ hypothesis, viruses appeared early in the history of life, even before the three domains diverged.

28. According to the _____ hypothesis, viruses evolved from small independent cells that were parasites in larger cells.

SUBVIRAL AGENTS

Satellites depend on helper viruses

29. Satellites are _____ agents that depend on co-infection of a host cell with a helper virus.

Viroids are the smallest known pathogens

30. Viroids are hardy and resist heat and ultraviolet radiation because of the condensed folding of their _____.

31. Viroids are generally found within the host cell (a) _____ and appear to interfere with (b) _____.

Prions are protein particles

32. Prions are found in the brains of patients with fatal degenerative brain diseases called _____.

BUILDING WORDS

Use combinations of prefixes and suffixes to build words for the definitions that follow.

Prefixes	The Meaning		Suffixes	The Meaning
			-ate	characterized by having
			-gen	production of
			-id	tending to
			-oid	resembling
bacterio-	bacteria		-phage	eat, devour
caps-	box		-retro	backward
oblig-	obliged			
patho-	suffering, disease, feeling			
vir-	poisonous slime			

Prefix	Suffix	Definition
_____	_____	1. A virus that infects and destroys bacteria.
_____	_____	2. A short, circular, single strand of naked RNA.
_____	_____	3. A protein coat surrounding the nuclear core of a virus.
_____	_____	4. An intracellular parasite that can only survive using the resources of a host cell.
_____	-virus	5. Viruses that transcribe the RNA genome into a DNA intermediate..

_____ _____   6. Any disease-producing organism.

MATCHING

Terms:

a. Attachment
b. Coevolution hypothesis
c. Capsid
d. Satillite
e. Retrovirus
f. Lysogenic conversion
g. Tamiflu
h. Regressive hypothesis
i. Envelope
j. Temperate
k. Virulent
l. Penetration
m. Viroid
n. Prion

For each of these definitions, select the correct matching term from the list above.

_____ 1. States that viruses evolved from small independent cells that were parasites in larger cells.

_____ 2. Found in the host cell nucleus and appear to interfere with replication.

_____ 3. Viruses that cause disease and often death.

_____ 4. Bacterial cells containing temperate viruses may exhibit new properties.

_____ 5. Subviral agent dependent upon co-infection of a host cell with a helper virus.

_____ 6. The protein covering of a virus.

_____ 7. Outer membranous covering surrounding a capsid.

_____ 8. An antiviral drug.

_____ 9. Viruses that do not always destroy their host.

_____ 10. Phage DNA enters the bacterial cell.

MAKING COMPARISONS

Fill in the blanks.

Characteristic	Viruses	Prions	Viroids
May contain RNA	yes	no	#1
May be covered by an envelope			
Can arise spontaneously			
Pathogenic			

Characteristic	Viruses	Prions	Viroids
Infect plants			

MAKING CHOICES

Place your answer(s) in the space provided. Some questions may have more than one correct answer.

_____ 1. Viruses
    a. are acellular.
    b. do not carry on metabolic activities.
    c. can only reproduce when occupying a cell.
    d. do not produce rRNA.
    e. are not included in any of the three domains.

_____ 2. A virulent virus
    a. has a lytic cycle.
    b. can reproduce outside of cells.
    c. causes disease.
    d. degrades its host cell's nucleic acids.
    e. invade but do not destroy their host cell.

_____ 3. Viruses
    a. can be cultured in the laboratory using bacteria.
    b. gave rise to prions.
    c. respond to treatment with antibiotics.
    d. are called prophages when their genome is integrated into the host DNA.
    e. may have an envelope made of the host-cell plasma membrane.

_____ 4. When a host bacterium exhibits new properties because of a prophage, the phenomenon is called
    a. lysogenic conversion.
    b. assembly.
    c. transduction.
    d. viroid induction.
    e. reverse transcription.

_____ 5. Viruses are usually grouped (classified)
    a. from species to orders.
    b. based on their host range
    c. into cellular and acellular groups.
    d. into families with the suffix _viridae_.
    e. in the traditional Linnean system.

_____ 6. The characteristic(s) of life that viruses do not exhibit is/are
    a. presence of nucleic acids.
    b. independent movement.
    c. reproduction.
    d. a cellular structure.
    e. independent metabolism.

_____ 7. A small, circular piece of DNA that is separate from the main chromosome is
    a. called a plasmid.
    b. a mesosome.
    c. responsible for photosynthesis.
    d. found in archaebacteria and cyanobacteria, but not eubacteria.
    e. found in gram-negative, but not gram-positive bacteria.

_____ 8. Prions
    a. are protein particles
    b. can arise spontaneously as the result of a mutation
    c. induce PrP proteins to fold abnormally
    d. have either a DNA or RNA core.
    e. can cause cells to malfunction

_____9. The coat surrounding the nucleic acid core of a virus is
   a. a capsid.
   b. made of protein.
   c. a virion.
   d. a viral envelope.
   e. composed of capsomeres.

_____10. An important difference between virulent and temperate viruses is that only the temperate virus
   a. destroys the host cell.
   b. lyses the host cell.
   c. does not always lyse host cells in the lysogenic cycle.
   d. contains DNA.
   e. infects eukaryotic cells.

_____11. The virus that causes AIDS and some types of cancer is a/an
   a. paramyxovirus.
   b. herpesvirus.
   c. adenovirus.
   d. retrovirus.
   e. RNA virus.

VISUAL FOUNDATIONS

Color the parts of the illustration below as indicated. Also label helical shape, polyhedral shape, and combination (helical and polyhedral) shape.

RED   ☐ DNA
GREEN ☐ RNA
VIOLET ☐ capsid
ORANGE ☐ fibers
BROWN ☐ tail

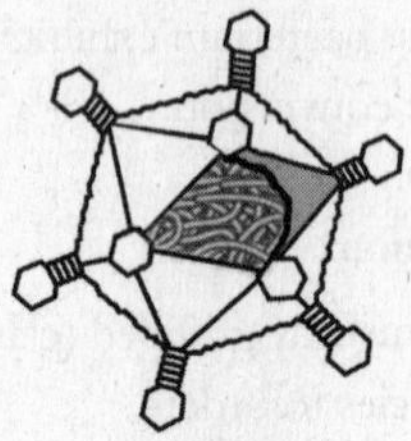

Color the parts of the illustration below as indicated. Also label these components inside of the nucleus, mRNA viral proteins, ribosomes, nucleus, glycoproteins

RED ☐ envelope

GREEN ☐ capsid (everyplace it is present)

VIOLET ☐ mRNA

ORANGE ☐ endoplasmic reticulum

BROWN ☐ viral proteins

BLUE ☐ viral nucleic acid

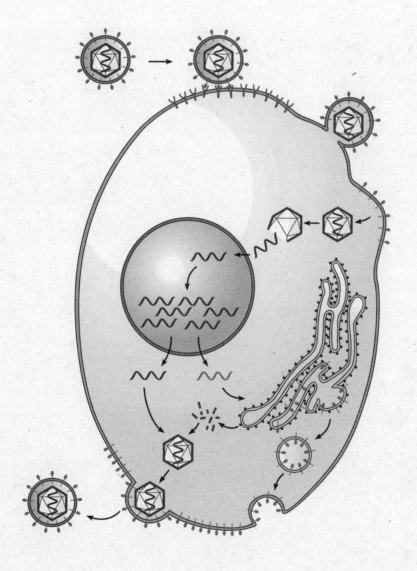

Bacteria and Archaea

This chapter examines two of the three domains, Bacteria and Archaea, often referred to as prokaryotes. These organisms are mostly unicellular, but some form colonies or filaments. They have ribosomes, but lack membrane-bounded organelles. In most the genetic material is a circular DNA molecule, but in some bacteria it is linear. Some have flagella. Whereas most prokaryotes get their nourishment either from dead organic matter or symbiotically from other organisms, some manufacture their own organic molecules. Although most prokaryotes reproduce asexually, some exchange genetic material. The former are thought to be the original prokaryotes from which all cellular life descended. Although most prokaryotes are harmless, a few are notorious for the diseases they cause. Prokaryotes have important medical and industrial uses.

REVIEWING CONCEPTS
Fill in the blanks.

INTRODUCTION

1. Pathogens are microorganisms such as bacteria, fungi, and protozoa that cause _____.

2. Prokaryotes and fungi account for approximately (a) _____% of the Earth's biomass while plants account for (b) _____% and animals (c)_____%.

THE STRUCTURE OF BACTERIA AND ARCHEA

Prokaryotes have several common shapes

3. Spherical prokaryotes are called (a)_____. They may occur in pairs called (b)_____, in chains called (c)_____, or in clumps or bunches called (d)_____.

4. Rod-shaped prokaryotes are known as _____.

5. A rigid spiral-shaped prokaryote is known as a (a)_____, and a flexible spiral-shaped prokaryote is called a (b)_____.

6. A spirillum shaped like a comma is called a _____.

Prokaryotic cells do not have membrane-enclosed organelles

7. In prokaryotes, there is a nuclear area that contains DNA and resembles the nucleus of an eukaryote; it is referred to as the _____.

A cell wall protects most prokaryotes

8. The Eubacterial cell wall includes _____, a polymer that consists of two unusual types of sugars linked with short polypeptides.

9. Bacteria that absorb and retain crystal violet stain are referred to as _____.

10. Penicillin works most effectively against _____ bacteria.

Some bacteria produce capsules or slime layers

11. In free-living species of bacteria, the outer covering may provide the cell with protection against phagocytosis by _____.

12. In disease-causing bacteria, outer covering may protect against phagocytosis by _____.

Some prokaryotes have fimbriae or pili

13. Fimbriae are made of (a) _____ and are shorter than (b) _____.

Some bacteria survive unfavorable conditions by forming endospores

14. Some bacteria form dormant, durable cells called endospores. However, endospores are not a form of _____.

Many types of prokaryotes are motile

15. Prokaryotes have two main methods of motility: (a) _____, consisting of a basal body, a hook, and a single filament, and (b) _____, movement in response to chemicals in the environment.

PROKARYOTE REPRODUCTION AND EVOLUTION

16. In addition to their genomic DNA, many bacteria have small amounts of genetic information present as one or more _____, or circular fragments of DNA.

Rapid reproduction contributes to prokaryote success

17. The three different types of reproduction in prokaryotes are

(a) _____

(b) _____

(c) _____.

Prokaryotes transfer genetic information

18. Prokaryotes usually reproduce asexually by simple transverse binary fission, but sometimes genetic material is exchanged in one of several ways. In _____, fragments of DNA released by a cell are taken in by another bacterial cell.

19. In _____, a phage carries bacterial genes from one bacterial cell into another.

20. Prokaryotes of two different "mating types" may exchange genetic material during _____.

Evolution proceeds rapidly in bacterial populations

21. Because bacteria reproduce by binary fission, _____ are quickly passed on to new generations, and natural selection effects are quickly evident.

NUTRITIONAL AND METABOLIC ADAPTATIONS

22. Organisms that are able to use inorganic compounds as a source of carbon for making organic compounds are called _____.

23. (a) _____ obtain energy from chemical compounds and
(b) _____ obtain energy from sunlight.
24. _____ obtain their energy from dead organic matter.

Most propkaryotes require oxygen

25. Most bacterial cells are (a)_____, or require oxygen for cellular respiration. Some other bacteria, however, are
(b)_____ that use oxygen if it's available, or (c)_____ that carry out anaerobic respiration.

Some prokaryotes fix and metabolize nitrogen

26. Organisms require nitrogen for the synthesis of (a) _____ and (b) _____.
27. Animals obtain nitrogen when they eat _____.

THE PHYLOGENY OF THE TWO PROKARYOTE DOMAINS

28. Under a microscope, most prokaryotes appear to be similar in (a)_____ and (b) _____.
29. The two domains of prokaryotes are (a) _____ and
(b) _____

Key characteristics distinguish the three domains

30. When compared to bacteria, archaea to not have _____ in their cell walls.
31. In some ways, Archaea are actually more like _____ than bacteria.

Taxonomy of archaea and bactria continuously changes

32. Prokaryote taxonomy is now based largely on molecular dta, RNA sequencing and more recently on _____.

Many archaea inhabit harsh environments

33. Archaea that require either a very high or a very low temperature for growth are considered to be _____.
34. _____ are a group of archaea that inhabit oxygen-free environments such as sewage or swamps.

IMPACT ON ECOLOGY, TECHNOLOGY, AND COMMERCE

Prokaryotes form intimate relationships with other organisms

35. The three types of symbiotic relationships that can be observed when prokaryotes interact with other organisms are
(a)_____ when both partners benefit from the relationship,
(b) _____ when one partner benefits and the other is neither harmed nor helped,
(c) _____ when one benefits and the other is harmed in some way as a result of the relationship.

36. Many bacteria that inhabit watery environments form dense films called

(a) _____ that attach to solid surfaces.

(b) _____ that forms in the mouth is one
    example.

Prokaryotes play key ecological roles

37. The most numerous of the prokaryotes found in soil are in the
    Domain_____.

38. Plant growth is dependent upon (a) _____ which must be
    continually added to the soil. (b) _____ bacteria that form
    mutualistic relationships with legumes are involved in this process.

Prokaryotes are important in many commercial processes and in technology

39. Compounds that either inhibit the growth of or destroy other microorganisms
    are called _____.

40. _____ uses microorganisms to detoxify, the
    process of removing toxic chemicals, from the environment.

BACTERIA AND DISEASE

41. Harmless symbionts that inhabit the surface of the human skin are called
    _____.

Many scientists have contributed tour understanding of infectious disease

42. The set of guidelines known as _____
    are used to demonstrate that a specific pathogen causes specific disease
    symptoms.

Many adaptations contribute to pathogen success

43. To cause a disease, a pathogen must do the following three things

(a) _____

(b) _____

(c) _____.

Antibiotic resistance is a major public health problem

44. Plasmids that have genes for antibiotic resistance to drugs are called
    _____.

45. A high percentage of infections acquired in hospitals involve bacteria
    inhabiting _____.

BUILDING WORDS

Use combinations of prefixes and suffixes to build words for the definitions that follow.

Prefixes	The Meaning	Suffixes	The Meaning
an-	without, lacking	-ate	characterized by having
chem.(o)	chemical	-gen	production of
endo-	within	-ive	tending to
eu-	good, well, "true"	-karyo(te)	nucleus
exo-	outside, outer, external	-phil(e)	love
facult	faculty	-taxis	ordered movement
halo-	salt		
oblig-	obliged		

patho-   suffering, disease, feeling

Prefix	Suffix	Definition
_____	_____	1. Movement toward or away from chemicals in the environment.
_____	-toxin	2. A poison that is secreted by bacterial cells into the external environment.
_____	-toxin	3. A poison that is a component "within" the cell wall of most gram-negative bacteria.
methano-	_____	4. A bacterium that produces methane from carbon dioxide and water.
_____	-spore	5. A thick-walled spore that forms within a bacterium in response to adverse conditions.
_____	_____	6. Any disease-producing organism.
_____	-aerobe	7. An organism that metabolizes only in the absence of (without) molecular oxygen.
pro-	_____	8. An organism that lacks a nuclear membrane.
_____	_____	9. An organism with a distinct nucleus surrounded by nuclear membranes.
_____ _____		10. Anaerobic organisms that cannot survive in an environment containing oxygen.
_____ _____		11. Organisms that survive in environments that are high in salt.

MATCHING

Terms:

a. Bacillus
b. Binary fission
c. Capsule
d. Conjugation
e. Flagellum
f. F factor

g. Lysogenic conversion
h. Mycoplasma
i. Nucleoid
j. Peptidoglycan
k. Pilus
l. Plasmid

m. Saprobe
n. Transduction
o

For each of these definitions, select the correct matching term from the list above.

_____ 1. An area of a prokaryote that contains DNA.

_____ 2. A rod-shaped bacterium.

_____ 3. An eubacterium that lacks cell walls.

_____ 4. A structure used for locomotion in some bacteria.

_____ 5. A process whereby one cell divides into two similar cells.

_____ 6. A layer covering the cell wall of many prokaryotes..

_____ 7. A component of eubacterial cell walls.

_____ 8. A method of reproduction in single-celled organisms in which two cells link and exchange nuclear material.

_____ 9. A type of genetic recombination resulting from transfer of genes from one organism to another by a virus.

_____10. A hair-like structure associated with prokaryotes.

_____11. A small, circular fragment of DNA

_____12. A DNA sequence that is necessary for a bacterium to serve as a donor during conjugation.

MAKING COMPARISONS

Fill in the blanks.

Characteristic	Bacteria	Archaea	Eukarya
Peptidoglycan in cell wall	Present	Absent	#1
70S ribosomes	#2	#3	#4
Nuclear envelope	#5	#6	#7
Simple RNA polymerase	#8	#9	#10
Membrane-bounded organelles	#11	#12	#13

MAKING CHOICES

Place your answer(s) in the space provided. Some questions may have more than one correct answer.

_____ 1. Prokaryotes

    a. are cellular organisms.   d. are divided into three domains.

    b. may have finmraie   e. may be gram-positive organisms.

    c. may transfer genetic information horizontally.

_____ 2. Conjugation

    a. is a form of asexual reproduction.   d. is an example of horizontal gene transfer.

    b. occurs in bacteria.   e. contributes to genetic variation.

    c. is the exchange of plasmids.

_____ 3. Autotrophs include

    a. the majority of prokaryotes.   d. decomposers

    b. organisms relying on organic compounds as a carbon source.   e. cyanobacteria

    c. most bacterial pathogens.

_____ 4. Prokaryote taxonomy relies on

    a. genomic sequencing.   d. molecular data.

    b. RNA sequencing   e. nutritional source.

    c. the use of bioflims.

_____ 5. The pathogen responsible for causing stomach ulcers is

    a. _Bordetella pertussis_   d. _Yersinia pestis_

    b. _Bacillus anthracis_   e. _Helicobacter pylori_

    c. _Salmonella typhi_

_____ 6. The disease MRSA is caused by

    a. *Staphylococcus aureus*

    b. plasmid transfer between gram negative bacteria.
pneumonia and

    c. normal microbiota

    d. a drug resistant bacterium.

    e. a symbiotic relationship between *S.*

        E. coli

_____ 7. The possible shape for a prokaryote organism is

    a. bacillus

    b. vibrio

    c. spirochete

    d. coccus

    e. spirillum

_____ 8. The majority of bacteria are

    a. autotrophs.

    b. heterotrophs.

    c. pathogens.

    d. saprobes.

    e. photosynthesizers.

_____ 9. Exchange of genetic material between bacteria sometimes occurs by

    a. fusion of gametes.

    b. conjugation.

    d. transfer of genes in bacteriophages.

    e. transduction.

    c. incorporation by a bacterium of DNA fragments from another bacterium.

_____10. A small, circular piece of DNA that is separate from the main chromosome is

    a. called a plasmid.

    b. a mesosome.

    c. responsible for photosynthesis.

    d. found in archaebacteria and cyanobacteria, but not eubacteria.

    e. found in gram-negative, but not gram-positive bacteria.

_____11. Bacteria that retain crystal violet stain differ from those that do not retain the stain in having a

    a. thick lipoprotein layer.

    b. thick lipopolysaccharide layer.

    c. thicker peptidoglycan layer.

    d. cell wall.

    e. mesosome.

_____12. Prokaryotic cells are distinguished from eukaryotic cells by

    a. absence of a nuclear envelope.

    b. absence of mitochondria.

    c. absence of DNA in the genetic material.

    d. absence of a plasma membrane.

    e. bacteriorhodopsin in cell walls.

VISUAL FOUNDATIONS

Color the parts of the illustration below as indicated. Also label cell wall, flagellum, and pili.

RED	☐	plasma membrane
BROWN	☐	outer membrane
YELLOW	☐	peptidoglycan layer
BLUE	☐	DNA
TAN	☐	capsule
GREEN	☐	plasmid

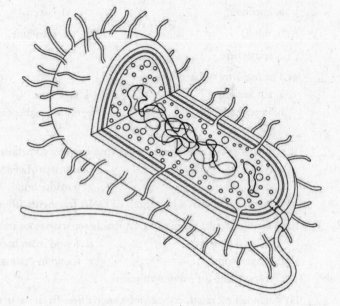

Color the parts of the illustration below as indicated.

RED ☐ bacterial chromosome

GREEN ☐ F plasmid

VIOLET ☐ outline of the F⁺ donor cell

ORANGE ☐ outline of the F⁻ recipient cell

BROWN ☐ the conjugation tube

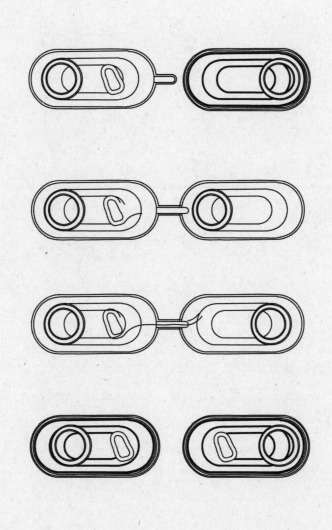

CHAPTER 26

❑

Protists

The protist kingdom consists of a vast assortment of primarily aquatic eukaryotic organisms with diverse body forms, types of reproduction, modes of nutrition, and lifestyles. Biologists currently recognize dozens of protist taxa which are included in domain Eukarya and are part of the five "supergroups" of eukaryotes. Protists range in size from microscopic single cells to meters-long multicellular organisms. They do not have specialized tissues. Protists obtain their nutrients autotrophically or heterotrophically. Whereas some protists are free-living, others form symbiotic associations ranging from mutualism to parasitism. Reproductive strategies vary considerably; most reproduce both sexually and asexually, others only asexually. Whereas some are nonmotile, most move by flagella, pseudopodia, or cilia. Protists are believed to be the first eukaryotes. Evidence suggests that mitochondria and chloroplasts may have originated as prokaryotes that became incorporated into larger cells.

REVIEWING CONCEPTS
Fill in the blanks.

INTRODUCTION

1. Protists are included in domain (a) _____ which means that protists have a (b) _____.

DIVERSITY IN THE PROTISTS

2. Most protists have a body plan that is (a) _____ and many exist as (b)_____ organisms.

3. Most protists are aquatic, and many of them are floating microscopic forms called _____.

4. Protists may form _____ relationships with other, unrelated organisms

HOW DID EUKARYOTES EVOLVE

Mitochondria and chloroplasts probably originated from endosymbionts

5. In the _____hypothesis, eukaryotic organelles such as mitochondria and chloroplasts arose from a symbiotic relationship between larger cells and smaller prokaryotes that were incorporated and lived in them.

6. Protists in the group _____have a nonfunctional chloroplast surrounded by four membranes. This group includes *Plasmodium* which causes malaria.

A consensus in eukaryote classification is beginning to emerge

7. In some cases electron microscopy supports molecular data suggesting that certain protist taxa are (a)_____ because they evolved from a common ancestor. However, the data show that other protists are actually a (b) _____ group.

231

EXCAVATES

8. Unlike other protists, evacuates have atypical or greatly modified
_____.

9. Excavates are so named because many have a deep, or excavated
_____.

Diplomonads are small, mostly parasitic flagellates

10. Diplomonads are characterized as having one or two nuclei,
no (a)_____,
no (b)_____,
and up to eight flagella.

Parabasilids are anaerobic endosymbionts that live in animals

11. Trichonymphs live in the guts of (a) _____
and (b) _____.

Euglenoids and trypanosomes have both free-living species and parasites

12. About one-third of all euglenoids are _____.

13. Some euglenoids have a flexible outer covering called a
_____.

14. Autotrophic euglenoids are single-celled, flagellated protists containing the
same _____ pigments as those found in green algae
and higher plants.

15. Trypanosomes have a single _____ that contains
a deposit of DNA called a kinetoplastid.

16. The genus and species of the parasitic flagellate that causes African sleeping
sickness is _____.

CHROMALVEOLATES

17. Unifying features of alveolates include similar ribosomal DNA sequences and
alveoli which are _____ located just inside
the plasma membrane.

Most dinoflagellates are a part of marine plankton

18. A typical dinoflagellate has two (a) _____ and alveoli
containing interlocking (b) _____ plates.

19. Dinoflagellates that live in the bodies of marine invertebrates are called
(a) _____. Using photosynthesis they
produce (b) _____ for the invertebrate hosts.
These dinoflagellates are particularly important to the survival of
(c) _____ found in tropical shallow waters.

Apicomplexans are spore-forming parasites of animals

20. Apicomplexans lack specific structures such as flagella which are necessary for
_____ and instead flex.

21. Apicomplexans have a specialized _____ that
is used to attach to its host cell.

22. At some point in their life cycle, apicomplexans produce
_____ which are small infective agents that are transmitted
to a host.

23. An apicomplexan in the genus _____ can enter human RBCs and cause malaria

Ciliates use cilia for locomotion

24. Ciliates are distinct in having one or more small (a)_____ that function during the sexual process and a larger (b)_____ that regulates other cell functions.

25. "Cross fertilization" occurs among many ciliates during the sexual process known as _____.

Water molds produce biflagellate reproductive cells

26 The water molds, like the fungi, have a body called the _____.

27. Water molds reproduce asexually by biflagellated (a)_____ and sexually by means of (b)_____.

Diatoms have shells composed of two parts

28. Diatoms are mostly single-celled, with _____ impregnated in the cell walls that form the shell

29. Diatoms may have either (a)_____ or (b)_____ symmetry.

30. When diatoms die, their shells accumulate over time, forming sediment called _____ when it becomes exposed on land.

Brown algae are multicellular stramenopiles

31. The largest and most complex of all algae are commonly called _____.

32. Brown algae reproduce using biflagellated cells. When reproducing sexually the cells are (a)_____ and when reproducing asexually the cells are (b)_____.

33. Brown algae spend part of their life as multicellular haploid organisms and part as multicellular diploid organisms. This type of a life cycle is an example of _____.

Most golden algae are unicellular biflagellates

34. Most golden algae are flagellated, _____ organisms.

35. Golden algae may covered in tiny scales of either (a)_____ or (b)_____.

RHIZARIANS

36. Rhizarians often have a hard outer shell which is also called a _____.

Forams extend cytoplasmic projections that form a threadlike, interconnected net

37. Almost all foraminiferans live in a _____ environment.

38. The dead bodies of foraminiferans are eventually transformed into _____.

39. Foram fossils are used by as _____ to help identify rock layers.

Actinopods project slender axopods

40. The cytoplasmic extensions of actinopods are strengthened by _____.

ARCHAEPLASTIDS

Red algae do not produce motile cells

41. Most multicellular red algae attach to surfaces by means of a structure called the
_____.

42. Two extracts of commercial value which prepared from red algae are
(a)_____, a polysaccharide which can be used as a
food thickener, and (b) _____ another
polysaccharide which is used as a food stabilizer in soft processed foods such as
ice cream and chocolate milk.

Green algae share many similarities with land plants

43. In single celled green algae, asexual reproduction occurs by
(a) _____ followed by (b)_____.

44. In multicellular green algae, asexual reproduction occurs by
_____.

45. Green algae produce zoospores which are
_____.

UNIKONTS

46. Unikonts are a supergroup of organisms where the flagellated cells have a singular
_____.

47. Bikonts are characterized as having _____.

Amoebozoa are unikonts with lobose pseudopodia

48. Because they have an extremely flexible outer membrane, amoebas have an
_____ body form.

49. Amoebas glide along surfaces by flowing into cytoplasmic extensions called
_____, meaning "false feet."

50. The parasitic species _____ causes amoebic
dysentery.

51. Plasmodial slime molds form intricate stalked reproductive structures called
_____, within which meiosis occurs.

52. When environmental conditions are suitable, spores open and one-celled haploid gametes
of two types, one a flagellated form called a (a)_____, the other an ameboid
(b)_____, emerge and fuse.

53. Slime molds have close affinities to two other organisms
(a) _____ and (b) _____.

54. The single flagellum of the choanoflagellates is surrounded at the base by a
delicate collar of _____.

BUILDING WORDS

Use combinations of prefixes and suffixes to build words for the definitions that follow.

Prefixes	**The Meaning**	**Suffixes**	**The Meaning**
Conju-	to join together	-ation	the process of
cyto-	cell	-cle	little
iso-	equal, "same"	-ile	having the character of
macro-	large, long, great, excessive	-pod(ium)	foot, footed
micro-	small	-zoa	animal
pell(i)	skin		
proto-	first, earliest form of		
pseudo-	false		
sess-	without a stem		

Prefix	**Suffix**	**Definition**
_____	_____	1. Single-celled, animal-like protists including amoebae, ciliates, flagellates, and sporozoans; members of this group are believed to have given rise to the earliest form of animal life.
_____	_____	2. .An outer flexible covering associated with euglenoids.
_____	_____	3. A temporary protrusion of the cytoplasm of an ameboid cell that the cell uses for feeding and locomotion; a "false foot."
_____	-nucleus	4. A small nucleus found in ciliates.
_____	-nucleus	5. A large nucleus found in ciliates.
_____	_____	6. A sexual union in which genetic material is exchanged between cells.
_____	_____	7. Species that are attached to the substrate as opposed to being free living.

MATCHING

Terms:

a.	Actinopod	f.	Hypha	k.	Zoosporangium
b.	Axopod	g.	Mycelium	l.	Sporozoite
c.	Conjugation	h.	Oospore	m.	Excavates
d.	Foraminiferan	i.	Plankton	n.	Trichonympha
e.	Holdfast	j.	Plasmodium	o.	Giardia

For each of these definitions, select the correct matching term from the list above.

_____ 1. One of the long, filamentous cytoplasmic projections that protrude through pores in the skeletons of actinopods.

_____ 2. A structure that produces tiny biflagellate zoospores.

_____ 3. The vegetative body of fungi; consists of a mass of hyphae.

_____ 4. Free-floating, mainly microscopic aquatic organisms found in the upper layers of the water.

_____ 5. Marine protozoans that produce chalky, many-chambered shells or tests.

_____ 6. A sexual process in which two individuals come together and exchange genetic material.

_____ 7. One of the filaments composing the mycelium of a fungus.

_____ 8. The basal root-like structure in multicellular algae that anchors the organism to a solid surface.

_____ 9. Multinucleate mass of cytoplasm that constitutes the feeding stage in the life cycle of slime molds.

_____10. Small infective agents produced by apicomplexans that are transmitted to the next host.

_____11. A diverse group of unicellular protists with flagella.

_____12. A major cause of water-borne diarrhea around the world.

MAKING CHOICES

Place your answer(s) in the space provided. Some questions may have more than one correct answer.

_____ 1. Protists may be
   a. unicellular.   d. eukayotic.
   b. colonial.   e. Archaea.
   c. simple multicellular organisms.

_____ 2. Mitochondria and chloroplasts are thought to have been derived from
   a. complex viruses.   d. intrasymbiotic eukaryotes.
   b. protozoa.   e. aerobic bacteria.
   c. endosymbionts.

_____ 3. Escavates
   a. have ciliate.   d. obtain energy through fermentation.
   b. are colonial organisms.   e. are flagellated.
   c. are generally endosymbionts.

_____ 4. The *Paramecium*
   a. is a ciliate.   d. possesses a pellicle.
   b. is eukaryotic.   e. has multiple nuclei.
   c. is colonial, sometimes multicellular.

_____ 5. The parasite that causes malaria is in the phylum
   a. Ciliophora.   d. Dinoflagellata.
   b. Apicomplexa.   e. Euglenophyta.
   c. Sarcomastigophora.

_____ 6. A protozoan whose cells bear a striking resemblance to specialized cells in sponges is the
   a. foraminiferan.   d. ameba.
   b. diatom.   e. radiolarian.
   c. choanoflagellate.

_____ 7. One of the reasons that protists are placed in a separate kingdom is that many of them possess both plantlike and animal-like characteristics, which is particularly well-illustrated in the genus
   a. *Paramecium*.   d. *Ameba*.
   b. *Euglena*.   e. *Didinium*.
   c. *Plasmodium*.

_____ 8. Members of the genus *Euglena*
   a. are flagellated.   d. are heterotropic.
   b. do not possess a pellicle.   e. are autotrophic.
   c. possess chlorophyll.

_____ 9. A symbiotic relationship in which both "partners" benefit is known as

a. commensalism.

b. mutualism.

c. parasitism.

d. endosymbiosis.

e. fraternism.

_____10. A symbiotic relationship in which one "partner" benefits and the other is unaffected is known as

a. commensalism.

b. mutualism.

c. parasitism.

d. endosymbiosis.

e. fraternism.

_____11. Protists can be found in the supergroup

a. Chromalveolates

b. Fungi

c. Amoebozoas

d. Rhizarians

e. Ciliates

_____ 12. Red tide

a. is caused by a toxin produced by dinoflagellates.

b. is caused by an organism in the supergroup excavates.

marine c. is responsible for large fish kills.

d. occurs in nutrient-rich, warm waters.

e. occurs when apicomplexans infect invertebrates.

_____13. The diploid stage of a *Plasmodium* life cycle occurs in

a. humans

b. red blood cells.

c. liver cells.

d. mice

e. mosquitoes

_____ 14. Because of their similarity to other organisms, water molds were once classified as

a. bread molds.

b. unikonts

c. fungi

d. algae

e. Rhizarians.

_____15. Cellular slime molds

a. are unikonts.

b. reproduces by meiosis.

c. reproduce using spores.

d. are amoebozoa.

e. form a slug when conditions are not optimal.

VISUAL FOUNDATIONS

Color the parts of the illustration below as indicated. Also label cilia.

RED   ☐ micronucleus

ORANGE   ☐ oral groove

BLUE   ☐ contractile vacuole

YELLOW   ☐ food vacuole

TAN   ☐ anal pore

VIOLET   ☐ macronucleus

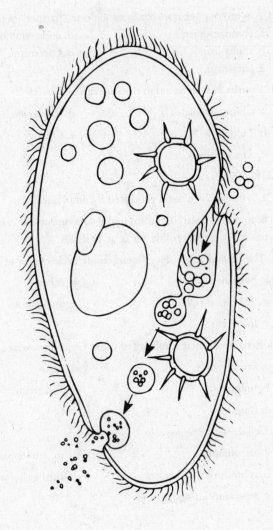

Color the parts of the illustration below as indicated. Label the portions of the illustration depicting sexual reproduction, and the portions depicting asexual reproduction. Also label diploid generation, haploid generation, mitosis, meiosis, and fertilization.

RED ☐ positive strain

GREEN ☐ negative strain

YELLOW ☐ zygote

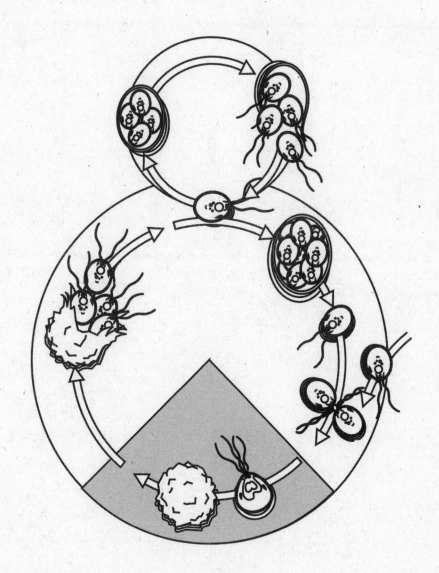

CHAPTER 27

Seedless Plants

Plants, believed to have evolved from green algae, are multicellular and exhibit great diversity in size, habitat, and form. They photosynthesize to obtain their energy. To survive on land, plants evolved 1) a waxy cuticle to protect against water loss, 2) openings in the surface layer that allow for gas exchange, 3) a strengthening polymer in the cell walls that enables them to grow tall, and 4) multicellular sex organs that protect the developing embryo. The plant kingdom consists of four major groups — bryophytes, seedless vascular plants, gymnosperms, and angiosperms. Except for the bryophytes, all plants have a vascular system. A vascular system allows plants to achieve larger sizes because water and nutrients can be transported over great distances to all parts of the plant. The focus of this chapter is on the seedless plants, i.e., the bryophytes and the ferns and their allies. Unlike the bryophytes, the ferns and their allies also have a larger and more dominant diploid stage, which is a trend in the evolution of land plants. Most of the seedless plants produce only one kind of spore as a result of meiosis. A few, however, produce two types of spores, an evolutionary development that led to the evolution of seed plants.

REVIEWING CONCEPTS

Fill in the blanks.

INTRODUCTION

1. Because of the numerous characteristics that they share, plants are thought to have evolved from ancient forms of _____.

2. Plants and the group from which they are thought to be derived have the same photosynthetic pigments, which are
   (a)_____
   and they both store excess carbohydrates as (b)_____ and
   have (c)_____ as a major component of their cell walls.

3. Plants have a character that distinguishes them from green algae. That characteristic is development from multicellular embryos enclosed in
   _____.

ADAPTATIONS OF PLANTS TO LIFE ON LAND

4. The adaptation of plants to land environments involved a number of important adaptations. Among them was the (a)_____ that covers aerial parts, protecting the plant against water loss. Openings in this covering also developed. These (b)_____ facilitate the gas exchange that is necessary for photosynthesis.

5. The carbon used by plants to make sugars and other organic molecules is obtained from _____ present in the atmosphere.

6. The sex organs of plants, or _____, as they are called, are multicellular structures containing gametes.

The plant life cycle alternates haploid and diploid generations

7. Plants live one part of their lives in a multicellular haploid stage and another part in a multicellular diploid stage. Having two stages to the life cycle is referred to as

   _____.

8. The haploid part of the life cycle is called the

   (a) _____ and the

   diploid part of the life cycle is called the

   (b) _____.

9. In plants, the male gametes form in the (a)_____, and the female gametes in the (b)_____.

10. In the plant life cycle, the first stage in the sporophyte generation is the

    (a)_____, and the haploid

    (b)_____ are the first stage of the gametophyte generation.

Four major groups of plants exist today

11. Molecular data and structural indicate that land plants probably descended from a group of green algae known as _____.

12. Of the four major groups of plants only the

    (a)_____ lack a vascular

    (conducting) system. The two groups that reproduce and disperse

    primarily via spores are the

    (b)_____. The

    (c)_____ have exposed seeds

    on a stem or in a cone, while the

    (d)_____ produce seeds within a

    fruit.

13. _____ is a polymer in the cell walls of large, vascular plants that strengthens and supports the plant and its conducting tissues.

BRYOPHYTES

14. Bryophytes can be separated into three phyla. The phyla are

    _____

    _____

    _____.

Moss gametophytes are differentiated into "leaves" and "stems"

15. Each moss plant has tiny, hairlike absorptive structures called

    _____.

16. In mosses, fertilization occurs in the

    _____.

17. The diploid zygote of mosses develops into a mature

    _____.

18. After it germinates, a moss spore grows into a filament of cells called a
_____.

19. Among the most important mosses are those with large water storage
vacuoles in their cells. These mosses are in the genus
(a)_____, and are known by their common name,
the (b)_____.

Liverwort gametophytes are either thalloid or leafy

20. The body of a liverwort is a flattened, lobed structure called a
_____.

21. Liverworts reproduce sexually and asexually. In sexual reproduction, the
haploid gametangia are called _____.

22. During asexual reproduction in liverworts, tiny balls of tissue
called_____
form on the thallus, and after they are dispersed by raindrops they will
develop into a new liverwort.

Hornwort gametophytes are inconspicuous thalloid plants

23. Hornworts belong to the phylum _____ and
live in disturbed habitats such as fallow fields and roadsides.

24. A unique feature of hornworts is that the sporophytes continue to grow
form their gases for the remainder of the gametophyte's life. Botanists
refer to this characteristic
as_____.

Bryophytes are used for experimental studies

25. Botanists use certain bryophytes as experimental models to study
_____, i.e., plant responses to varying periods of
night and day length.

Recap: details of bryophyte evolution are based on fossils and on structural and molecular evidence

26. Plants are considered to be a _____
group.

27. Fossil evidence indicates that _____ are
probably the first group of plants to arise from the common plant ancestor.

SEEDLESS VASCULAR PLANTS

28. There are two basic types of true leaves: the small
(a)_____ with its single vascular strand and the larger
(b)_____ with multiple vascular strands. Of these two
types of leaves, the (c)_____ represent the
majority of leaves we see today.

29. The two main clades of seedless vascular plants are
(a) _____ and (b)
_____.

Club mosses are small plants with rhizomes and short, erect branches

30. Club mosses were major contributors to our present-day

_____.

Ferns are a diverse group of spore-forming vascular plants

31. In ferns, the sporophyte is composed mainly of an underground stem called the (a) _____, from which extend roots and leaves called (b)_____.

32. (a)_____, the spore cases on the leaves of ferns, often occur in clusters called (b)_____. The mature fern gametophyte is a tiny, green, often heart-shaped structure called a (c)_____.

33. The main organs of photosynthesis in Psilotum are the

(a) _____ that exhibit (b) _____ branching (dividing into two equal halves).

34. The small haploid gametophyte of the whisk fern is a non-photosynthetic subterranean plant that apparently obtains nourishment through a symbiotic relationship with a _____.

35. Horsetails are found on every continent except _____.

36. The stems of horsetails are impregnated with _____.

Some ferns and club mosses are heterosporous

37. Bryophytes, horsetails, whisk ferns, and most ferns and club mosses produce only one type of spore, a condition known as (a)_____. Some ferns and club mosses exhibit (b)_____, the production of two different types of spores. The spores are either (c)_____ which develop into male gametophytes or (d) _____ which develop into female gametophytes.

Seedless vascular plants are used for experimental studies

38. The _____ is the area at the tip of the root or shoot where growth occurs.

Seedless vascular plants arose more than 420 mya

39. Microscopic spores of early vascular plants appear in the fossil record earlier than _____, suggesting that even older megafossils of simple vascular plants may be discovered.

BUILDING WORDS

Use combinations of prefixes and suffixes to build words for the definitions that follow.

Prefixes	The Meaning
arch(e)-	primitive
bryo-	moss
gamet(o)-	sex cells, eggs and sperm
hetero-	different, other
homo-	same
mega-	large, great
micro-	small
spor(o)-	spore
strobil-	a cone
thall-	a young shoot
xantho-	yellow

Suffixes	The Meaning
-angi(o)(um)	vessel, container
-gon(o)(ium)	sexual, reproductive
-(o)logy	study of
-phyte	plant
-spor(e)(y)	spore
-us	thing

Prefix	Suffix	Definition
_____	-phyll	1. Yellow plant pigment.
_____	_____	2. Special structure (container) of plants, protists, and fungi in which gametes are formed.
_____	_____	3. The female reproductive organ in primitive land plants.
_____	_____	4. The gamete-producing stage in the life cycle of a plant.
_____	_____	5. The spore-producing stage in the life cycle of a plant.
_____	-phyll	6. A small leaf that contains one vascular strand.
_____	-phyll	7. A large leaf that contains multiple vascular strands.
_____	_____	8. Special structure (container) of certain plants and protists in which spores and sporelike bodies are produced.
_____	_____	9. Production of one type of spore in plants (all spores are the same type).
_____	_____	10. Production of two different types of spores in plants, microspores and megaspores.
_____	_____	11. Large spore formed in a megasporangium.
_____	_____	12. Small spore formed in a microsporangium.
_____	_____	13. Mosses, liverworts, and their relatives.
_____	_____	14. The study of mosses.
_____	_____	15. A plant body not developed into true roots, stems, and leaves.
_____	_____	16. An elongated, cone-like reproductive structure.

MATCHING

Terms:

a. Antheridium
b. Dichotomous
c. Gametangia
d. Lignin
e. Liverwort
f. Megaphylls
g. Phloem
h. Prothallus
i. Protonema
j. Rhizoid
k. Rhizome
l. Sorus
m. Stoma
n. Strobilus
o. Thallus
p. Xylem

For each of these definitions, select the correct matching term from the list above.

_____ 1. Vascular tissue that conducts dissolved organic molecules in plants.

_____ 2. A strengthening polymer found in the walls of cells that function for support and conduction..

_____ 3. Hair-like absorptive structures similar in function to roots that extend from the base of the stem of mosses, liverworts, and fern prothallia.

_____ 4. A cluster of sporangia (in the ferns).

_____ 5. A type of branching in which the branches or veins always branch into two more or less equal parts.

_____ 6. The heart-shaped, haploid gametophyte plant found in ferns, whisk ferns, club mosses, and horsetails.

_____ 7. The male gametangium in certain plants.

_____ 8. The sex organs of plants.

_____ 9. Vascular tissue that conducts water and dissolved minerals in plants.

_____10. A horizontal underground stem.

_____11. A bryophyte that looks like an internal human organ.

_____12. Evolved from stem branches that filled in to eventually form leaves.

MAKING COMPARISONS

Fill in the blanks.

Plant	Nonvascular or Vascular	Method of Reproduction	Dominant Generation
Hornworts	Nonvascular	Seedless, reproduce by spores	Gametophyte
Angiosperms	#1	Seeds enclosed within a fruit	#2
Ferns	#3	Seedless, reproduce by spores	#4
Club mosses	#5	Seedless, reproduce by spores	#6
Gymnosperms	#7	#8	Sporophyte

Plant	Nonvascular or Vascular	Method of Reproduction	Dominant Generation
Mosses	#9	#10	#11
Horsetails	Vascular	#12	Sporophyte
Whisk ferns	Vascular	#13	#14

MAKING CHOICES

Place your answer(s) in the space provided. Some questions may have more than one correct answer.

_____ 1. Spores grow
   a. into gametophyte plants.
   b. into sporophyte plants.
   c. into a haploid plant.
   d. to form a plant body by meiosis.
   e. to form a plant body by mitosis.

_____ 2. Plants all have
   a. chlorophyll a.
   b. chlorophyll b.
   c. xanthophyll.
   d. carotene.
   e. yellow pigments.

_____ 3. A strobilus is
   a. on a diploid plant.
   b. on a haploid plant.
   c. on a vascular plant.
   d. found on horsetails.
   e. found on plants that do not have true leaves.

_____ 4. The gametophyte generation of a plant
   a. is diploid.
   b. is haploid.
   c. produces haploid spores.
   d. produces haploid gametes by mitosis.
   e. produces haploid gametes by meiosis.

_____ 5. When a gamete produced by an archegonium fuses with a gamete produced by a male gametophyte plant, the result is
   a. an anomaly.
   b. a diploid zygote.
   c. the first stage in the sporophyte generation.
   d. the first stage in the gametophyte generation.
   e. called fertilization.

_____ 6. Plant sperm cells form in
   a. diploid gametophyte plants.
   b. haploid sporophyte plants.
   c. haploid gametophyte plants.
   d. antheridia.
   e. archegonia.

_____ 7. The leafy green part of a moss is the
   a. sporophyte generation.
   b. gametophyte generation.
   c. rhizoid.
   d. product of buds from a protonema.
   e. thallus.

_____ 8. The sporophyte generation of a plant
   a. is diploid.
   b. is haploid.
   c. produces haploid gametes.
   d. produces haploid spores by mitosis.
   e. produces haploid spores by meiosis.

_____ 9. Liverworts

    a. are bryophytes.   d. contain a medicine that cures liver disease.

    b. are vascular plants.   e. are in the same class as hornworts.

    c. can produce archegonia and antheridia on a haploid gametophyte.

_____10. The spore cases on a fern are

    a. usually on the fronds.   d. often arranged in a sorus.

    b. formed by the haploid generation.   e. precursors to the fiddlehead.

    c. called sporangia.

_____11. The leaves of vascular plants that evolved from stem branches

    a. are megaphylls.   d. contain one vascular strand.

    b. are microphylls.   e. contain more than one vascular strand.

    c. evolutionarily derived from stem tissue.

_____12. Land plants are thought to have evolved from

    a. fungi.   d. charophytes.

    b. green algae.   e. mosses.

    c. bryophytes.

_____13. All plants

    a. have a cuticle covering the aerial portion of the plant.   d. store starch.

    b. develop from multicellular embryos.   e. evolved from a green algae.

    c. are a monophyletic group.

_____14. The water conducting portion of the plant vasculature is the

    a. phloem.   d. xylem.

    b. protonema.   e. gemmae.

    c. microphyll.

_____15. Red algae, green algae, and land plants are collectively classified as

    a. charophytes.   d. bryophytes.

    b. archaeplastids.   e. Hepatophyta.

    c. Anthocerophyta.

_____16. Whisk ferns

    a. are found mainly in the tropics and subtropics.   d. are in phhlum Pteridophyta.

    b. are extinct.   e. have vascularized stems.

    c. should be classified as reduced ferns.

VISUAL FOUNDATIONS

Color the parts of the illustration below as indicated. Also label haploid gametophyte generation, diploid sporophyte generation, fertilization, and meiosis.

RED	☐ zygote	BROWN	☐ capsule
GREEN	☐ archegonium	TAN	☐ gametophyte plants
YELLOW	☐ egg	PINK	☐ sporophyte
BLUE	☐ sperm	VIOLET	☐ spore
ORANGE	☐ antheridium		

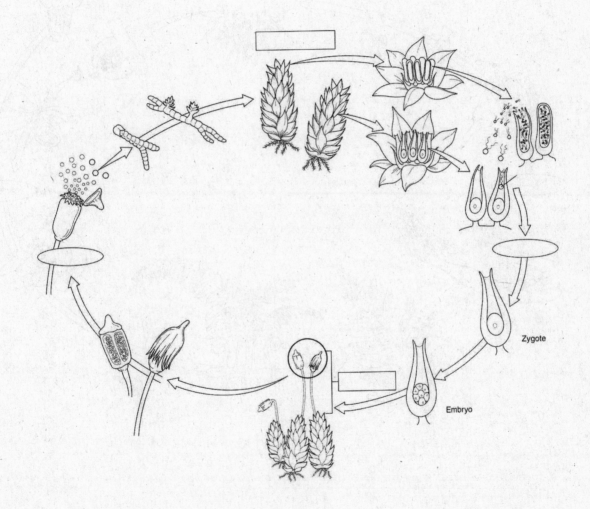

Color the parts of the illustration below as indicated.

RED	☐	zygote
GREEN	☐	archegonium
YELLOW	☐	egg
BLUE	☐	sperm

BROWN	☐	sporangium
TAN	☐	gametophyte
PINK	☐	mature sporophyte
VIOLET	☐	spores

Seed Plants

The most successful plants on earth are the seed plants, also known as the gymnosperms and angiosperms (flowering plants). Whereas gymnosperm seeds are totally exposed or borne on the scales of cones, those of angiosperms are encased within a fruit. All seeds are multicellular, consisting of an embryonic root, stem, and leaves. They are protected by a seed coat that allows them to survive until conditions are favorable for germination, and they contain a food supply that nourishes the seed until it becomes self-sufficient. Seed plants have vascular tissue, and they exhibit alternation of generations in which the gametophyte generation is significantly reduced. They produce both microspores and megaspores. There are four phyla of gymnosperms and one of angiosperms. Seeds and seed plants have been intimately connected with the development of human civilization. Human survival, in fact, is dependent on angiosperms, for they are a major source of human food. Angiosperm products play a major role in world economics. The organ of sexual reproduction in angiosperms is the flower. Flowering plants have several evolutionary advancements that account for their success. The fossil record indicates that seed plants evolved from seedless vascular plants.

REVIEWING CONCEPTS

Fill in the blanks.

INTRODUCTION

1. The primary means of reproduction in plants is by _____.
2. Although most of the seeds eaten by humans are produced by flowering plants, the seeds of the _____, a gymnosperm, may also be eaten.

AN INTRODUCTION TO SEED PLANTS

3. The two groups of seed plants are the (a)_____, from the Greek "naked seed," and the (b)_____, meaning "seed enclosed in a vessel or case."
4. Seed plants all have two types of vascular tissues: (a)_____ for conducting water and minerals and (b)_____ for conducting dissolved sugar.

GYMNOSPERMS

5. The four phyla of plants with "naked seeds" are: _____
_____.

Conifers are woody plants that produce seeds in cones

6. Many conifers produce a clear substance called _____ which helps to protect the plants from attack by insects and fungi.
7. The conifers are the largest group of gymnosperms. Most of them are _____, which means that the male and female reproductive organs are in different places on the same plant.

8. The largest genus (a) _____ contains the largest number of conifers and pine trees are mature (b) _____.

9. The leaf-like structures that bear sporangia on male cones of conifers are (a)_____, at the base of which are two microsporangia containing many "microspore mother cells," also called (b)_____. These cells develop into the (c)_____.

10. Female cones have (a)_____ on the upper surface of each cone scale, within which are megasporocytes. Each of these cells produces four haploid (b)_____, three of which disintegrate, while the fourth develops into the egg producing (c)_____.

11. One of the pollen grains that adheres to the sticky surfaced of the ovule grows to form a _____ which digests its way through the megasporangium until it reaches the egg where two sperm cells will be released.

12. A major adaptation of the pine life cycle was the elimination of (a) _____ and the use of (b) _____ as a means of dispersing the sperm.

13. Reproduction in conifers is totally adapted for life _____.

Cycads have seed cones and compound leaves

14. Cycads are in the phylum _____.

15. The cycads reproduce in a manner similar to pines, except cycads are _____, meaning that male cones and female cones are on *separate* plants.

Ginkgo biloba is the only living species in its phylum

16. *Ginko biloba* is in the phylum _____.

Gnetophytes include three unusual genera

17. Gnetophytes belong to the phylum _____.

18. The gnetophytes share a number of advances over the rest of the gymnosperms, one of which is the presence of distinct _____ in their xylem tissues.

FLOWERING PLANTS

19. Flowering plants are (a) _____ plants that reproduce (b) _____ by forming flowers.

20. Flowering plants have efficient water-conducting cells called _____ in their xylem.

Monocots and eudicots are the two largest classes of flowering plants

21. A few examples of monocots include _____ _____.

22. A few examples of eudicots include _____ _____.

23. The monocots have floral parts in multiples of (a)_____, and their seeds contain (b)_____(#?) cotyledon(s). The nutritive tissue in their mature seeds is the (c)_____.

24. The eudicots have floral parts in multiples of (a)_____, and their seeds contain (b)_____(#?) cotyledons. The nutritive tissue in their mature seeds is usually in the (c)_____.

Sexual reproduction takes place in flowers

25. The four main organs in flowers are the (a)_____ _____, each of them arranged in whorls. The "male" organs are the (b)_____ and the "female" organs are the (c)_____. Flowers with both "male" and "female" parts are said to be (d)_____.

26. The (a)_____ are the lowermost and outermost whorl on the floral shoot and they are green and (b)_____ in their appearance.

27. Pollen forms in the (a)_____, a saclike structure on the tip of a stamen, and ovules form within the (b)_____.

The life cycle of flowering plants includes double fertilization

28. The megasporocyte in an ovule produces four haploid (a)_____, three of which disintegrate while the remaining one develops into the female gametophyte or (b)_____, as it is also called.

29. Microsporocytes in the anther produce four haploid (a)_____, each of which develops into a (b)_____.

30. Double fertilization is a phenomenon that is unique to flowering plants. It results in the formation of two structures the (a)_____ and the (b) _____.

Seeds and fruits develop after fertilization

31. As a seed develops from an ovule following fertilization, the ovary wall surrounding it enlarges dramatically and develops into a _____.

Flowering plants have many adaptations that account for their success

32. Pollen transfer results in (a)_____, which mixes the genetic material and promotes (b)_____ among the offspring.

33. The stems and roots of flowering plants are often modified for (a) _____ and (b) _____ features that help flowering plants survive in severe environments.

Floral structure provides insights into the evolutionary process

34. Many botanists have concluded that stamens and carpels are probably derived from _____.

THE EVOLUTION OF SEED PLANTS

35. (a)_____ is an extinct group of plants descended from ancestral seedless vascular plants. This group probably gave rise to conifers and to another group of extinct plants called the (b)_____, which in turn are thought to be the ancestors of cycads and possibly ginkgo.

Our understanding of the evolution of flowering plants has made great progress in recent years

36. Flowering plants evolved from _____.

37. Currently, many botanists think that the closest living gymnosperms related to flowering plants are _____.

BUILDING WORDS

Use combinations of prefixes and suffixes to build words for the definitions that follow.

Prefixes	The Meaning	Suffix	The Meaning
angio-	vessel, container	-sperm	seed
di-	two, twice, double	-zation	the process of
endo-	within		
fertil-	fertile		
gymn(o)-	naked, bare, exposed		
mon(o)-	alone, single, one		

Prefix	Suffix	Definition
_____	_____	1. A plant that produces seeds that are totally exposed or on the scales of cones.
_____	_____	2. A plant that produces their seeds within a fruit.
_____	-oecious	3. Pertains to plant species in which the male and female reproductive parts are in different locations on one plant.
_____	-oecious	4. Pertains to plant species in which the male and female reproductive parts are on two different plants.
_____	_____	5. The fusion of two gametes producing a zygote.
_____	_____	6. The nutritive tissue that results from double fertilization and is found in a mature seed.

MATCHING

Terms:

a. Anther
b. Calyx
c. Carpel
d. Corolla
e. Cotyledon

f. Integument
g. Ovary
h. Ovule
i. Petals
j. Pistil

k. Pollination
l. Sepals
m. Style
n. Tracheid

For each of these definitions, select the correct matching term from the list above.

____ 1. The part of the stamen in flowers that produces microspores and, ultimately, pollen.

____ 2. An embryonic seed leaf.

____ 3. One of the layers of sporophyte tissue that surrounds the megasporangium.

____ 4. The colored cluster of modified leaves that constitute the next-to-outermost portion of a flower.

____ 5. In seed plants, the transfer of pollen from the male to the female part of the plant.

____ 6. The neck connecting the stigma to the ovary of a carpel.

____ 7. The part of a plant that develops into seed after fertilization.

____ 8. The outermost parts of a flower, usually leaflike in appearance, that protect the flower as a bud.

____ 9. The collective term for the sepals of a flower.

____10. The female reproductive unit of a flower that bears the ovules.

____11. A long, tapering cell with pits through which water flows.

MAKING COMPARISONS

Fill in the blanks.

Plant Features	Monocot	Eudicot
Vascular bundles in stems	Scattered vascular bundles	Vascular bundles arranged in a circle
Growth tissue type	Herbaceous	#1
Nutritive material in mature seeds	#2	#3
Leaf shape	Long, narrow leaves	#4
Seeds	#5	2 cotyledons
Flowers	Floral parts in multiples of 3	#6
Leaf venation	#7	#8

MAKING CHOICES

Place your answer(s) in the space provided. Some questions may have more than one correct answer.

_____ 1. The major difference between gymnosperms and angiosperms is that in gymnosperms
   a. self fertilization occurs.   d. flowers generally have 4 or 5 sepals and petals.
   b. seeds are exposed ("naked").   e. the mature ovary is a fruit.
   c. there is no ovary wall surrounding ovules.

_____ 2. The petals of a flower are collectively known as the
   a. corolla.   d. androecium.
   b. calyx.   e. florus perfecti.
   c. gynoecium.

_____ 3. The sepals of a flower are collectively known as the
   a. corolla.   d. androecium.
   b. calyx.   e. florus perfecti.
   c. gynoecium.

_____ 4. The triploid endosperm of angiosperms develops from fusion of
   a. two sperm and one polar nucleus.   d. one diploid ovule and one haploid sperm.
   b. two polar nuclei and one sperm.   e. one haploid egg and one diploid sperm.
   c. three polar nuclei.

_____ 5. Which of the following is/are correct about a pine tree?
   a. Sporophyte generation is dominant.   d. Nutritive tissue in seed is gametophyte tissue.
   b. Male gametophyte produces an antheridium.   e. Pollen grain is an immature male gametophyte.
   c. Gametophyte is dependent on sporophyte for nourishment.

_____ 6. A perfect flower has
   a. stamens only.   d. anthers.
   b. carpels only.   e. ovules.
   c. both stamens and carpels.

_____ 7. In gymnosperms, the pollen grain develops from
   a. microspore cells.   d. the gametophyte generation.
   b. spores.   e. meiosis of cells in the microsporangium.
   c. the male gametophyte.

_____ 8. Dioecious plants with naked seeds, motile sperm, and whose pollen is carried by air or insects are
   a. ginkgoes.   d. gnetophytes.
   b. cycads.   e. gymnosperms.
   c. extinct.

_____ 9. A pine tree has
   a. sporophylls.   d. two sizes of spores in separate cones.
   b. megasporangia on the female cones. e. female cones that are larger than male cones.
   c. separate male and female parts on the same tree.

_____ 10. In angiosperms, of the four haploid megaspores that result from meiosis of the megaspore mother cell, one of them becomes the
   a. male gametophyte generation.   d. archegonium.
   b. female gametophyte generation.   e. antheridium.
   c. embryo sac.

_____11. Seeds are reproductively superior to spores because seeds

      a. are single cells with stored nutrients.   d. are protected by a seed coat.

      b. seeds live for an extended period of time.   e. a seed is further developed than a spore.

      c. seeds have a reduced rate of metabolism.

_____12. Like bryophytes and seedless plants, seed plants

      a. have a life cycle with an alternation of generation.   d. have a dominate sporophyte generation.

      b. have free living gametophytes.   e. are heterosporous.

      c. a female gametophyte attached to the sporophyte generation.

_____13. Humans use conifers

      a. for their wood.   d. for medical purposes.

      b. for the production of chemicals.   e. for landscaping.

      c. in some cases as a source of food.

_____14. Many botanists think that female cones are modified

      a. megasporangia.   d. branch systems.

      b. sporophylls.   e. cycads

      c. sieve tubes.

_____15. The ancestors of all other flowering plants are thought to be

      a. seed ferns   d. core angiosperms

      b. progymnosperms.   e. magnoliids

      c. basal angiosperms.

VISUAL FOUNDATIONS

Color the parts of the illustration below as indicated. Also label haploid gametophyte generation, diploid sporophyte generation, fertilization, and meiosis.

RED ☐ zygote

GREEN ☐ female cone

YELLOW ☐ megasporangium, megaspore,

BLUE ☐ pollen grain

ORANGE ☐ microsporangium, microspore

BROWN ☐ male cone

TAN ☐ sporophyte

PINK ☐ gametophyte

VIOLET ☐ embryo

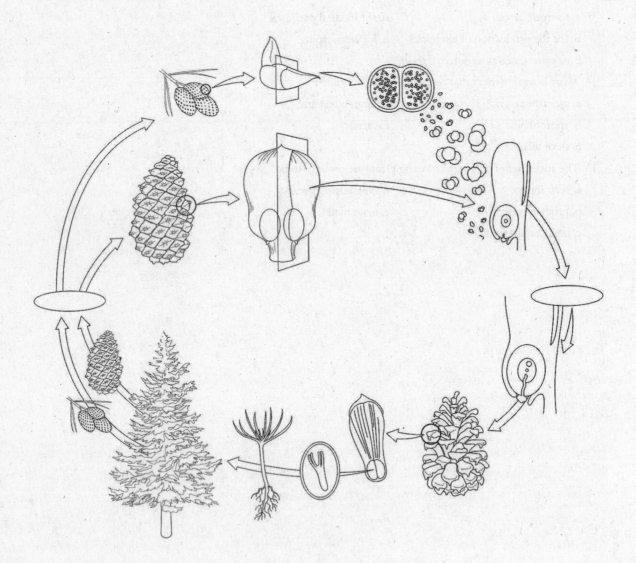

Color the parts of the illustration below as indicated. Also label haploid gametophyte generation, diploid sporophyte generation, double fertilization, and meiosis.

RED ☐ zygote

GREEN ☐ female floral part

YELLOW ☐ megasporangium, megaspore

BLUE ☐ pollen tube

ORANGE ☐ microsporangium, microspore

BROWN ☐ male floral part

TAN ☐ male and female sporophytes

PINK ☐ male and female gametophytes

VIOLET ☐ embryo

The Fungi

Fungi are eukaryotes with cell walls. They digest their food outside their body by secreting digestive enzymes onto the food source. Fungi then absorb the nutrients into their own body. Fungi are nonmotile and reproduce by means of spores that are formed either sexually or asexually. Some fungi are unicellular (the yeasts) but most are multicellular, having a filamentous body plan. Fungi are classified into five phyla. Most fungi are decomposers, and as such play a critical role in the cycle of life. Others, however, form symbiotic relationships with other organisms. Fungi are of both positive and negative economic importance. Some damage stored goods and building materials. Some provide food for humans. Some function in brewing beer and in baking. Some produce antibiotics and other drugs and chemicals of economic importance, and some cause serious diseases in humans and economically important animals and plants.

REVIEWING CONCEPTS

Fill in the blanks.

INTRODUCTION

1. _____ are biologists who study fungi.
2. Like prokaryotes, most fungi are _____ that obtain nutrients from dead organic matter.

CHARACTERISTICS OF FUNGI

3. Fungi share key characteristics including being _____ (heterotrophs or autotrophs?).
4. Many fungi are _____(more or less?) sensitive to osmotic pressures than are bacteria.

Fungi absorb food from the environment

5. Digestion in fungi occurs _____ (intracellularly or extracellulary?).
6. Fungi store excessive amounts of nutrients as _____.

Fungi have cell walls that contain chitin

7. At some stage in their life, fungi have a feature present in bacteria, certain protists, and plants. This feature is _____.
8. Cell walls of most fungi contain (a) _____, which is highly resistant to microbial breakdown. This same material is also part of the protective covering of organisms in phylum (b) _____.
9. The simplest fungi are the _____.
10. Multicellular fungi have a body consisting of threadlike filaments called _____.

11. Molds form a tissuelike network called a (a) _____.
    Fungi that form these networks are called (b) _____.

FUNGAL REPRODUCTION

12. Must fungi reproduce by microscopic structures called _____.

13. The large part of the common mushroom is called a (a) _____ in
    which (b) _____ are produced.

Many Fungi reproduce asexually

14. Yeasts reproduce asexually by forming (a) _____. Many species of
    multicellular fungi reproduce asexually by producing (b) _____.

Most fungi reproduce sexually

15. Most fungal cells contain a _____ nucleus.

16. Fungi communicate chemically by secreting signaling molecules called
    _____.

17. When the cells of two different mating types of fungi fuse a hyphal cell is
    produced which has two distinctly different nuclei. These cells are said to be
    (a) _____, whereas cells containing only one haploid
    nucleus are said to be (b) _____.

FUNGAL DIVERSITY

18. Unlike plants, the walls of fungal cells do not contain _____.

Fungi are assigned to the opisthokont clade

19. It is hypothesized that the common ancestor of all plants, fungi, and
    animals was a _____.

20. Based upon their characteristics, fungi are considered to be more closely
    related to _____ (plants or animals?).

Diverse groups of fungi have evolved

21. Based on the characteristics of sexual spores, fruiting bodies, and
    molecular data, fungi are generally assigned to the five phyla

    _____

    _____

    _____

22. When individuals within a given taxa share a common ancestor they are
    said to be _____.

23. Fungi have been classified as deuteromycetes simply because no one has
    observed them to have a _____ in their life cycles.

Chytrids have flagellate spores

24. Most chytrids are unicellular or composed of a few cells that form a
    simple body called a

    (a)_____, which may have slender extensions called
    (b)_____.

25. Chytrids are the only group of fungi to produce a _____
    cell.

Zygomycetes reproduce sexually by forming zygospores

26. The zygomycete species _____ is the well-known black bread mold.

27. Black bread mold is _____, meaning that an individual fungal hypha is self-sterile and mates only with a hypha of a different mating type.

Microsporidia have been a taxonomic mystery

28. Microsporidia are (a) _____ pathogens. They infect people with (b) _____.

29. Gene sequence data from microsporidia indicate that they originally had _____.

Glomeromycetes are symbionts with plant roots

30. The symbiotic relationships between fungi and the roots of plants are called _____.

31. Glomeromycetes and plants exchange nutrients in structures called _____.

Ascomycetes reproduce sexually by forming ascospores

32. Ascomycetes produce sexual spores in sacs called (a)_____, and asexual spores called

(b)_____ at the tips of (c)_____.

33. As the asci develop, they are surrounded by intertwining hyphae that develop into a fruiting body known as an _____.

34. Asexual reproduction in yeast occurs primarily by _____.

Basidiomycetes reproduce sexually by forming basidiospores

35. Basidiomycetes develop an enlarged, club-shaped hyphal cell called a (a)_____, on the surface of which four spores called (b)_____ develop.

36. What we call a mushroom grows from a compact mass of hyphae along the mycelium called a (a)_____. A mushroom is more formally referred to as a (b)_____.

ECOLOGICAL IMPORTANCE OF FUNGI

37. Most fungi are free-living decomposers that absorb nutrients from organic wastes and dead organisms. In the process, they release (a)_____ to the atmosphere and return (b)_____ to the soil.

Fungi form symbiotic relationships with some animals

38. Symbiotic relationships have formed between certain fungi and animals; the fungi break down (a)_____and (b)_____ that the animals ingest but are incapable of digesting.

Mycorrhizae are symbiotic relationships between fungi and plant roots

39. Glomeromycetes form (a)_____ connections inside the roots of plants, but species of acomycetes and basidiomycetes form

(b)_____ that coat the surface of roots.

A lichen consists of two components: a fungus and a photoautotroph

40. The phototroph portion of the lichen is usually a
(a)_____, or (b) _____, or
(c)_____. The fungus is sometimes a basidiomycete, but usually an
(d)_____.

41. Lichens typically assume one of three forms: (a)_____, a
low, flat growth; (b)_____, a flat growth with leaf-like lobes;
or (c)_____, an erect and branching growth.

42. Lichens reproduce asexually by fragmentation, wherein pieces of lichen
called _____ break off and begin growing after landing on a
suitable substrate.

ECONOMIC, BIOLOGICAL, AND MEDICAL IMPACT OF FUNGI

43. Cellulose is the most abundant organic compound on earth. The second most
abundant is _____.

Fungi provide beverages and food

44. (a)_____ are the fungi used to make wine and beer, and to
produce baked goods. Wine is produced when the fungus ferments
(b)_____, beer results from fermentation of
(c)_____, and
(d)_____ bubbles cause bread to rise.

45. Roquefort and Camembert cheeses are produced using the genus
(a)_____, and the genus
(b)_____ is used to make soy sauce from
soybeans.

46. Some of the most toxic mushrooms, such as the "destroying angel" and
"death cap," belong to the genus _____.

47. The chemical _____ found in some mushrooms causes
hallucinations and intoxication.

Fungi are important to modern biology and medicine

48. Alexander Fleming noticed that bacterial growth is inhibited by the mold
(a)_____, and this discovery eventually led to the
development of the most widely used of all antibiotics, (b)_____.

49. An ascomycete that infects cereal plant flowers produces a structure called
an _____ where seeds would normally form. When livestock or
humans eat grain or grain products contaminated with this fungus, they
may be poisoned by the extremely toxic substances contained therein.

Fungi are used in bioremediation and to biologically control pests

50. Some fungi can biodegrade pesticides, herbicides, (a)_____
and (b) _____.

51. Scientists are investigating the use of micropsoridia to control the spread
of (a) _____ a disease that is transmitted by the
(b) _____ mosquito.

Some fungi cause diseases in humans and other animals

52. Some species of Aspergillus produce potent mycotoxins called _____ that harm the liver and are known carcinogens.

Fungi cause many important plant diseases

53. All plants are susceptible to fungal diseases. Fungi enter plants through stomata, or wounds, or by dissolving a portion of cuticle with the enzyme _____.

54. Parasitic fungi often extend specialized hyphae called _____ into host cells in order to obtain nutrients from host cell cytoplasm.

BUILDING WORDS

Use combinations of prefixes and suffixes to build words for the definitions that follow.

Prefixes	The Meaning		Suffixes	The Meaning
Basid(io)-	pedestal		-carp	fruiting body
coeno-	common		-cyt(ic)	cell
			-gam(y)	union
			-(el)ium	region
hetero-	different		-karyo(tic)	nucleus
			-oid	resembling
homo-	same		-phore	bearer
mono-	alone, single, one		-troph	nutrition, growth, "eat"
myc-	fungus			
pher-	to carry			
plasm-	formed			
rhiz-	root			
sapro-	rotten			

Prefix	Suffix	Definition
_____	_____	1. Pertains to an organism made up of a multinucleate, continuous mass of cytoplasm enclosed by one cell wall (all nuclei share a common cell).
_____	_____	2. Pertains to hyphae that contain only one nucleus per cell.
_____	-thallic	3. Pertains to an organism that has two different mating types.
conidio-	_____	4. Hyphae in the ascomycetes that bear conidia.
_____	-thallic	5. Pertains to an organism that can mate with itself.
_____	_____	6. All the hyphae of a fungus.
_____	_____	7. The fusion of two mating type cells forming one larger cell.
_____	____omone____	8. Signaling molecule used by fungi to communicate chemically.
_____	_____	9. A hypha used by the fungus as an anchor and to absorb nutrients.
_____	_____	10. The basic mushroom structure consisting of a base and a cap.

MATCHING

Terms:

a.   Ascocarp
b.   Ascospore
c.   Basidiocarp
d.   Basidium
e.   Budding
f.   Chitin
g.   Conidium
h.   Dikaryotic
i.   Haustorium
j.   Lichen
k.   Mycelium
l.   Mycorrhizae
m.   Mycotoxin
n.   Psilocybin
o.   Soredium
p.   Zygospore

For each of these definitions, select the correct matching term from the list above.

_____ 1. Compound organism composed of a photosynthetic green alga or cyanobacterium and a fungus.

_____ 2. The vegetative body of fungi; consists of a mass of hyphae.

_____ 3. The sexual spores produced by an ascomycete.

_____ 4. The fruiting body of a basidiomycete.

_____ 5. Asexual reproductive body produced by lichens.

_____ 6. Mutualistic associations of fungi and plant roots that aid in the absorption of materials.

_____ 7. Asexual reproduction in which a small part of the parent's body separates from the rest and develops into a new individual.

_____ 8. The club-shaped spore-producing organ of certain fungi.

_____ 9. Cells having two nuclei.

_____10. A component of fungal cell walls; a polymer that consists of subunits of a nitrogen-containing-sugar.

_____ 11. Poisonous compound produced by some fungal species.

MAKING COMPARISONS

Fill in the blanks.

Phylum	Mode of Sexual Reproduction	Mode of Asexual Reproduction	Representative(s) of the Group
#1	Zygospores	Haploid spores	Black bread mold
#2	Ascospores	Conidia	Yeasts, cup fungi, morels, truffles, and blue-green and pink molds
#3	Basidiospores	Uncommon	Mushrooms, puffballs, rusts, and smuts
#4	Unknown	Blastospores	Symbiotic fungi

MAKING CHOICES
Place your answer(s) in the space provided. Some questions may have more than one correct answer.

_____ 1. Fungi
    a. are heterotrophs.
    b. are photoautotrophs.
    c. are eukaryotes.
    d. possess cell walls.
    e. digest food outside their bodies.

_____ 2. Fungal hyphae that contain two genetically distinct nuclei within each cell are known as
    a. dihaploid.
    b. disomic.
    c. dikaryotic.
    d. $2n$.
    e. $n + n$.

_____ 3. A lichen can be composed of a/an
    a. alga and fungus.
    b. photoautotroph and fungus.
    c. cyanobacterium and ascomycete.
    d. alga and basidiomycete.
    e. alga and ascomycete.

_____ 4. A common fungal infection of the mucous membranes of the mouth, throat, or vagina is
    a. an autoimmune response.
    b. St. Anthony's fire.
    c. histoplasmosis.
    d. ergotism.
    e. candidiasis.

_____ 5. Fungi can reproduce
    a. sexually.
    b. asexually.
    c. by spore formation.
    d. by simple division.
    e. by budding.

_____ 6. The black fungus growing on a piece of bread
    a. is heterothallic.
    b. has male and female strains.
    c. has coenocytic hyphae.
    d. has sporangia on the tips of stolons.
    e. is in the phylum Zygomycota.

_____ 7. The genus *Penicillium*
    a. is an ascomycete.
    b. is a sac fungus.
    c. produces the flavor in Roquefort cheese.
    d. produces the antibiotic penicillin.
    e. produces the flavor in Brie cheese.

_____ 8. Lichens can be used as indicators of air pollution because they
    a. cannot excrete absorbed elements.
    b. tolerate sulfur dioxide.
    c. can endure large quantities of toxins.
    d. do not grow well in polluted areas.
    e. overgrow polluted areas.

_____ 9. A mass of filamentous hyphae is called a
    a. hypha.
    b. mycelium.
    c. conidium.
    d. thallus.
    e. ascocarp.

_____ 10. Yeast participates in the brewing of beer by
    a. adding vital amino acids.
    b. fermenting grain sugars.
    c. fermenting fruit sugars.
    d. producing ethyl alcohol.
    e. converting barley to hops.

_____11. If you eat just any mushroom that you find in the wild, there's a chance that you will
   a. die.
   b. become intoxicated.
   c. see colors that aren't really there.
   d. not become nauseated or die.
   e. ingest the hallucinogenic drug psilocybin.

_____12. A fungus infection throughout the body obtained by exposure to bird droppings is likely
   a. an autoimmune response.
   b. St. Anthony's fire.
   c. histoplasmosis.
   d. ergotism.
   e. candidiasis.

_____13. Chytrids (aka chytridiomycetes)
   a. are fungi.
   b. are funguslike protists.
   c. inhabit damp or wet environments.
   d. have flagellated spores.
   e. are among the latest in their kingdom to evolve.

_____14. Fungi can be found growing
   a. in moist habitats.
   b. in tree and plant roots.
   c. in the grasses of geothermal hot springs.
   d. where organic material is available.
   e. in the wood of buildings.

_____15. The cell walls of fungi contain
   a. carbohydrates.
   b. lignin
   c. cellulose.
   d. cutinase.
   e. chitin.

_____16. Coenocytic fungi
   a. are elongated.
   b. multinucleated.
   c. lack septa.
   d. round.
   e. have a single nucleus.

_____17. The clade opisthokonts includes
   a. plants.
   b. fungi.
   c. amoebozoa
   d. animals.
   e. choanoflagellates.

_____18. Terms associated with sexual reproduction in fungi include
   a. plasmogamy.
   b. karyogamy.
   c. conidiophores.
   d. zygote nucleus.
   e. conidia.

_____19. Features some fungi share with some plants include
   a. a cell wall.
   b. growth to enormous size.
   c. asexual reproduction.
   d. active growth in dry conditions.
   e. alternation of generation.

_____20. Microsporidia
   a. infect eukaryotic cells.
   b. were once assigned to protozoa.
   c. produce a polar tube.
   d. are opportunistic pathogens.
   e. have two developmental stages.

_____ 21. Species of yeast that are self-fertile are referred to as being

a. heterothallic.

b. polythallic.

c. unithallic.

e. multithallic.

e. homothallic.

_____ 22. Terms associated with basidiomycetes include

a. primary mycelium.

b. cytoplasmic streaming.

c. karyogamy.

d. gills.

e. "fairy rings"

_____ 23. Mycorrhizae are symbiotic associations between

a. fungi and bacteria.

b. fungi and grasses.

c. fungi and fruits.

d. fungi and plant roots.

e. fungi and grains.

VISUAL FOUNDATIONS

Color the parts of the illustration below as indicated.

RED ☐ haustoria
GREEN ☐ hypha
YELLOW ☐ epidermal cells
BROWN ☐ spore
VIOLET ☐ stoma

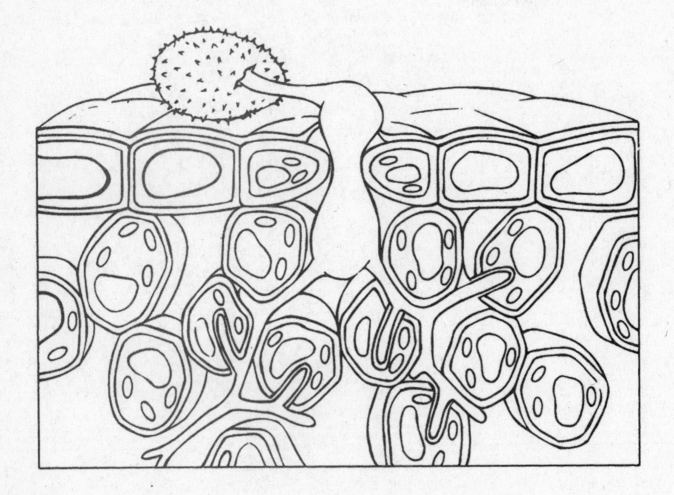

Color the parts of the illustration below as indicated. Also label asexual reproduction, sexual reproduction, and the diploid and haploid generations.

RED ☐ gamete type A
GREEN ☐ gamete type B
YELLOW ☐ haploid zoospore
BLUE ☐ diploid thallus
ORANGE ☐ haploid thallus
BROWN ☐ motile zygote
TAN ☐ resting sporangium
VIOLET ☐ diploid zoospore

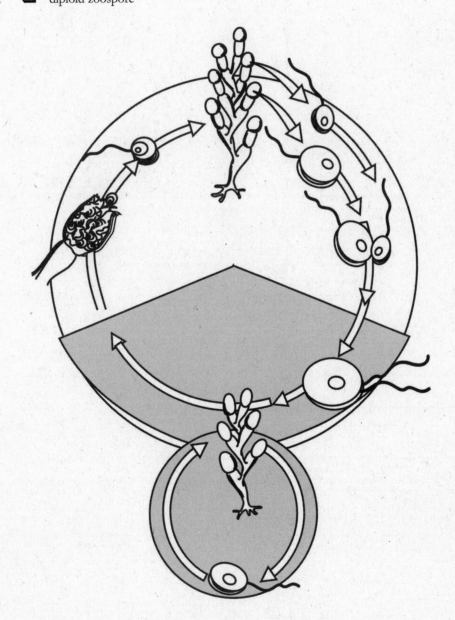

An Introduction to Animal Diversity

This is the first of three chapters discussing the animal kingdom. Animals are eukaryotic, multicellular, heterotrophic organisms. They are made up of localized groupings of cells with specialized functions. One grouping, for example, might function in locomotion, another might respond to external stimuli, and another might function in sexual reproduction. Animals inhabit virtually every environment on earth. As discussed in this chapter, animals have adapted to living in ocean, freshwater, and terrestrial habitats. Animals may be classified based upon the characters they posses such as radial or bilaterally symmetry; the absence of a body cavity (the acoelomates); the presence of a body cavity between mesoderm and endoderm (the pseudocoelomates); the presence of a body cavity within mesoderm (the coelomates); those in which the first opening that forms in the embryonic gut becomes either the mouth (the protostomes); or those in which the first opening in the embryonic gut forms the anus (the deuterostomes). This chapter discusses all of these characters.

REVIEWING CONCEPTS

Fill in the blanks.

INTRODUCTION

1. An estimated _____% of all animal species that ever inhabited our planet are extinct.

2. Molecular data indicate that many groups of animals are _____.

ANIMAL CHARACTERS

3. Animals lack cell walls and instead rely on _____ for structural support.

4. Animals are _____, or consumers that depend on producers for their raw materials and energy.

ADAPTATIONS TO OCEAN, FRESHWATER, AND TERRESTRAIL HABITATS

Marine habitats offer many advantages

5. (a) _____ and (b) _____ balance are more easily maintained in a salt water environment than a freshwater environment.

6. The freshwater environment is _____ than that of seawater.

Some animals are adapted to freshwater habitats

7. Freshwater environments are _____ to the tissue fluids of animals living there.

Terrestrial living requires major adaptations

8. Land environments tend to _____ animals.

9. Many land animals have evolved _____ which allows them to reproduce in a dry environment.

ANIMAL EVOLUTION

10. The common ancestor of animals was a _____.

11. According to the principle of _____ complex molecules such as RNA are likely to have evolved only once.

Molecular systematics helps biologists interpret the fossil record

12. The rapid appearance of an amazing variety of body plans in the fossil record is known as the _____.

Biologists develop hypotheses about the evolution of development

13. Mutations in _____ genes could have resulted in rapid changes in body plans.

RECONSTRUCTING ANIMAL PHYLOGENY

14. Biologists use (a) _____ and differences in

    (b) _____ to infer relationships between animals.

Animals exhibit two main types of body symmetry

15. (a) _____ and (b) _____ are two adaptations for locomotion.

16. There are many anatomical terms that are used to describe the locations of body parts. Among them are: (a)_____ for front and (b)_____ for rear; (c)_____ for a back surface and (d)_____ for the under or "belly" side.

17. Other locations of body parts include (a)_____ for parts closer to the midline and (b)_____ for parts farther from the midline, towards the sides; (c)_____ toward the head end; and (d)_____ when away from the head, toward the tail.

Animal body plans are linked to the level of tissue development

18. In the early development of all animals except sponges, cells form layers called _____ that give rise to linings, tissues, and body structures.

Most bilateral animals have a body cavity lined with mesoderm

19. Animals can also be grouped according to the presence or absence of a body cavity and, when present, the type of body cavity. For example, the flatworms are called _____ because they do not have a body cavity.

20. Animals with body cavities are either (a)_____, possessing a body cavity between the mesoderm and endoderm, or (b)_____ with a true body cavity within the mesoderm.

Bilateral animals form two main clades based on differences in development

21. Coelomate animals can be grouped by shared developmental characteristics. In mollusks, annelids, and arthropods — all protostomes — the blastopore develops into the (a)_____, whereas in echinoderms and chordates — both deuterostomes — the blastopore becomes the (b)_____.

22. Early cell divisions in deuterostomes are either parallel to or at right angles to the polar axis, a pattern of division known as

(a)_____; in protostomes, early cell divisions are diagonal to the polar axis, a type of division referred to as

(b)_____.

Biologists have identified major animal clades based on structure, development, and molecular data

23. Animals are referred to as _____.

24. Animals that have germ layers are classified as _____.

25. The two major clades of protostomes are

(a)_____

and (b) _____.

26. One important innovation of body forms has been _____, a body plan in which certain structures are repeated producing body compartments.

BUILDING WORDS

Use combinations of prefixes and suffixes to build words for the definitions that follow.

Prefixes	The Meaning		Suffixes	The Meaning
blast-	bud, sprout		-age	collection of
cleav-	divide		-coel(om)(y)	cavity
deutero-	second		-derm	skin
ecto-	outer, outside, external		-stome	mouth
entero-	intestine		-ula	little
gastro-	stomach			
meso-	middle			
proto-	first, earliest form of			
pseudo-	false			
schizo-	split			

Prefix	Suffix	Definition
_____	_____	1. A body cavity between the mesoderm and endoderm; not a true coelom.
_____	_____	2. The outermost germ layer.
_____	_____	3. The middle layer of the three basic germ layers.
_____	_____	4. Major division of the animal kingdom in which the mouth forms from the first opening in the embryonic gut (the blastopore).
_____	_____	5. Major division of the animal kingdom in which the mouth forms from the second opening in the embryonic gut.
_____	_____	6. The process of coelom formation in which the mesoderm splits into two layers.
_____	_____	7. The process of coelom formation in which the mesoderm forms as "outpocketings" of the developing intestine.

_____ _____ 8. The hollow ball of cells that forms during early embryonic development.

_____ _____ 9. The tissue lining the gut cavity ("stomach") that functions in digestion in certain phyla.

_____ _____ 10. A series of mitotic cell divisions.

MATCHING

Terms:

a. Acoelomate
b. Autotroph
c. Cephalization
d. Cleavage
e. Coelom
f. Cuticle

g. Dorsal
h. Eumetazoa
i. Gastrulation
j. Heterotroph
k. Invertebrate
l. Metazoa

m. Pseudocoelom
n. Segmentation
o. Sessile
p. Ventral

For each of these definitions, select the correct matching term from the list above.

_____ 1. A body cavity not completely lined with mesoderm.

_____ 2. Depends upon a producer to provide molecules for energy and raw materials.

_____ 3. Without a body cavity (coelom).

_____ 4. The main body cavity of most animals.

_____ 5. Describes animals with either two or three germ cell layers.

_____ 6. The clustering of neural tissues at the anterior (leading) end of an animal.

_____ 7. Referring to the belly aspect of an animal's body.

_____ 8. A body plan which results in a series of body compartments.

_____ 9. Permanently attached to one location.

_____ 10. The process that forms and segregates the three germ layers during embryonic development.

MAKING COMPARISONS

Fill in the blanks.

Deuterostome	Protostome	Character
#1	#2	Pattern of cleavage
#3 cleavage	#4 cleavage	Developmental fate of the early cells
#5	#6	Developmental fate of the blastopore
#7	#8	Method of coelom formation

MAKING CHOICES

Place your answer(s) in the space provided. Some questions may have more than one correct answer.

_____ 1. Eumetazoans that have bilateral symmetry and are in the clade Lophotrochozoa include

    a. echinoderms.

    b. arthropods.

    c. flatworms.

    d. nematodes.

    e. None of these choices.

_____ 2. Features of arthropods include

    a. radial symmetry.

    b. a protostome pattern of development.

    c. a coelom.

    d. molting.

    e. a deuterostome pattern of development.

_____ 3. Examples of acoelomate eumetazoans include

    a. sponges.

    b. insects.

    c. flatworms.

    d. those groups whose coelom developed from a blastocoel.

    e. nematodes.

_____ 4. The clade Bilateria includes

    a. sponges.

    b. cnidarians.

    c. ctenophores.

    d. arthropods

    e. mollusks.

_____ 5. When a group of cells moves inward to form a sac in early embryonic development, and that pore ultimately develops into a mouth, the group of animals is considered to be a

    a. protostome.

    b. deuterostome.

    c. pseudocoelomate.

    d. parazoan.

    e. schizocoelomate.

_____ 6. The one invertebrate phylum with radial symmetry and distinctly different tissues is

    a. Porifera.

    b. Cnidaria.

    c. Platyhelminthes.

    d. Nemertea.

    e. Echinodermata.

_____ 7. The roundworms

    a. are pseudocoelomates.

    b. have a mastax.

    c. are in the phylum Platyhelminthes.

    d. include hookworms.

    e. are nematodes.

_____ 8. Animals with a water vascular system include

    a. hemichordates.

    b. chordates.

    c. sponges.

    d. nematodes.

    e. echinoderms.

_____ 9. The taxon that has epidermal stinging cells is

    a. Scyphozoa.

    b. Hydrozoa.

    c. Porifera.

    d. Cnidaria.

    e. Ctenophora.

_____ 10. Segmentation occurs in

    a. arthropods.

    b. nematodes.

    c. vertebrates.

    d. chordates.

    e. earthworms.

_____ 11. Animals with bilateral symmetry include

    a. ecydyzoans.

    b. protostomes.

    c. sponges.

    d. deuterostomes.

    e. lophotrochozoans.

_____12. Animls with a muscle layer derived from mesoderm include
    a. coelomates.   d. nematodes.
    b. flatworms.   e. pseudocoelomates.
    c. vertebrates.

_____13. A monophyletic grouping
    a. can be used to describe the protists.   d. includes all the descendents of a common ancestor.
    b. can be used to describe the clade Deuterostomia.   e. is supported using molecular data.
    c. can be used to describe the clades Prostomia and Bilateria.

_____14. Clade Lophotrochozoa includes
    a. the nematodes and arthropods.   d. animals that molt.
    b. platyhelminths.   e. animals with radial symmetry.
    c. protostomes.

VISUAL FOUNDATIONS

Color the parts of the illustration below as indicated. Label the animals depicted as acoelomate, pseudocoelomate, or coelomate.

RED ☐ mesoderm
GREEN ☐ endoderm
YELLOW ☐ coelom
BROWN ☐ pseudocoelom
BLUE ☐ ectoderm
ORANGE ☐ mesenchyme

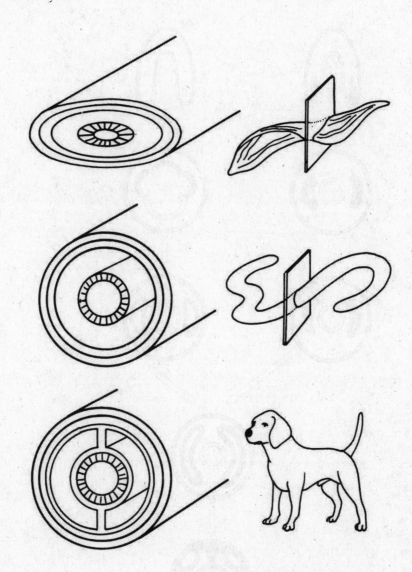

Color the parts of the illustration below as indicated. Label the ectoderm, mesoderm, endoderm, gut, and coelom in the last figure.

BLUE ☐ Ectoderm

RED ☐ Mesoderm

YELLOW ☐ Endoderm

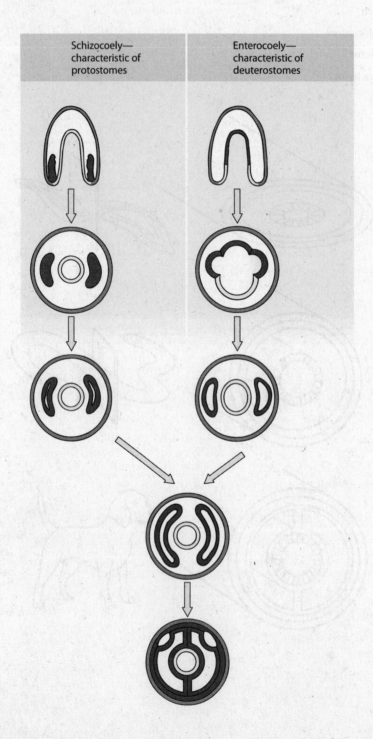

Sponges, Cnidarians, Ctenophores, and Protostomes

This chapter discusses the three phyla that diverged early in the evolutionary history of the animal kingdom: the Parazoa (sponges), the Cnidaria (hydras, jelly fish, sea anemones, and corals), and the Ctenophora (comb jellies), and it concludes with a discussion of the protostomes. The sponges are at the cellular level of organization and lack true tissues. The cnidarians and ctenoporans have tissues and a body that is radially symmetrical. The protostomes are animals with coeloms in which the first opening that forms in the embryonic gut gives rise to the mouth, a complete digestive tract, and most have well-developed circulatory, excretory, and nervous systems. The coelom separates muscles of the body wall from those of the digestive tract, permitting movement of food independent of body movements, and it serves as a space in which organs develop and function. The fluid it contains helps transport food, oxygen, and wastes. Additionally, it can be used as a hydrostatic skeleton, permitting a greater range of movement than in acoelomate animals and conferring shape to the body of soft animals. The coelomate protostomes include the clades Lophotrochozoa (platyhelminthes, nemerteans, mollusks, annelids, and the lophophorate phyla) and Ecdysozoa (nematodes and arthropods).

REVIEWING CONCEPTS

Fill in the blanks.

INTRODUCTION

1. An estimated _____% of all animal species belong to clade Bilateria.
2. The success of bilateral animals can be attributed to three major things -
   (a) _____,
   (b) _____,
   (c) _____.

SPONGES, CNIDARIANS, AND CTERNOPHORES: ANIMALS WITH ASYMMETRY, RADIAL, OR BILATERAL SYMMETRY

3. Sponges which have been considered to be paraozaons because the cells of sponges _____.

Sponges have collar cells and other specialized cells

4. Based upon the structure of their collar cells, sponges are thought to have evolved from _____.
5. Sponges have three types of canals systems through which water flows to enter the sponge.
   In the (a) _____ system the choanocytes lining the internal cavity pull water in through tiny pores, ostia.
   In the (b) _____ system the choanocytes line channels through which the water flows into the sponge.

In the (c)_____ system the wall of the sponge is highly folded and the increased system of canals lined with choanocytes provides more surface area for capturing food.

6. In asconoid sponges, water enters the internal cavity or (a)_____ through the porocytes and then exits through the (b)_____.

7. Between the outer and inner cell layers of the sponge body is a gelatin-like layer called the (a) _____ supported by slender skeletal spikes called (b) _____.

Cnidarians have unique stinging cells

8. The phylum Cnidaria is divided into four classes: Hydras and hydroids are in the class (a)_____, jellyfish are in class (b)_____, sea anemones and corals are in class (c)_____, and the "box jellyfish" are in class (d) _____.

9. The epidermis of cnidarians contains specialized "stinging cells" called _____.

10. Stinging "thread capsules" in the stinging cells of cnidarians are called _____.

11. Cnidarians have two definite tissue layers, the outer (a)_____ and the inner (b)_____ separated by a gelatinous region called the (c)_____.

12. Hydra can be found living in (a) _____ and during the fall they reproduce (b) _____.

13. Among the jellyfish, members of the Scyphozoa, the _____ stage is the dominant body form.

14. Cubozoans are food for _____.

15. Anthozoan polyps produce eggs and sperm, and the fertilized egg develops into a small, ciliated larva called a _____.

Comb Jellies have adhesive glue cells that trap prey

16. Ctenophores have _____ symmetry.

17. Ctenophores are similar to cnidarians in that they have a thick jellylike layer, the mesoglea. However they differ from cnidarians because they lack (a) _____ in the epithelium and they have a digestive system with (b) _____ and (c) _____.

THE LOPHOTROCHOZOA

18. Most lophotrochozoans have (a) _____ symmetry and have (b) _____ primary germ layers.

19. Flatworms are in phylum (a) _____ which is further divided into four classes. These classes are (a)_____, the free-living flatworms; two classes of flukes (b)_____ and (c) _____ and (d)_____, the tapeworms.

20. The organs of chemoreception on planarians that resemble "ears" are called _____.

21. Planarians use an extendable _____ to suck prey into their digestive tube.

22. Most tapeworms have suckers and/or hooks on the _____ for attachment to their hosts.

23. In tapeworms, each segment, or _____, contains both male and female reproductive organs.

Nemerteans are characterized by their proboscis

24. The proboscis organ is a long, hollow, muscular tube that can be rapidly _____.

Mollusks have a muscular foot, visceral mass, and mantle

25. Mollusks possess a ventral foot for locomotion and a (a)_____ that covers the visceral mass.

26. Most mollusks have a (a) _____ located in the mouth region, and an open circulatory system, meaning the blood flows through a network of open sinuses called the blood cavity or (b) _____.

27. Most marine mollusks have larval stages, the first a free-swimming, ciliated form called a (a) _____ larva. Some, the snails and bivalves go through a second stage a (b) _____ larva with a shell, foot, and mantle.

28. A distinctive feature of chitons is their shell, which is composed of _____(#?) overlapping plates.

29. Chitons lack (a) _____ and (b) _____.

30. Class Gastropoda, the largest and most successful group of mollusks, includes the snails, slugs, and their relatives. It is a class second in size only to the _____.

31. The gastropod shell (when present) is coiled, and the visceral mass is twisted, a phenomenon known as _____.

32. Foreign matter lodged between the (a) _____ and mantle of bivalves may cause the secretion of (b)_____ by the epithelial cells of the mantel. The result is the formation of a pearl..

33. Class Cephalopoda includes the squids and octopods, which are active predatory animals. The foot is divided into (a) _____ that surround the mouth, (b)_____(#?) of them in squids, (c)_____(#?) of them in octopods.

34. The term Cephalopoda literally means _____.

Annelids are segmented worms

35. In polychaetes and earthworms, body segments are separated from one another by partitions called (a)_____. Also, each body segment bears bristle-like structures called (b)_____ that function to anchor the body during locomotion.

36. Most polychaetes are (a) _____ worms and each body segment bears a pair of paddle-shaped appendages called (b) _____ used for locomotion and gas exchange.

37. The two parts of the earthworm's "stomach" are the (a)_____, where food is stored, and the (b)_____, where ingested materials are ground into bits.

38. The respiratory pigment in earthworms is (a)_____; the excretory system consists of paired (b)_____ in most segments; and the gas exchange surface is the (c)_____.

The lophophorates are distinguished by a ciliated ring of tentacles

39. There are three lophophorate phyla: (a) _____, or lampshells; (b)_____, wormlike sessile animals; and (c)_____, or microscopic aquatic animals.

Rotifers have a crown of cilia

40. Rotifers have a nervous system which includes a (a) _____ and (b) _____. They have protonephridia with (c) _____ that remove excess water from the body and may also excrete metabolic wastes.

41. Rotifers grow through an _____.

THE ECDYSOZOA

Roundworms are of great ecological importance

42. Members of the phylum Nematoda play key ecological roles as

(a) _____ and

(b) _____.

Arthropods are characterized by jointed appendages and an exoskeleton of chitin

43. More than ____% of all known animals are arthropods.

44. Arthropods are characterized as having an armor-like body covering, (a)_____, which is composed of (c)_____.

45. Gas exchange in terrestrial arthropods occurs in a system of thin, branching tubes called _____.

46. Among the earliest arthropods were the _____ which inhabited seas and are now extinct.

47. The members of the class (a)_____, or "centipedes," have (b)_____ pair(s) of legs per body segment.

48. The members of the class (a)_____, or "millipedes," have (b)_____ pair(s) of legs on most body segments.

49. The subphylum of arthropods that lacks antennae and includes horseshoe crabs and arachnids is _____.

50. The arachnid body consists of a (a)_____ and abdomen. They have (b)_____(#?) pairs of jointed appendages which include _____(#?) pairs of legs used for walking.

51. Gas exchange in arachnids occurs in tracheae and/or across thin, vascularized plates called _____.

52. Silk glands in spiders secrete an elastic protein that is spun into web fibers by organs called _____.

53. The subphylum (a) _____ includes lobsters, crabs, shrimp, and barnacles. This subphylum uses its (b) _____ for biting food and its (c) _____ for manipulating and holding food, and it has (d)_____ appendages and (e)_____(#?) pairs of antennae.

54. _____ are the only sessile crustaceans.

55. Lobsters, crayfish, crabs, and shrimp are members of the largest crustacean order, the _____.

56. Among the many highly specialized appendages in decapods are the (a)_____, two pairs of feeding appendages just behind the mandibles; the (b)_____ that are used to chop food and pass it to the mouth; the (c)_____ or pinching claws on the fourth thoracic segments; and the pairs of (d)_____ on the last four thoracic segments.

57. Appendages on the first abdominal segment of decapods are part of the (a)_____ system. (b)_____ on the next four abdominal segments are paddle-like structures adapted for swimming and holding eggs.

58. In terms of numbers of individuals, number of species, and geographic distribution, animals in the class (a) _____ and subphylum (b) _____ are considered to be the most successful group of animals on the planet.

59. Adult insects typically have (a)_____(#?) pairs of legs, (b)_____(#?) pairs of wings, and (c)_____(#?) pair(s) of antennae. The excretory organs are called (d)_____.

BUILDING WORDS

Use combinations of prefixes and suffixes to build words for the definitions that follow.

Prefixes	The Meaning	Suffixes	The Meaning
arthro-	joint, jointed	-coel	abdominal cavity, hollow
aur-	ear	-cyte(s)	cell
bi-	twice, two	-icle	little
cephalo-	head	-pod	foot, footed
choan(o)-	funnel	-stat(ic)	to control
exo-	outside, outer, external		
hexa-	six		
hydr(o)-	water		
tri-	three		
uni-	one		

Prefix	Suffix	Definition
_____	-skeleton	1. An external skeleton, such as the shell of arthropods.
_____	-valve	2. A mollusk that has two shells (valves) hinged together.
_____	-ramous	3. Consisting of or divided into two branches.
_____	-ramous	4. Unbranched; consisting of only one branch.
_____	-lobite	5. An extinct marine arthropod characterized by two dorsal grooves that divide the body into three longitudinal parts (or lobes).
_____	-thorax	6. Anterior part of the body in certain arthropods consisting of the fused head and thorax.
_____	_____	7. Having six feet; an insect.
_____	_____	8. An animal with paired, jointed legs.
spongo-	_____	9. The central cavity in the body of a sponge.
_____	_____	10. Cells used by sponges to trap food particles.
_____	_____	11. The type of skeleton used by cnidarians to shorten and elongate.
_____	_____	12. A flap-like extension on the head of a planarian that is used to sense the environment.

MATCHING

Terms:

a. Acoelomate
b. Auricle
c. Cephalization
d. Choanocyte
e. Cnidocyte
f. Coelom
g. Cuticle
h. Dorsal
i. Ecdysis
j. Hermaphroditic
k. Invertebrate
l. Mandible
m. Mesohyl
n. Metamorphosis
o. Nematocyst
p. Pseudocoelom
q. Radula
r. Sessile
s. Spongin
t. Trachea
u. Trochophore
v. Ventral

For each of these definitions, select the correct matching term from the list above.

_____ 1. The fibrous part of a sponge skeleton.

_____ 2. Possessing sex organs of both the male and the female.

_____ 3. A rasplike structure in the mouth region of certain mollusks.

_____ 4. A fluid-filled space lined by mesoderm that lies between the digestive tube and the outer body wall.

_____ 5. A unique cell having a flagellum surrounded by a collar of microvilli.

_____ 6. The clustering of neural tissues at the anterior (leading) end of an animal.

_____ 7. The first pair of appendages in arthropods.

_____ 8. A stinging structure in cnidarians.

_____ 9. Permanently attached to one location.

_____10. The shedding and replacement of the arthropod exoskeleton.

_____11. A larval form found in mollusks and many polychaetes.

_____12. Transition from one developmental stage to another, such as from a larva to an adult.

_____13. Internal branching air tubes throughout the body of many terrestrial arthropods and some terrestrial mollusks.

MAKING COMPARISONS

Fill in the blanks.

Phylum	Examples	Body Plan	Key Characteristics
Porifera	Sponges	Asymmetrical; sac-like body with pores, central cavity, and osculum	Choanocytes
Cnidaria	#1	Radial symmetry; diploblastic; gastrovascular cavity with one opening	Tentacles with cnidocytes (stinging cells); polyp and medusa body forms;
Ctenophora	Comb jellies	#2	Eight rows of cilia; tentacles with adhesive glue cells
#3	Flatworms, tape worms, planarians, flukes	#4	No body cavity; some cephalization
#5	Clams, snails, squids	Biradial symmetry; triploblastic; organ systems; complete digestive tube	#6
Annelida	#7	Biradial symmetry; triploblastic; organ systems; complete digestive tube	#8
#9	Wheel animals	Biradial symmetry; triploblastic; organ systems; complete digestive tube	#10
Nematoda	Roundworms: Ascaris, hookworms, trichina worms	Biradial symmetry; triploblastic; organ systems; complete digestive tube	#11
Arthropoda	#12	Biradial symmetry; triploblastic; organ systems; complete digestive tube	#13
#14	Starfish, sea urchins, sand dollars	#15	Endoskeleton; water vascular system; tube feet

MAKING CHOICES

Place your answer(s) in the space provided. Some questions may have more than one correct answer.

_____ 1. Marine carnivores with a proboscis that can be everted to capture prey are

      a. members of Nemertea.   d. known as ribbon worms.

      b. are protostomes.   e. members of Mollusca.

      c. are deuterostomes.

_____ 2. The taxon that has biradial symmetry, tentacles with adhesive glue cells, and eight rows of cilia is

      a. Scyphozoa.   d. Cnidaria.

      b. Hydrozoa.   e. Porifera.

      c. Ctenophora.

_____ 3. Examples of acoelomate eumetazoans include

      a. sponges.   d. those groups whose coelom developed from a blastocoel.

      b. insects.   e. cnidarians.

      c. flatworms.

_____ 4. The phylum/phyla that does/do not have specialized nerve cells is/are

      a. Echinodermata.   d. Nemertea.

      b. Cnidaria.   e. Porifera.

      c. Platyhelminthes.

_____ 5. When a group of cells moves inward to form a sac in early embryonic development, and that pore ultimately develops into a mouth, the group of animals is considered to be a

      a. protostome.   d. parazoan.

      b. deuterostome.   e. schizocoelomate.

      c. pseudocoelomate.

_____ 6. The one invertebrate phylum with radial symmetry and distinctly different tissues is

      a. Porifera.   d. Nemertea.

      b. Cnidaria.   e. Echinodermata.

      c. Platyhelminthes.

_____ 7. The roundworms

      a. are pseudocoelomates.   d. include hookworms.

      b. have a mastax.   e. are nematodes.

      c. are in the phylum Platyhelminthes.

_____ 8. Sessile, marine animals with no medusa stage and a partitioned gastrovascular cavity include

      a. hydrozoans.   d. jellyfish.

      b. scyphozoans.   e. corals, sea anemones, and their close relatives.

      c. anthozoans.

_____ 9. The taxon that has "thread capsules" in epidermal stinging cells is

      a. Scyphozoa.   d. Cnidaria.

      b. Hydrozoa.   e. Ctenophora.

      c. Porifera.

_____10. Deuterostomes characteristically have

      a. radial cleavage.   d. schizocoely.

      b. spiral cleavage.   e. enterocoely.

      c. indeterminate cleavage.

_____11. The excretory organ(s) in earthworms is/are the

    a. parapodia.

    b. kidneys.

    c. metanephridia.

    d. lophophores.

    e. crop.

_____12. The so-called "wheel animals"

    a. are unicellular.

    b. are generally aquatic.

    c. have no cell mitosis after completing embryonic development.

    d. move by spinning head over tail, hence the name "wheel animals".

    e. are rotifers.

_____13. Roundworms are

    a. nematodes.

    b. mollusks.

    c. chilopods.

    d. annelids.

    e. cephalopods.

_____14. Earthworms are

    a. nematodes.

    b. mollusks.

    c. chilopods.

    d. annelids.

    e. cephalopods.

_____15. The animal that has a shell consisting of eight separate, overlapping dorsal plates is

    a. an insect.

    b. a chiton.

    c. in the same class as squids.

    d. in the class Polyplacophora.

    e. a mollusk.

_____16. Earthworms are held together in copulation by mucous secretions from the

    a. clitellum.

    b. typhlosole.

    c. seminal receptacles.

    d. epidermis.

    e. prostomium.

_____17. Earthworms are

    a. annelids.

    b. oligochaets.

    c. in the phylum Polychaeta.

    d. protostomes.

    e. acoelomate animals.

_____18. Butterflies, termites, and black widows all

    a. are insects.

    b. are arthropods.

    c. have paired, jointed appendages.

    d. have closed circulatory systems.

    e. have an abdomen.

_____19. A marine animal with parapodia and a trochophore larva might very well be

    a. a tubeworm.

    b. an oligochaet.

    c. bilaterally symmetrical.

    d. a leech.

    e. a polychaet.

_____20. An octopus is a

    a. vertebrate.

    b. mollusk.

    c. chilopod.

    d. protostome.

    e. cephalopod.

_____21. Animals with most of the organs in a visceral mass that is covered by a mantle are

    a. annelids.

    b. oligochaets.

    c. onychophorans.

    d. mollusks.

    e. cheliceratans.

_____22. Clams, scallops, and oysters are

a. nematodes.

b. mollusks.

c. chilopods.

d. annelids.

e. cephalopods.

_____23. Insects are

a. hexapods.

b. arthropods.

c. chelicerates.

d. in the same subphylum as lobsters and crabs.

e. in the same subphylum as spiders.

_____24. Spiders are

a. hexapods.

b. arthropods.

c. chelicerates.

d. in the same subphylum as lobsters and crabs.

e. in the same subphylum as insects.

VISUAL FOUNDATIONS

Color the parts of the illustration below as indicated.

RED ☐ digestive tract

GREEN ☐ shell

YELLOW ☐ foot

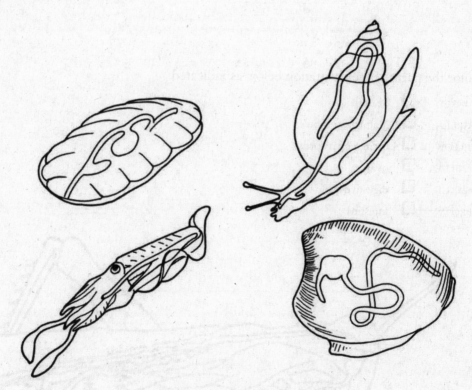

Color the parts of the illustration below as indicated.

RED	☐	dorsal and ventral blood vessels
GREEN	☐	metanephridium
YELLOW	☐	nerve cord
BLUE	☐	cerebral ganglia
ORANGE	☐	pharynx
BROWN	☐	intestine
TAN	☐	coelom

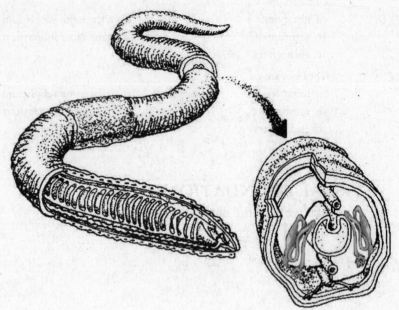

Color the parts of the illustration below as indicated.

RED	☐	heart
GREEN	☐	Malpighian tubules
YELLOW	☐	brain, nerve cord
BLUE	☐	ovary
ORANGE	☐	digestive gland
BROWN	☐	intestine

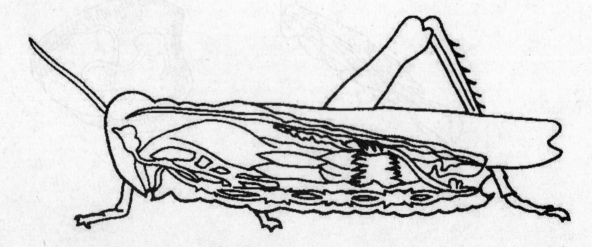

The Deuterostomes

This chapter discusses the deuterostomes, animals with coeloms in which the second opening that develops in the embryo becomes the mouth, and the first becomes the anus. The deuterostomes include the echinoderms, hemichordates, and the chordates. The echinoderms and the chordates are by far the more ubiquitous and dominant deuterostomes. The echinoderms all live in the sea. They have spiny skins, a water vascular system, and tube feet. The chordates include animals that live both in the sea and on land. Chordates share four derived characters which distinguish them from other animals. They are a notochord, a dorsal tubular nerve cord, a post anal tail, and an endostyle or thyroid gland, at least at some time in their life cycle. At some point in their life cycle, chordates also have pharyngeal gill slits. There are three groups of chordates: the tunicates, the lancelets, and the vertebrates. The vertebrates are characterized by a vertebral column, a cranium, pronounced cephalization, a differentiated brain, muscles attached to an endoskeleton, and two pairs of appendages. There are 10 classes of living vertebrates: 6 of fishes and 4 of animals with four limbs. The fishes include the jawless, cartilaginous, and bony fishes, and, the four-limbed vertebrates include the amphibians, reptiles, birds, and mammals. Mammals are classified first on the basis of whether embryonic development occurs within an egg, a maternal pouch, or utilizes an organ of exchange, i.e., a placenta.

REVIEWING CONCEPTS

Fill in the blanks.

INTRODUCTION

1. Scientists speculate that the last common ancestor of the deuterostomes was an animal that was a _____ feeder.

2. The two major phyla assigned to the deuterostomes are the

   (a) _____ and

   (b) _____.

WHAT ARE DEUTEROSTOMES?

3. Deuterostomes evolved during the _____ eon.

4. Deuterostomes are characterized by (a) _____cleavage. The cleavage is (b) _____, which means the fate of their cells is fixed later in development than is the case in protostomes

ECHINODERMS

5. Echinoderm larvae have (a)_____ symmetry; most adults have

   (b)_____ symmetry. Echinoderms also have a true

   (c) _____ that houses the internal organs.

6. Echinoderms have (a) _____ skeleton and

   (b) _____ which keep the surface of the body clean.

7. Echinoderms are thought to have evolved during the
_____ period.

Feather stars and sea lilies are suspension feeders

8. Sea lilies and feather stars are placed in class (a) _____.
Unlike other echinoderms, the oral surface of this class is located on the
(b) _____ surface.

Many sea stars capture prey

9. Sea stars are in the class _____.

10. Sea stars have a _____ from which radiate five or more arms
or rays.

11. The arms or rays on sea stars contain hundreds of small _____, each
of which contains a canal that leads to a central canal.

12. As a means of obtaining food, most sea stars are
_____.

Basket stars and brittle stars makeup the largest group of echinoderms

13. Basket stars and brittle stars are placed in class _____.

14. This class uses their arms primarily for (a) _____ and their
tube feet for (a)_____.

Sea urchins and sand dollars have movable spines

15. Sea urchins and sand dollars are in the class (a) _____.
The arms in this class are (b) _____. They have a solid shell
called a
(c) _____, and their body is covered with (d) _____..

Sea cucumbers are elongated, sluggish animals

16. Sea cucumbers belong to class (a) _____. Their tentacles
are modified (b) _____.

THE CHORDATES: DEFINING CHARACTERISTICS

17. Chordates are in the phylum _____.

18. The four characteristics that distinguish chordates from all other groups are
(a) _____
(b) _____
(c) _____
(d) _____

INVERTEBRATE CHORDATES

Tunicates are common marine animals

19. Larval tunicates have typical chordate characteristics and superficially resemble
_____.

20. Adult tunicates develop a protective covering called a
_____.

Lancelets clearly exhibit chordate characteristics

21. Most cephalochordates belong to the genus _____.

22. Lancelets collect food by _____.

Systematists debate chordate phylogeny

23. The most recent common ancestor of cephalochordates and vertebrates may
   have resembled a _____.

24. _____ are a series of blocks of mesoderm that develop on
   each side of the notochord.

INTRODUCING THE VERTEBRATES

The vertebral column is a derived vertebrate character

25. Subphylum Vertebrata includes animals with a backbone or
   (a) _____, a braincase or (b)_____,
   and a concentration of nerve cells and sense organs in a definite head, a
   phenomenon known as (c)_____.

Vertebrate taxonomy is a work in progress

26. The extant vertebrates are currently assigned to (a)_____(#?) classes of
   fishes and (b)_____(#?) of tetrapods. The four classes of tetrapods are

   (c) _____

   (d) _____

   (e) _____

   (f) _____.

JAWLESS FISHES

27. Some of the earliest known vertebrates, the _____, consisted
   of several groups of small, armored, jawless fishes that lived on the ocean floor
   and strained their food from the water.

28. Hagfishes are in the class _____.

29. Class Petromyzontida contains the _____.

30. Contemporary hagfishes and lampreys have neither jaws nor paired _____.

EVOLUTION OF JAWS AND LIMBS: JAWED FISHES AND AMPHIBIANS

31. Jaws evolved from a portion of the _____.

Most cartilaginous fish inhabit marine environments

32. Cartilaginous fishes belong to class _____.

33. The skin of cartilaginous fish contains numerous _____,
   toothlike structures composed of layers of enamel and dentine.

34. (a)_____ are sensory grooves along the
   sides of all fish that contain cells capable of perceiving water movements.
   Similarly, the (b)_____ on the shark's head can
   sense very weak electrical currents, like those generated by another animal's
   muscle contractions.

35. Sharks may be (a)_____, that is, lay eggs; or
   (b)_____, that is the eggs are incubated and hatch
   internally; or (c)_____, nutrients are transferred between
   the maternal and fetal blood and there is a live birth.

The ray-finned fishes gave rise to modern bony fishes

36. In the Devonian period, bony fishes diverged into two major groups: the

    (a)_____ fish, or actinopterygians, and the

    (b)_____ with their fleshy, lobed fins and lungs.

37. The actinopterygians gave rise to the modern bony fish in which lungs became

    modified as _____.

Descendants of the lungfishes moved onto the land

38. Ancestors of the _____ gave rise to the tetrapods.

39. Molecular data support the hypothesis that tetrapod limbs evolved from

    _____.

40. (a) _____, which is considered to be a fish, is considered

    to be the transitional form between fishes and tetrapods. While it was

    considered a fish because it had scales and fins it also had the tetrapod features

    of (b) _____ and

    (c) _____.

Amphibians were the first successful land vertebrates

41. The three orders of modern amphibians are: (a)_____,

    amphibians with long tails, including salamanders, mud puppies, and newts; the

    (b)_____, tailless frogs and toads; and

    (c) _____, the wormlike caecilians.

42. Some salamanders retain larval characteristics even when sexually mature. This

    phenomenon is called _____.

43. Amphibians use lungs and their (a)_____ for gas exchange. They

    have a (b)_____(#?)-chambered heart with systemic and pulmonary

    circulations.

AMNIOTES

44. The evolution of the _____ allowed terrestrial vertebrates to

    complete their life cycle entirely on land.

Our understanding of amniote phylogeny is changing

45. Diapsid amniotes have (a)_____ pairs of openings in the temporal

    bones whereas synapsid amniotes have (b)_____ pair.

Reptiles have many terrestrial adaptations

46. Most reptiles have a (a)_____(#?)-chambered heart, and are

    (b)_____, which means they cannot regulate body temperature.

We can assign extant reptiles to five groups

47. Reptiles have been assigned to four orders or one clade. These are the

    (a) _____ which includes the turtles,

    terrapins and tortoises;

    (b) _____which includes lizards, snakes,

    and amphisbaenians;

    (c) _____ which includes the tuataras

(d) _____ which includes the crocodiles, alligators, caimans, and gavials; and the clade

(e) _____ which includes extinct flying reptiles and dinosaurs as well as the order including the crocodilians.

Turtles have protective shells

48. A turtle's jaws are _____ but they contain no teeth.

Lizards and snakes are the most common modern reptiles

49. The scales of animals in order Squamata _____ to form a continuous, flexible armor covering.

50. Vipers have a specialized _____ on each side of the head that is used to detect heat emitted by prey.

Tuataras superficially resemble lizards

51. In their outward appearance, tuataras look like _____.

Crocodilians have an elongated skull

52. Crocodiles are distinguished from alligators and caimans by two major features

(a) _____ and the outward appearance of the

(b) _____ in the bottom jaw when the mouth is closed.

How do we know that birds are really dinosaurs?

53. Modern birds use _____ feathers for flight.

54. Insulating feathers could have contributed to the evolution of _____ which would have allowed animals to be more active.

55. Some extinct theropods as well as modern birds have fucula (wishbone) which is formed when _____.

Modern birds are adapted for flight

56. Birds have a (a)_____(#?)-chambered heart, very efficient lungs, a high metabolic rate, and they are (b)_____, meaning they can maintain a constant body temperature. They excrete wastes as semi-solid (c)_____.

57. The saclike portion of a bird's digestive system that temporarily stores food is the (a)_____, whereas the (b)_____ portion of the stomach secretes gastric juices, and the (c)_____ grinds the food.

Mammals have hair and mammary glands

58. Derived characters of mammals include (a) _____, (b) _____, (c) _____ and (d) _____.

59. Fertilization is internal and the majority of mammals are (a) _____ and develop a (b) _____ which serves as an exchange organ between the mother and the fetus.

New fossil discoveries are changing our understanding of the early evolution of mammals

60. Mammals probably evolved from (a)_____, a reptilian group, about 200 million years ago in the (b)_____ period.

61. The fact that early mammals were (a) _____ (tree-dwelling) and (b) _____ (active at night) helped them to survive during the reign of the reptiles.

Modern mammals are assigned to three subclasses

62. _____ are mammals that lay eggs. They include the duck-billed platypus and the spiny anteater.

63. (a)_____ are pouched mammals. The young are born immature and complete development in the (b)_____, where they are nourished from the mammary glands.

64. Animals with a well-developed placenta are classified as _____.

BUILDING WORDS

Use combinations of prefixes and suffixes to build words for the definitions that follow.

Prefixes	The Meaning	Suffixes	The Meaning
a-	without	-chord	cord
Chondr(o)-	cartilage	-derm	skin
Cephal-	head	-gnath(an)	jaw
endo-	within	-ichthy(es)	fish
echino-	sea urchin, "spiny"	-ite	belonging to
noto-	back	-ization	process of
tetra-	four	-pod(al)	foot, footed
som-	body	-therm	heat
Uro-	tail	-ur(o)(a)	tail

Prefix	Suffix	Definition
_____	_____	1. A fish without jaws (jawless fish); a member of the class of vertebrates including lampreys and hagfishes.
An-	_____	2. An order of amphibia with legs but no tail; tailless frogs or toads.
_____	_____	3. Pertains to those amphibians with no feet, i.e., the wormlike caecilians.
_____	_____	4. The class comprising the cartilaginous fishes.
_____	_____	5. A vertebrate with four legs; a member of the superclass Tetrapoda.
_____	-dela	6. The order that comprises the amphibians with long tails, e.g., the salamanders, mudpuppies, and newts.
_____	_____	7. A spiny-skinned animal.

_____   -skeleton   8. Bony and cartilaginous-supporting structures within the body that provide support from within.

_____ _____   9. A dorsal, longitudinal rod of cartilages.

_____ _____   10. A paired, block-like segment of mesoderm.

_____ _____   11. A concentration of sensory structures on the anterior (front end) of an animal.

_____ _____   12. An animal that maintains its own body temperature through metabolism and behavior.

MATCHING

Terms:

a. Amnion
b. Amphibian
c. clasper
d. Ctylosaur
e. Cotylosaur
f. crop
g. gizzard

h. Labyrinthodont
i. Marsupium
j. Neoteny
k. Notochord
l. Oviparous
m. Ovoviviparous
n. Placenta

o. Placoderm
p. Squ;amata
q. Therapsid
r. Vertebrate

For each of these definitions, select the correct matching term from the list above.

_____ 1. The dorsal, longitudinal rod that serves as an internal skeleton in the embryos of all, and in the adults of some, chordates.

_____ 2. The pouch in which marsupial young develop.

_____ 3. Animals that are egg-layers.

_____ 4. An extra-embryonic membrane that forms a fluid-filled sac for the protection of the developing embryo.

_____ 5. An extinct jawed fish.

_____ 6. A group of mammal-like reptiles of the Permian period that gave rise to the mammals.

_____ 7. An organ of exchange between developing embryo and mother in eutherian mammals.

_____ 8. A chordate possessing a bony vertebral column.

_____ 9. Order containing lizards and snakes.

_____10. A sac-like structure in which food is temporarily stored.

_____11. Used by male sharks to transfer sperm to a female.

MAKING COMPARISONS

Fill in the blanks.

Vertebrate Class	Representative Animals	Heart	Skeletal Material	Other Characteristics
Myxini	Hagfish	Two-chambered heart	Cartilage	Jawless, most primitive vertebrates, gills,
Chondrichthyes	#1	Two-chambered heart	#2	#3
#4	Salmon, tuna	Two-chambered heart	#5	Gills, swim bladder
Amphibia	#6	#7	Bone	Tetrapods, aquatic larva metamorphoses into terrestrial adult, moist skin and lungs for gas exchange, ectothermic
#8	Birds	#9	#10	Amniotes with feathers, many adaptations for flight, endothermic, high metabolic rate
#11	Monotremes, marsupials, placentals	Four-chambered heart	#12	#13

MAKING CHOICES

Place your answer(s) in the space provided. Some questions may have more than one correct answer.

_____ 1. Humans are members of the taxon(a)

a. Lagomorpha.

b. Rodentia.

c. that has chisel-like incisors that grow continually.

d. that includes placental mammals.

e. Cetacea.

_____ 2. The phylum(a) that many biologists believe had a common ancestry with our own phylum is/are

a. Mollusca.

b. Annelida.

c. Arthropoda.

d. Echinodermata.

e. Chordata.

_____ 3. The group(s) of animals that maintain a constant internal body temperature is/are
   a. Reptilia.
   b. Mammalia.
   c. Aves.
   d. Amphibia.
   e. Pisces.

_____ 4. The group(s) of animals with a part of the stomach that secretes gastric juices and a separate part of the stomach that grinds food is/are
   a. Reptilia.
   b. Mammalia.
   c. Aves.
   d. Amphibia.
   e. Pisces.

_____ 5. A turtle is a member of the taxon(a)
   a. Reptilia.
   b. Chelonia.
   c. Urodela.
   d. Amphibia.
   e. Anura.

_____ 6. Mammals probably evolved from
   a. amphibians.
   b. a type of fish.
   c. reptiles.
   d. monotremes.
   e. therapsids.

_____ 7. The lateral line organ is found
   a. only in sharks.
   b. only in bony fish.
   c. only in agnathans.
   d. in all fishes.
   e. in all fish except placoderms.

_____ 8. Members of the phylum Chordata all have
   a. an embryonic notochord.
   b. embryonic pharyngeal gill slits.
   c. a dorsal, tubular nerve cord.
   d. well-developed germ layers.
   e. bilateral symmetry.

_____ 9. The "spiny-skinned" animals include
   a. sea stars.
   b. sand dollars.
   c. snails.
   d. lancets.
   e. marine worms.

_____ 10. The acorn worm
   a. has a notochord.
   b. is a deuterostome.
   c. is in the same subphylum as *Amphioxus*.
   d. is in the phylum Chordata.
   e. is a hemichordate.

_____ 11. The group(s) of animals with both four-chambered hearts and a double circuit of blood flow is/are the
   a. earthworms.
   b. birds.
   c. mammals.
   d. class Aves.
   e. amphibians.

_____ 12. An animal that traps food in mucus secreted by cells of the endostyle is a/an
   a. holothuroidean.
   b. enchinoderm.
   c. urochordate.
   d. sea cucumber.
   e. tunicate.

_____ 13. The first successful land vertebrates were
   a. tetrapods.
   b. reptiles.
   c. chondrichthyes.
   d. amphibians.
   e. in the superclass Pisces.

_____14. Cats are members of the taxon(a)

    a. Lagomorpha.   d. that includes placental mammals.

    b. Rodentia.   e. Cetacea.

    c. that has chisel-like incisors that grow continually.

_____15. Unique features of the echinoderms include

    a. radial larvae.   d. gas exchange by diffusion.

    b. ciliated larvae.   e. water vascular system.

    c. endoskeleton composed of $CaCO_3$ plates.

_____16. A frog is a member of the taxon(a)

    a. Reptilia.   d. Amphibia.

    b. Chelonia.   e. Anura.

    c. Urodela.

_____ 17. Features of echinoderms that distinguish them from other animals include

    a. tube feet.   d. pedicellariae.

    b. radial symmetry.   e an endoskeleton.

    c. a coelom.

_____18. Included in the five extant classes of echinoderms are

    a. crinoids.   d. asteroids.

    b. sea lilies.   e. sand dollars.

    c. feather stars.

_____19. In a cladogram of deuterostomes, a unifying feature(s) of chordates would include

    a. pharyngeal gill slits.   d. radial cleavage.

    b. determinate cleavage   e. a notochord.

    c. a cranium.

_____20. Terms associated with tunicates include

    a. endostyle.   d. urochordates.

    b. siphon.   e. atrium.

    c. ganglion.

_____ 21. Vertebrates are distinguished from other chordates

    a. by their notochord.   d. by their pronounced cephalization.

    b. by having a postanal tail.   e. by the presence of a vertebral column.

    c. by having amniotic eggs.

_____ 22. Terms associated with lamprey include

    a. conodont.   d. myxini.

    b. ostracoderm.   e. parasite.

    c. jaws.

_____ 23. Reproductively, sharks may be

    a. hermaphrodites.   d. viviparous.

    b. egg layers.   e. ovoviviparous.

    c. monogamous.

VISUAL FOUNDATIONS

Color the parts of the illustration below as indicated. Also label the postanal tail.

RED	☐	heart
GREEN	☐	mouth
YELLOW	☐	brain and dorsal, hollow nerve tube
BLUE	☐	notochord
ORANGE	☐	pharyngeal gill slits
PINK	☐	pharynx, intestine
TAN	☐	muscular segments

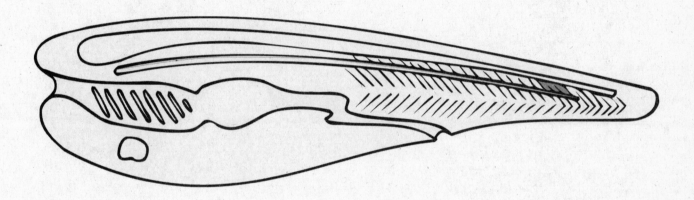

CHAPTER 33

❑

Plant Structure, Growth, and Development

This is the first of six chapters addressing the integration of plant structure and function. This chapter examines the external structure of the flowering plant body; the organization of its cells, tissues, and tissue systems; and its basic growth patterns. The plant body is organized into a root system and a shoot system. Plants are composed of cells that are organized into tissues, and tissues that are organized into organs. Each organ performs a single function or group of functions, and is dependent on other organs for its survival. All vascular plants have three tissue systems: the dermal tissue system, the vascular tissue system, and the ground tissue system. Plant tissues may be simple or complex. Plants grow by increasing both in girth and length. Unlike growth in animals, plant growth is localized in areas of unspecialized cells, and involves cell proliferation, elongation, and differentiation.

REVIEWING CONCEPTS
Fill in the blanks.

INTRODUCTION

1. (a) _____ plants grow, reproduce and die in a year or less;
   (b) _____ plants take two years to complete their life cycle;
   and (c) _____ plants can often live for more than two years.

2. (a) _____ perennials shed their lives in fall and are dormant over winter but (b) _____ shed their leaves over an extended period of time which means that some leaves are always present.

THE PLANT BODY

3. The body of vascular plants is organized into a (a) _____ system and a (b) _____ system.

The plant body consists of cells and tissues

4. A (a) _____ is a group of cells that form a structural and functional unit. (b) _____ are composed of only one kind of cell whereas (c) _____ have two or more kinds of cells.

5. Roots, stems, leaves, flower parts, and fruits are _____ because each is composed of all three tissue systems.

The ground tissue system is composed of three simple tissues

6. The three tissues composing the ground tissue system of herbaceous plants are
   (a) _____, (b) _____,
   and (c) _____.

7. The three primary functions of parenchyma tissue include
   (a) _____, (b) _____,
   and (c) _____.

8. Parenchyma cells have the ability to _____ into other kinds of cells, for example when a plant has been injured.

9. Collenchyma tissue is composed of living cells that function primarily to _____ the plant.

10. The two types of sclerenchyma cells are (a) _____ and (b) _____.

11. Cellulose microfibrils are cemented together by a matrix of _____ and _____.

12. An important component of the cell walls of wood is _____ which may comprise up to 35% of the dry weight of the secondary cell wall.

The vascular tissue system consists of two complex tissues

13. Xylem conducts (a) _____ from roots to stems and leaves. Xylem also contains (b)_____ for storage and (c)_____ for support.

14. Having undergone (a) _____, mature tracheids and vessel cells are dead. They are also (b) _____ which contributes to their conducting properties.

15. Phloem is a complex tissue that functions to conduct(a) _____ throughout the plant through the (b) _____ which are among the most specialized plant cells.

16. _____ are cytoplasmic connections through which cytoplasm extends from one cell to another.

The dermal tissue system consists of two complex tissues

17. The dermal tissue system, the (a) _____and (b) _____, provides a protective covering over plant parts. Epidermal cells secrete a waxy layer called the (b) _____ that restricts water loss. (c)_____ are openings in this layer through which gases diffuse.

18. The epidermis contains special outgrowths, or hairs, called _____, which occur in many sizes and shapes and have a variety of functions.

19. The periderm replaces the epidermis in the stems and roots of older woody plants. Its primary function is for (a) _____. It is composed mainly of (b) _____ cells whose walls are coated with suberin, a waterproofing substance.

PLANT MERISTEMS

20. Plant growth is localized in regions called (a)_____.

21. Meristematic cells do not differentiate which means that plants and not most animals are able to _____ throughout their lifespan.

22. Plants have two kinds of meristematic growth: (a)_____ growth, an increase in length; and (b)_____ growth, an increase in the girth of the plant.

Primary growth takes place at apical meristems

23. The root apical meristem consists of three zones (areas), which, in order from the root cap inward, are the a)_____,
    (b)_____, and
    (c)_____.

24. (a)_____(developing leaves) and
    (b)_____ (developing buds) arise from the shoot apical meristem.

Secondary growth takes place at lateral meristems

25. There are two lateral meristems responsible for secondary growth (increase in girth), the (a)_____, a layer of meristematic cells that forms a long, thin, continuous cylinder within the stem and root, and the
    (b)_____, a thin cylinder or irregular arrangement of meristematic cell in the outer bark.

26. The outermost covering over woody stems and roots is the _____.

DEVELOPMENT OF FORM

The plane and symmetry of cell division affect plant form

27. The _____ appears just prior to mitosis and determines the plane in which the cells will divide.

The orientation of cellulose microfibrils affects the direction of cell expansion

28. Growth occurs in plants when cells (a) _____
    and (b) _____.

Cell differentiation depends in part on a cell's location

29. _____ is responsible for variations in chemistry, behavior, and structure among plant cells.

Morphogenesis occurs through pattern formation

30. Depending upon their location, cells are exposed to
    (a) _____ that specify positional information.

BUILDING WORDS

Use combinations of prefixes and suffixes to build words for the definitions that follow.

Prefixes	**The Meaning**		**Suffixes**	**The Meaning**
bi-	two		-derm(is)	skin
cut-	skin		-(i)cle	little
decidu-	falling off		-ome	mass
epi-	on, upon		-ous	pertaining to
stom-	mouth		-ennial	pertaining to a year
trich-	hair			

Prefix	**Suffix**	**Definition**
_____	_____	1. A hair or other special outgrowth growing out from the epidermis of plants.
_____	_____	2. A plant that takes two years to complete its life cycle.
_____	_____	3. Along with the periderm, provides a protective covering on the surface of plants.
_____	-a	4. A small pore ("mouth") in the epidermis of plants.
_____	_____	5. The shedding of leaves at the end of the growing season.
_____	_____	6. A waxy covering of plant epidermal cells that prevents water loss.

MATCHING

Terms:

a. Apical meristem
b. Companion cell
c. Dermal tissue
d. Ground tissue
e. Lateral meristem

f. Meristem
g. Parenchyma
h. Perennial
i. Periderm
j. Root system

k. Sclerenchyma
l. Shoot system
m. Tracheid
n. Vascular cambium
o. Xylem

For each of these definitions, select the correct matching term from the list above.

_____ 1. Vascular tissue that conducts water and dissolved minerals through the plant.

_____ 2. The chief type of water-conducting cell in the xylem of gymnosperms.

_____ 3. An area of dividing tissue located at the tips of roots and shoots in plants.

_____ 4. Secondary meristem that produces the secondary xylem and secondary phloem.

_____ 5. The type of tissue system that gives rise to the epidermis.

_____ 6. Plant cell that has thick secondary walls, is dead at maturity and functions in support.

_____ 7. A nucleated cell in the phloem responsible for loading and unloading sugar into the sieve tube member.

_____ 8. A plant that lives longer than two years.

_____ 9. General term for all localized areas of mitosis and growth in the plant body.

_____10. Plant cells that are relatively unspecialized, are thin walled, and function in photosynthesis and in the storage of nutrients.

MAKING COMPARISONS

Fill in the blanks.

Tissue	Tissue System	Function	Location in Plant Body
Parenchyma	Ground tissue	Photosynthesis, storage, secretion	Throughout plant body
Collenchyma	#1	Flexible structural support	#2
#3	Ground tissue	#4	Throughout plant body; common in stems and certain leaves, some nuts and pits of stone fruit
#5	#6	Conducts water, dissolved minerals	Extends throughout plant body
Phloem	Vascular tissue	#7	#8
Epidermis	#9	#10	#11
#12	Dermal tissue	#13	Covers body of woody plants

MAKING CHOICES

Place your answer(s) in the space provided. Some questions may have more than one correct answer.

_____ 1. Plants with the potential for living more than two years are called

    a. annuals.   d. perennials.

    b. biennials.   e. polyannuals.

    c. triennials.

_____ 2. The most common type of cell and tissue found throughout the plant body is the

    a. parenchyma.   d. vascular tissue.

    b. sclerenchyma.   e. xylem.

    c. collenchyma.

_____ 3. The kind of growth that results in an increase in the girth of the plant is known as

    a. differentiation.   d. primary growth.

    b. elongation.   e. secondary growth.

    c. apical meristem growth.

_____ 4. All plant cells have

    a. cell walls.   d. primary cell walls.

    b. secondary growth.   e. secondary cell walls.

    c. the capacity to form a complete plant.

_____ 5. Hairlike outgrowths of plant epidermis are called

    a. periderm.   d. trichomes.

    b. fiber elements.   e. companion cells.

    c. lateral buds.

_____ 6. The types of cells in phloem include
   a. tracheids.
   b. vessel elements.
   c. parenchyma.
   d. sieve tube elements.
   e. fiber cells.

_____ 7. The cell type found throughout the plant body that often functions in photosynthesis, secretion, and storage is called
   a. sclerenchyma.
   b. collenchyma.
   c. parenchyma.
   d. companion cells.
   e. a tracheid.

_____ 8. Increase in the girth of a plant is due to growth of the
   a. vascular cambium.
   b. cork cambium.
   c. area of cell maturation.
   d. area of cell elongation.
   e. lateral meristems.

_____ 9. Localized areas of cell division resulting in plant growth are called
   a. primordia.
   b. mitotic zones.
   c. meristems.
   d. primary growth centers.
   e. secondary growth centers.

_____ 10. The types of cells in xylem include
   a. tracheids.
   b. vessel elements.
   c. parenchyma.
   d. sieve tube members.
   e. fiber cells.

_____ 11. The soft tissue part of a plant and those that are often edible consist of
   a. sclerenchyma.
   b. parenchyma.
   c. periderm.
   d. collenchyma.
   e. lignin.

_____ 12. Terms associated with collenchymal cells include
   a. storage.
   b. dead.
   c. secretion.
   d. support.
   e. thick primary cell walls.

_____ 13. The most abundant polymer in the world is
   a. lignin.
   b. cellulose.
   c. suberin.
   d. pectin.
   e. lectin.

_____ 14. In plants, the vascular tissue system
   a. is embedded in the ground tissue.
   b. vessel elements.
   c. transports carbohydrates.
   d. includes the xylem and phloem.
   e. has parts composed of dead cells.

_____ 15. Plasmodesmota are
   a. connect cells of the phloem.
   b. are cytoplasmic connections.
   c. are located in the periderm.
   d. connect cells of the xylem.
   e. companion cells.

_____ 16. Stomata allow the passage of

    a. carbohydrates.   d. carbon dioxide.

    b. water vapor.   e. oxygen

    c. glucose.

_____ 17. Cork cambium

    a. is composed of meristematic cells.   d. is located in the outer bark.

    b. is part of the vascular cambium.   e. is part of the periderm.

    c. contains storage cells.

VISUAL FOUNDATIONS

Color the parts of the illustration below as indicated. Also label the area of cell division and the area of cell maturation.

RED   ❑   apical meristem

GREEN   ❑   area of cell elongation

YELLOW   ❑   root cap

ORANGE   ❑   root hairs

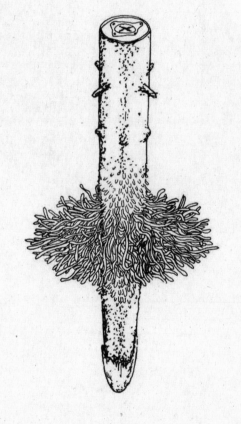

Color the parts of the illustration below as indicated.

RED   ☐   outer bark (periderm)

GREEN   ☐   inner bark (secondary phloem)

YELLOW   ☐   vascular cambium

Brown   ☐   wood (secondary xylem)

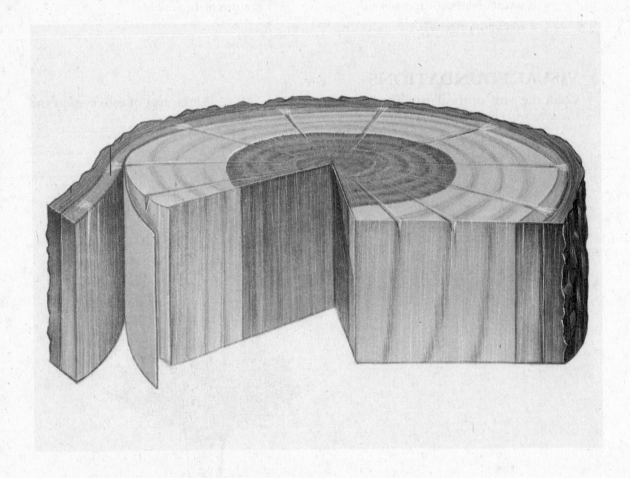

Leaf Structure and Function

This chapter, the second of six addressing the integration of plant structure and function, discusses the structural and physiological adaptations of leaves. Leaves, as the principal photosynthetic organs in plants, are highly adapted to collect radiant energy, convert it into the chemical bonds of carbohydrates, and transport the carbohydrates to the rest of the plant. They are also adapted to permit gas exchange, and to receive water and minerals transported into them by the plant vascular system. Water loss from leaves is controlled in part by the cell walls of the epidermal cells, by presence of a surface waxy layer secreted by epidermal cells, and by tiny pores in the leaf epidermis that open and close in response to various environmental factors. Most of the water absorbed by land plants is lost, either in the form of water vapor or as liquid water. Survival at low temperatures is facilitated in some plants by the loss of leaves. The leaves of many plants are modified for functions other than photosynthesis.

REVIEWING CONCEPTS

Fill in the blanks.

INTRODUCTION

LEAF FORM AND STRUCTURE

1. The broad, flat portion of a leaf is the (a)_____; the stalk that attaches the blade to the stem is the (b)_____. Some leaves also have (c)_____, which are leaflike outgrowths usually present in pairs at the base of the petiole.

2. Leaves may be (a)_____, having a single blade, or (b)_____, having a blade divided into two or more leaflets.

3. Leaves are arranged on a stem in one of three possible ways. These are (a)_____, with one leaf at each node; (b)_____, with two leaves at each node; and (c)_____, with three or more leaves at each node.

4. Leaf blades may have (a)_____ venation in which the strands of vascular tissue run parallel to one another. This venation is generally found in (b)_____. Venation may be (c)_____ in which the veins form a meshwork. This venation is generally found in (d)_____.

Leaf structure is adapted for maximum light absorption

5. The _____ forms the surface tissue of the leaf.

6. The surface tissue of the leaf secretes a waxy (a)_____ composed of (b)_____ which helps to reduce water loss.

7. Guard cells change shape to regulate the stomatal opening when water and ions flow into them from the _____ cells.

8. The photosynthetic ground tissue of the leaf is called the
   (a)_____. When this tissue is divided into two regions, the upper layer, nearest to the upper epidermis, is called the
   (b)_____, and the lower portion is called the (c)_____.

9. (a)_____ in veins of a leaf conduct water and essential minerals, while (b)_____ in veins conducts sugar produced by photosynthesis. Veins may be surrounded by a
   (c)_____, consisting of parenchyma or sclerenchyma cells.

10. Light can penetrate into the body of the leaf because the leaf epidermis is
    (a)_____. Light penetration is necessary for
    (b)_____ to occur in complexes present in the mesophyll.

11. (a)_____, a raw material of photosynthesis, diffuses into the leaf through stomata, and the (b)_____ produced during photosynthesis diffuses rapidly out of the leaf through stomata.

STOMATAL OPENING AND CLOSING

12. A guard cell's shape is determined by the amount of water it contains which contributes to its (a)_____. When water leaves a guard cell is becomes (b)_____.

Blue light triggers stomatal opening

13. Any plant response to light must involve a _____ molecule.

14. The pigment involved in the opening and closing of stomata is _____.

15. Blue light triggers the synthesis of (a)_____ and the hydrolysis of (b)_____.

16. Blue light also triggers the activation of (a)_____ located in the plasma membrane of guard cells. An electrochemical gradient is formed when (b)_____ is pumped out of the guard cells.

17. As evening approaches, the concentration of (a)_____ in the guard cell declines as it is converted to (b)_____ which is a less osmotically active molecule.

Additional factors affect stomatal opening and closing

18. The opening and closing of stomata is triggered by _____.

19. Other environmental factors that affect stomatal opening and closing include (a) _____, (b) _____, and (c) _____.

TRANSPIRATION AND GUTTATION

20. Transporation helps to move water from (a) _____ and throughout the plant.

21. Two benefits of transpiration are (a) _____ and (b)_____.

22. Transportation is like _____ in humans.

Some plants exude liquid water

23. Loss of liquid water from leaves by force is known as _____.

LEAF ABSCISSION

24. Abscission is the process of _____.

25. Leaf abscission is a complex process that involves many physiological changes, all initiated and orchestrated by changing levels of plant hormones, particularly _____.

26. The brilliant colors found in autumn landscapes in temperate climates are due to the various combinations of (a)_____ and (b)_____.

In many leaves, abscission occurs at an abscission zone near the base of the petiole

27. The area where a petiole detaches from the stem is a structurally distinct area called the (a) _____; it is a weak area because it contains relatively few strengthening (b)_____.

28. The "cement" that holds the primary cell walls of adjacent cells together is called the _____.

MODIFIED LEAVES

29. Leaves are variously modified for a plethora of specialized functions. For example, the hard, pointed _____ on a cactus are leaves modified for protection.

30. _____ are specialized leaves that anchor long, climbing vines to the supporting structures on which they are growing.

Modified leaves of carnivorous plants capture insects

31. The traps of the pitcher plant are (a)_____, whereas those of the Venus flytrap are (b)_____.

BUILDING WORDS

Use combinations of prefixes and suffixes to build words for the definitions that follow.

Prefixes	The Meaning		Prefixes	The Meaning
abscis-	cut off		-ation	the process of
circ-	around		-ion	process of
gutt-	tear		-phyll	leaf
mes(o)-	middle			
trans-	across, beyond			

Prefix	Suffix	Definition
_____	_____	1. The separation and falling away from the plant stem of leaves, fruit and flowers.
_____	-adian	2. Pertains to something that cycles at approximately 24-hour intervals.
_____	_____	3. The photosynthetic tissue of the leaf sandwiched between (in the middle of) the upper and lower epidermis.
_____	-piration	4. The loss of water vapor from the plant body across leaf surfaces.
_____	_____	5. The loss of liquid water which is forced out at specialized regions on leaves.

MATCHING

Terms:

a. Blade
b. Bundle sheath
c. Cuticle
d. Guard cell
e. Guttation
f. Petiole
g. Phloem
h. Spine
i. Stipule
j. Tendril
k. Trichome
l. Vascular bundle
m. Vein
n. Xylem

For each of these definitions, select the correct matching term from the list above.

____ 1. A ring of cells surrounding the vascular bundle in monocot and dicot leaves.

____ 2. A leaf or stem that is modified for holding or attaching to objects.

____ 3. Vascular tissue that transports sugars produced by photosynthesis.

____ 4. A waxy covering over the epidermis of the above-ground portion of plants.

____ 5. A hard, pointed leaf that is modified for protection.

____ 6. The part of a leaf that attaches to a stem.

____ 7. One of two cells that collectively form a stoma.

_____ 8. In plants, vascular bundles in leaves.

_____ 9. Leaflike outgrowth at the base of the peticule.

_____ 10. The flat portion of a leaf.

MAKING COMPARISONS

Fill in the blanks.

Structure of Leaf	Function of Leaf Structure
Thin, flat shape	Maximizes light absorption, efficient gas diffusion
Ordered arrangement on stem	#1
Waxy cuticle	#2
#3	Allow gas exchange between plant and atmosphere
Relatively transparent epidermis	#4
#5	Allows for rapid diffusion of CO_2 to mesophyll cell surface
#6	Provide support to prevent leaf from collapsing
Xylem in veins	#7
#8	Transports sugar to other plant parts

MAKING CHOICES

Place your answer(s) in the space provided. Some questions may have more than one correct answer.

_____ 1. Loss of water by evaporation from aerial plant parts is called
   a. transpiration.
   b. guttation.
   c. abscission.
   d. activation.
   e. aspiration.

_____ 2. The "cement" that holds the primary cell walls of adjacent cells together is called the
   a. abscission zone.
   b. leaf scar.
   c. adhesion layer.
   d. middle lamella.
   e. glial substance.

_____ 3. The portion of mesophyll that is usually composed of loosely and irregularly arranged cells is
   a. the palisade layer.
   b. the spongy layer.
   c. toward the leaf's underside.
   d. toward the leaf's upperside.
   e. an area of photosynthesis.

_____ 4. The process by which plants secrete water as a liquid is
   a. evaporation.
   b. transpiration.
   c. guttation.
   d. the potassium ion mechanism.
   e. found only in monocots.

_____ 5. The leaves of eudicots usually have
   a. bulliform cells.
   b. netted venation.
   c. differentiated palisade and spongy tissues.
   d. special subsidiary cells.
   e. bean-shaped guard cells.

_____ 6. Trichomes are found on/in the
   a. palisade layer.
   b. spongy layer.
   c. entire mesophyll.
   d. epidermis.
   e. guard cells.

_____ 7. Facilitated diffusion of potassium ions into guard cells
   a. requires ATP.
   b. closes stomates.
   c. causes guard cells to shrink and collapse.
   d. occurs more in daylight than at night.
   e. indirectly causes pores to open.

_____ 8. In general, leaf epidermal cells
   a. are living.
   b. lack chloroplasts.
   c. protect mesophyll cells from sunlight.
   d. are absent on the lower leaf surface.
   e. have a cuticle.

_____ 9. Factors that tend to affect the amount of transpiration include
   a. the cuticle.
   b. wind velocity.
   c. ambient temperature.
   d. relative humidity.
   e. amount of light.

_____ 10. Guard cells generally
   a. have chloroplasts.
   b. form a pore.
   c. are found in the epidermis.
   d. are found only in monocots.
   e. are found only in dicots.

_____ 11. Red water-soluble pigments in leaves are
   a. carotenoids.
   b. xanthophylls.
   c. anthocyanins.
   d. rhodophylls.
   e. found only in monocots.

_____ 12. Proton pumps in the plasma membranes of guard cells are activated by
   a. blue light.
   b. ultraviolet light.
   c. red light.
   d. wavelengths in the 800-1,000 nm range.
   e. wavelengths in the 400-500 nm range.

_____ 13. Water diffuses into guard cells by means of
   a. voltage-activated ion channels.
   b. proton pumps.
   c. active transport.
   d. facilitated diffusion.
   e. osmosis.

_____ 14. Monocot leaves
   a. a palisade layer.
   b. have parallel veins.
   c. have evenly spaced veins on cross section.
   d. have a spongy layer.
   e. may have guard cells shaped like dumbbells.

_____ 15. Terms associated with a leaf include

    a. axillary bud.   d. stipules.

    b. blade.   e. veins.

    c. petiole.

_____ 16. Venation patterns may be

    a. simple.   d. alternate.

    b. opposite.   e. pinnately netted.

    c. parallel.

_____ 17. The bundle sheath is composed of

    a. sclerenchyma cells.   d. mesohyl

    b. cambium.   e. parenchyma cells.

    c. cork cambium.

_____ 18. Water required for photosynthesis is obtained from

    a. reverse transpiration.   d. subsidiary cells.

    b. soil.   e. the interstitial space.

    c. guard cells.

VISUAL FOUNDATIONS

Color the parts of the illustration below as indicated. Also label the location of stomata. Using brackets, indicate the structures on the list below that would be components of the dermal tissue system, the ground tissue system, and the vascular tissue system.

RED   ❑   stoma

YELLOW   ❑   upper epidermis

ORANGE   ❑   lower epidermis

VIOLET   ❑   cuticle

BLUE   ❑   palisade mesophyll

GREEN   ❑   spongy mesophyll

PINK   ❑   xylem

PURPLE   ❑   phloem

BROWN   ❑   bundle sheath

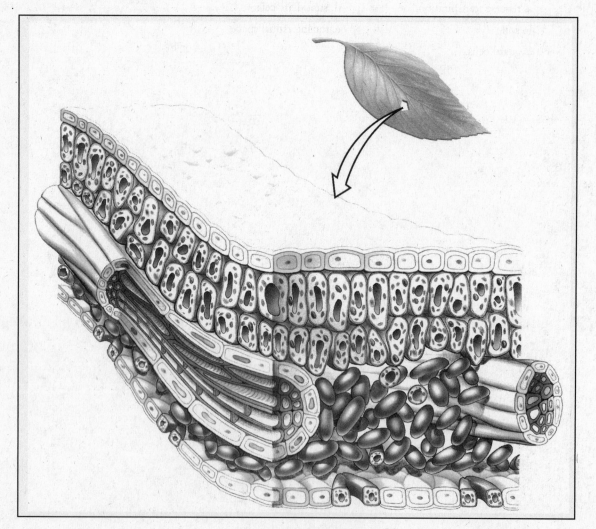

❑

Stem Structure and Transport

This chapter, the third of six addressing the integration of plant structure and function, discusses the structural and physiological adaptations of stems. Stems function to support leaves and reproductive structures, conduct materials absorbed by the roots and produced in the leaves to other parts of the plant, and produce new tissue throughout the life of the plant. Specialized stems have other functions as well. Although all herbaceous, or nonwoody, stems have the same basic tissues, the arrangement of the tissues in the monocot stem differs from that in the eudicot stem. The stems of herbaceous plants increase only in length. The stems of all gymnosperms and some eudicots increase in both length and girth. Increases in the length of plants result from mitotic activity in apical meristems, and increases in the girth of plants result from mitotic activity in lateral meristems. Simple physical forces are responsible for the movement of food, water, and minerals in multicellular plants.

REVIEWING CONCEPTS

Fill in the blanks.

INTRODUCTION

1. The three main parts of a vegetative vascular plant are
   (a) _____, (b) _____, and
   (c)_____.
2. The primary functions of stems are to
   (a) _____,
   (b) _____, and
   (c) _____.

STEM GROWTH AND STRUCTURE

3. All plants have (a) _____ growth.

Herbaceous eudicot and monocot stems differ in internal structure

4. A sunflower is a representative _____ stem.
5. Vascular tissues in herbaceous eudicots are located in _____.
6. Each vascular bundle contains two vascular tissues, the
   (a)_____, and in some herbaceous stems a single layer of
   cells, the (b)_____, sandwiched between them.
7. A cross section the vascular bundle of a monocot stem would have
   (a) _____ toward the inside and
   (b) _____ toward the outside of the bundle.
8. Monocot stems do not possess _____ which give
   rise to secondary growth in woody stems.

Woody plants have stems with se

9. In woody plants the (a) _____, gives rise to secondary xylem and phloem, and the (b)_____, along with the tissues it produces, is called the (c)_____.

10. The woody stem increases in diameter when cells in the vascular cambium divide _____.

11. The vascular cambium produces secondary (a)_____ and secondary (b)_____ to replace the primary conducting and supporting tissues.

12. Secondary tissues eventually replaces the function of primary tissues as a result of the primary tissues being _____ by the mechanical pressure of secondary growth.

13. Lateral transport takes place in (a)_____, which are chains of (b)_____ cells that radiate out from the center of the woody stem or root.

14. Cork cambium produces (a) _____ which is the functional replacement for the (b) _____.

15. Variation in cork cambia and their rats of division explain the differences in the _____ of different tree species.

16. Cork cambium cells divide to form new tissues toward the inside and the outside. The layer toward the outside consists of _____, heavily waterproofed cells that protect the plant.

17. To the inside, cork cambium forms the _____ that stores water and starch.

18. In a woody twig the apical meristem is dormant and covered by an outer layer of (a) _____ which are actually modified (b)_____.

19. The number of _____ on a termperate-zone woody twig indicates its age.

20. When looking at a woody stem during the winter the pattern of _____ indicates the leaf arrangement on the stem during the summer season.

21. The older wood in the center of a tree is called (a)_____, and the younger, lighter-colored, more peripheral wood is (b)_____.

22. Botanically speaking (a) _____ wood is the wood of flowering plants and (b) _____ wood is the wood of conifers.

23. The wood of conifers typically lacks (a) _____ and (b) _____ which accounts for the major differences between the wood of flowering plants and conifers.

24. Annual rings are composed of two types of cells arranged in alternating concentric circles, each layer appropriately named for the season in which it developed. The (a) _____ has large-diameter conducting cells and few fibers, whereas the (b)_____ has narrower conducting cells and numerous fibers.

WATER TRANSPORT

25. In contrast to internal circulation in animals, in plants movement of materials throughout the organism is driven largely by _____.

Water and minerals are transported in xylem

26. Water moves in plants as result of being either (a) _____ or (b) _____. Current evidence indicates that most water is transported by being (c) _____.

Water movement can be explained by a difference in water potential

27. One of the principal forces behind water movements through plants is a function of a cell's ability to absorb water by osmosis, also known as the "free energy of water," or the (a) _____. Water containing solutes has (b) _____ (more or less?) free energy than pure water. Water moves from a region of (c) _____ water potential to a region of (d)_____ water potential.

28. Under normal conditions, the water potential of the root is more _____ (negative or positive?) than the water potential of the soil. Thus water moves by osmosis from the soil into the root.

According to the tension-cohesion model, water is pulled up a stem

29. Water is pulled up the plant as a result of (a) _____ occurring at the top of the plant.

30. This upward pull is possible only as long as the column of water in the xylem remains unbroken. The two forces working to maintain this unbroken column are (a) _____ and (b) _____.

Root pressure pushes water from the root up a stem

31. Root pressure occurs when (a) _____ are actively absorbed and pumped into the xylem. This movement causes a (b) _____ (decrease, increase) in the water potential of the xylem.

TRANSLOCATION OF SUGAR IN SOLUTION

The pressure-flow model explains translocation in phloem

32. The movement of sugar in the phloem occurs as the result of a pressure gradient which exists between the (a) _____ where sugar is loaded into the phloem and the (b) _____ where the sugar is removed from the phloem.

33. The accumulation of sugar in the sieve tube element causes a (a) _____ (decrease, increase) in the water potential which causes water to move by (b) _____ and (c) _____ (increase, decrease) in the turgor pressure inside the sieve tube elements.

BUILDING WORDS

Use combinations of prefixes and suffixes to build words for the definitions that follow.

Prefixes	The Meaning	Suffixes	The Meaning
adher(s)-	to stick to		
cohes-	to stick together		
inter-	between, among	-derm	skin
osmo-	pushing		
peri-	around	-ion	the process of
trans-	across, beyond	-sis	the process of

Prefix	Suffix	Definition
_____	-location	1. The movement of materials (across distances) in the vascular tissues of a plant.
_____ _____		2. Layers of cells covering the surface of woody stems and roots (i.e., the outer bark); the "skin" of woody dicots and cone-bearing gymnosperms.
_____	-node	3. The region of a stem between two successive nodes.
_____ _____		4. The binding of water molecules to the walls of the xylem cells.
_____ _____		5. The binding of water molecules to each other.
_____ _____		6. The diffusion of water from an area of higher to an area of lower concentration.

MATCHING

Terms:

a. Apical meristem
b. Cork cambium
c. Ground tissue
d. Lateral bud
e. Lateral meristem

f. Lenticels
g. Periderm
h. Pith
i. Ray
j. Root pressure

k. Terminal bud
l. Vascular cambium
m. Xylem

For each of these definitions, select the correct matching term from the list above.

_____ 1. An area of dividing tissue located at the tips of plant stems and roots.

_____ 2. Technical term for the outer bark of woody stems and roots.

_____ 3. Large, thin-walled parenchyma cells found in the innermost tissue in many plants.

_____ 4. A lateral meristem in plants that produces cork cells and cork parenchyma.

_____ 5. The cortex and the pith are parts of this tissue system.

_____ 6. Lateral meristem that gives rise to secondary vascular tissues.

_____ 7. The vascular tissue responsible for transporting water and dissolved minerals in plants.

_____ 8. The positive pressure in the root tissues of plants.

_____ 9. A chain of parenchyma cells that functions for lateral transport of food, water, and minerals in woody plants.

_____10. Another name for axillary bud.

_____ 11. The embryonic shoot located at the tip of the stem.

MAKING COMPARISONS

Fill in the blanks.

Tissue	Source	Location	Function
Secondary xylem	Produced by vascular cambium	Wood	Conducts water and dissolved minerals
#1	Produced by vascular cambium	Inner bark	#2
Cork parenchyma	#3	Periderm	#4
#5	Produced by cork cambium	#6	Replacement for epidermis
#7	A lateral meristem produced by procambium tissue	Between wood and inner bark in vascular bundles	#8
Cork cambium	#9	In epidermis or outer cortex	#10

MAKING CHOICES

Place your answer(s) in the space provided. Some questions may have more than one correct answer.

_____ 1. Secondary xylem and phloem are derived from
   a. cork parenchyma.   d. periderm.
   b. cork cambium.   e. epidermis.
   c. vascular cambium.

_____ 2. The outer portion of bark is formed primarily from
   a. cork parenchyma.   d. periderm.
   b. cork cambium.   e. epidermis.
   c. vascular cambium.

_____ 3. If soil water contains 0.1% dissolved materials and root water contains 0.2% dissolved materials, one would expect
   a. a negative water potential in soil water.   d. water to flow from soil into root.
   b. a negative water potential in root water.   e. water to flow from root into soil.
   c. less water potential in roots than in soil.

_____ 4. When water is plentiful, wood formed by the vascular cambium is
   a. springwood.   d. composed of large diameter conducting cells.
   b. summerwood.   e. composed of thick-walled vessels.
   c. late summerwood.

_____ 5. One daughter cell from a mother cell in the vascular cambium remains as part of the vascular cambium, the other divides to form
   a. secondary xylem or phloem.   d. secondary tissue.
   b. wood or inner bark.   e. primary xylem and phloem.
   c. outer bark.

_____ 6. If a plant is placed in a beaker of pure distilled water, then
    a. water will move out of the plant.   d. water potential in the beaker is zero.
    b. water will move into the plant.   e. water potential in the plant is less than zero.
    c. water pressure in the plant is positive relative to water in the beaker.

_____ 7. The pull of water up through a plant is due in part to
    a. cohesion.   d. transpiration.
    b. adhesion.   e. a water potential gradient between the soil and the root.
    c. the low water potential in the atmosphere.

_____ 8. Which of the following is/are true of monocot stems?
    a. stem covered with epidermis   d. vascular bundles scattered through the stem
    b. vascular tissues embedded in ground tissue   e. vascular bundles arranged in circles
    c. stem has distinct cortex and pith

_____ 9. Which of the following is/are true of eudicot stems?
    a. stem covered with epidermis   d. vascular bundles scattered through the stem
    b. vascular tissues embedded in ground tissue   e. vascular bundles arranged in circles
    c. stem has distinct cortex and pith

_____10. The area on a stem where each leaf is attached is called the
    a. bud scale.   d. leaf scar.
    b. node.   e. lenticel.
    c. lateral bud.

_____11. Sites of loosely arranged cells along the bark of a woody twig allow gases to diffuse into the stem. These are called
    a. bud scales.   d. leaf scars.
    b. nodes.   e. lenticels.
    c. lateral buds.

_____12. Terms associated with secondary growth include
    a. an increase in girth.   d. lateral meristem.
    b. apical meristerm.   e. woody stems.
    c. vascular cambium.

_____13. The ground tissue at the center of a herbaceous eudicot stem is called
    a. cortex.   d. epidermis.
    b. cork.   e. pith.
    c. periderm.

_____14. Monocot stems lack
    a. sclerencyma   d. lateral meristems.
    b. vascular cambium.   e. cork cambium.
    e. an epidermis.

_____15. As a woody stem increases in circumference, the number of cells in the vascular cambium
    a. increase.
    b. decrease.

_____16. The area of a leaf where the leaf is attached is called the
    a. internode.   d. lateral node.
    b. axillary node.   e. node
    c. bundle.

VISUAL FOUNDATIONS

Color the parts of the illustration below as indicated.

RED ☐ vascular cambium
GREEN ☐ secondary phloem
YELLOW ☐ pith
BLUE ☐ primary phloem
ORANGE ☐ secondary xylem
BROWN ☐ primary xylem
PINK ☐ epidermis
PURPLE ☐ cortex
TAN ☐ periderm

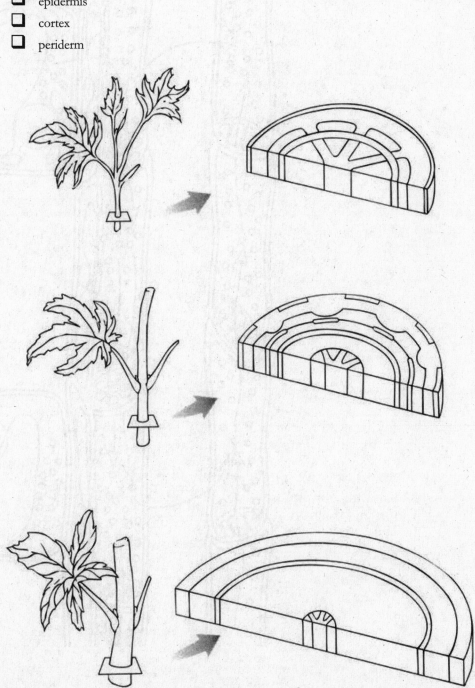

Color the parts of the illustration below as indicated. Also label xylem and phloem.

RED ☐ path of sucrose
YELLOW ☐ companion cell
BLUE ☐ path of water
ORANGE ☐ sink
BROWN ☐ sieve tube element
PINK ☐ source

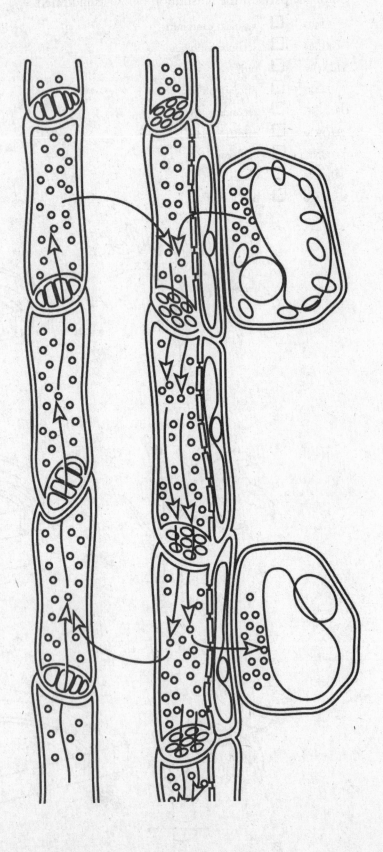

CHAPTER 36

Roots and Mineral Nutrition

The previous two chapters discussed the structural and physiological adaptations of the plant's shoot system, i.e., the leaves and stems. This chapter addresses the structural and physiological adaptations of the remaining portion of the plant, i.e., the root system. The roots of most plants function to anchor plants to the ground, to absorb water and dissolved minerals from the soil, and, in some cases, to store food. Additionally, the roots of some plants are modified for support, aeration, and/or photosynthesis. Although herbaceous roots have the same tissues and structures found in stems, they also have other tissues and structures as well. Soil is composed of inorganic minerals, organic matter, soil organisms, soil atmosphere, and soil water. Plants require at least 19 essential nutrients for normal growth, development, and reproduction. Some human practices, such as harvesting food crops, depletes the soil of certain essential elements, making it necessary to return these elements to the soil in the form of organic or inorganic fertilizer.

REVIEWING CONCEPTS

Fill in the blanks.

INTRODUCTION

1. The primary functions of roots include:

    (a) _____,

    (b) _____,

    and

    (c) _____.

ROOT STRUCTURE AND FUNCTION

2. Two types of root systems occur in plants; they are the

    (a) _____ root system and the (b) _____ root system.

3. The embryonic root in the seed gives rise to the (a) _____ root system and the (b) _____ root system develops from the stem tissue.

4. Roots that arise from the stem and not preexisting roots are called _____ roots.

Roots have root caps and root hairs

5. Each root tip is covered by a (a) _____, which protects the root

    (b) _____.

6. Short-lived tubular extensions of epidermal cells located just behind the growing root tip are called _____.

The arrangement of vascular tissues distinguishes the roots of herbaceous eudicots and monocots

7. The eudicot root is composed primarily of (a) _____ and lacks supporting (b) _____.

8. Cytoplasmic bridges connecting root cells are called
_____.

9. The inner layer of cortex in a eudicot root is the (a) _____, the cells of which possess a bandlike region on their radial and transverse walls. This "band," or (b) _____ as it's called, contains suberin.

10. _____ are integral membrane proteins that facilitate the rapid movement of water across membranes.

11. The stele at the center of the eudicot primary root is composed of a outer layer the (a) _____ and an inner region the (b) _____ in which are located patches of (c) _____.

12. Horizontal movement of water from the soil into the center of the root can be summarized as moving from root hair/epidermis → (a) _____ → (b) _____ → (c) _____ → root xylem.

13. Since secondary growth is absent in virtually all monocots, monocot roots lack a _____.

Woody plants have roots with secondary growth

14. Plants that produce stems with secondary growth also produce _____ with secondary growth.

15. As growth occurs in the roots of woody plants the epidermis is replaced by (a) _____ which is composed of cork cells and (b) _____.

Some roots are specialized for unusual functions

16. Adventitious roots often arise from the _____ of stems.

17. Roots that develop from branches or a vertical stem and provide support to the plants such as corn are called (a) _____. The swollen bases that help to support trees in an upright position are called (b) _____.

18. (a) _____ are plants that grow attached to other plants. They use (b) _____ roots to anchor themselves to bark, branches, and other surfaces.

ROOT ASSOCIATIONS AND INTERACTIONS

Mycorrhizae facilitate the uptake of essential minerals by roots

19. Subterranean associations between roots and soil fungi are known as
_____.

Rhizobial bacteria fix nitrogen in the roots of leguminous plants

20. Plants and nitrogen-fixing bacteria use _____ to establish contact and develop nodules.

THE SOIL ENVIRONMENT

21. Soils area generally formed from _____ continually being fragmented into smaller particles.

Soil is composed of inorganic minerals, organic matter, air, and water

22. _____ have the greatest surface area of all soil particles.

23. Roots secrete (a) _____ which are exchanged for positively charged mineral ions adhering to the surface of the soil molecules in a process called (b) _____.

24. Good, loamy agricultural soil contains about 40% each of (a) _____ and about 20% of (b) _____. The spaces between soil particles are filled with (c) _____.

25. Partly decayed organic matter called _____ contributes to the water holding ability of soils.

26. When water drains from large pores in the soil it draws in _____.

27. Water moving downward through soil spaces carries dissolved minerals with it in a process called (a) _____. The deposition of the dissolved minerals in lower layers of the soil is called (b) _____.

Soil organisms form a complex ecosystem

28. Bits of soil that have passed through the gut of an earthworm are called _____.

Soil pH affects soil characteristics and plant growth

29. Air pollution in which sulfuric and nitric acids produced by human activities fall to the ground as acid rain, sleet, snow, or fog is known as _____.

Soil provides most of the minerals found in plants

30. More than (a) _____(#) elements have been found on earth and over (b) _____(#) of them have been found in plant tissues.

31. Elements required in large quantities are called (a) _____ while those needed only in small amounts are called (b) _____.

32. In plants, potassium remains (a) _____ and plays an important role in maintaining (b) _____ of cells. Potassium is also involved in the opening and closing of (c) _____.

33. In order to identify elements that are essential for plant growth, investigators must reduce the number of variables to a minimum. To do this, since soil is very complex, plants are grown in aerated water containing known quantities of elements, a process known as _____.

Soil can be damaged by human mismanagement

34. The three elements that most often limit plant growth are
    (a) _____,
    (b) _____, and
    (c) _____.

35. The wearing away or removal of soil from the land is known as _____.

36. Irrigation sometimes causes salt to accumulate in the soil, a process called _____.

BUILDING WORDS

Use combinations of prefixes and suffixes to build words for the definitions that follow.

Prefixes	The Meaning	suffixes	The Meaning
endo-	within	-derm(is)	skin
epi-	upon	-phyt(e)	plant
hydro-	water	-rrhiz(a)	root
macro-	large, long, great, excessive	-ule	small
micro-	small		
myc(o)-	fungus		
nod-	knob		

Prefix	Suffix	Definition
_____	-ponics	1. Growing plants in water (not soil) containing dissolved inorganic minerals.
_____	-nutrient	2. An essential element that is required in fairly large amounts for normal plant growth.
_____	_____	3. Plants that grow attached to other plants.
_____	_____	4. The innermost layer of the cortex in the plant root.
_____	_____	5. An association between roots and certain fungi.
_____	_____	6. Swelling on roots that houses nitrogen-fixing bacteria.

MATCHING

Terms:

a.	Adventitious root	f.	Graft	k.	Prop root
b.	Apoplast	g.	Humus	l.	Root cap
c.	Casparian strip	h.	Leaching	m.	Root hair
d.	Endodermis	i.	Mycorrhizae	n.	Stele
e.	Fibrous root system	j.	Pneumatophore	o.	Taproot system

For each of these definitions, select the correct matching term from the list above.

_____ 1. Mutualistic associations of fungi and plant roots that aid in the plant's absorption of essential minerals from the soil.

_____ 2. Organic matter in various stages of decomposition in the soil.

_____ 3. The innermost layer of the cortex in the plant root.

_____ 4. A root that arises in an unusual position on a plant

_____ 5. A band of waterproof material around the radial and transverse walls of endodermal root cells.

_____ 6. An adventitious root that arises from the stem and provides additional support for plants.

_____ 7. Aerial "breathing roots" of some plants in swampy and tidal environments that assist in getting oxygen to submerged roots.

_____ 8. The type of root system present in a plant that has several roots of the same size developing from the end of the stem, with lateral roots of various sizes branching off these roots..

_____ 9. An extension of an epidermal cell in roots, which increases the absorptive capacity of the roots.

_____ 10. A covering of cells over the root tip that protects delicate meristematic tissue beneath it.

_____ 11. At the center of an eudicot primary root.

_____ 12. Root system formed from the seedling's enlarging radical.

_____ 13. The effect of water transporting dissolved minerals deeper into the soil.

MAKING COMPARISONS

Fill in the blanks.

Root Structure	Function of the Root Structure
Root cap	Protects the apical meristem, may orient the root downward
Root apical meristem	#1
#2	Absorption of water and minerals
Epidermis	#3
Cortex	#4
#5	Controls mineral uptake into root xylem
#6	Gives rise to lateral roots and lateral meristems
#7	Conducts water and dissolved minerals
Phloem	#8

MAKING CHOICES

Place your answer(s) in the space provided. Some questions may have more than one correct answer.

_____ 1. Structures found in primary roots that are also found in stems include the

    a. cortex.
    b. cuticle.
    c. vascular tissues.
    d. apical meristem cap.
    e. epidermis.

_____ 2. The three groups of organisms in soil that are the most important in decomposition are

    a. insects.
    b. fungi.
    c. soil protozoa.
    d. algae.
    e. bacteria.

_____ 3. The origin of multicellular branch roots is the

    a. cambium.
    b. Casparian strip.
    c. pericycle.
    d. parenchyma.
    e. cortex.

_____ 4. When water first enters a root, it usually

    a. enters parenchymal cells.
    b. is absorbed by cellulose.
    c. moves from a water negative potential to a positive water potential.
    d. moves along cell walls.
    e. enters cells in the Casparian strip.

_____ 5. The function(s) performed by all roots is/are

    a. absorption of water.
    b. absorption of minerals.
    c. food storage.
    d. anchorage.
    e. aeration.

_____ 6. The principal function(s) of the root cortex is/are

    a. conduction.
           d. water absorption.

    b. storage.
           e. mineral absorption.

    c. production of root hairs.

_____ 7. Essential macronutrients include

    a. phosphorus.
           d. hydrogen.

    b. potassium.
           e. magnesium.

    c. iron.

_____ 8. Essential micronutrients include

    a. phosphorus.
           d. hydrogen.

    b. potassium.
           e. magnesium.

    c. iron.

_____ 9. In general, the vascular tissues in monocot roots

    a. form a solid cylinder.
           d. contain a vascular cambium.

    b. are absent.
           e. continue into root hairs.

    c. are in bundles arranged around the central path.

_____ 10. The inorganic materials in soil come from

    a. fertilizers.
           d. the atmosphere.

    b. water runoff.
           e. percolating water.

    c. weathered rock.

_____ 11. Roots produced in unusual places on the plants, often as aerial roots, are called _____ roots.

    a. secondary
           d. contractile

    b. enhancement
           e. aerial water-absorbing

    c. adventitious

_____ 12. Sugar produced during photosynthesis are transported in the

    a. epidermis.
           d. radical.

    b. xylem
           e. phloem.

    c. cambium.

_____ 13. Root hairs are modified

    a. endodermis.
           d. parenchyma.

    b. sclerenchyma
           e. collenchymas.

    c. epidermis.

_____ 14. Oxygen needed by roots

    a. enters from spaces in the soil.
           d. is transported by the phloem.

    b. is produced by the breakdown of glucose.
           e. is less available when it rains.

    c. is transported by the xylem.

_____ 15. The Casparian strip

    a. contain suberin.
           d. is associated with the endodermis.

    b. is waterproof.
           e. forms when leaves prepare to fall from the plant.

    c. forms to prevent sugar loss from the root.

_____ 16. Aquaporins

    a. are intergral membrane proteins.
           d. can be found in roots.

    b. facilitate solute movement into roots.
           e. are located in the epidermis.

    c. contribute to the pericyle.

_____17. Aerial breathing roots are also known as

    a. prop roots.   d. arbuscules.

    b. nodules.   e. pneumatophores.

    c. rhizobia.

_____18. When soils are low in nitrogen, legume roots

    a. secrete flavinoids.   d. initiate cell signaling.

    b. secrete nitrogenase compounds.   e. enlarge their nodules.

    c. attract rhizobial bacteria.

_____19. The plant hormone _____ triggers cell division and is involved in root nodule formation.

    a. giberellin.   d. ethylene.

    b. cytokinin.   e. auxin.

    c. abscisic acid.

VISUAL FOUNDATIONS

Color the parts of the illustration below as indicated.

RED ☐ primary xylem
GREEN ☐ cortex
BLUE ☐ vascular cambium
ORANGE ☐ secondary xylem
YELLOW ☐ epidermis
BROWN ☐ periderm
TAN ☐ pericycle
PINK ☐ secondary phloem
VIOLET ☐ primary phloem

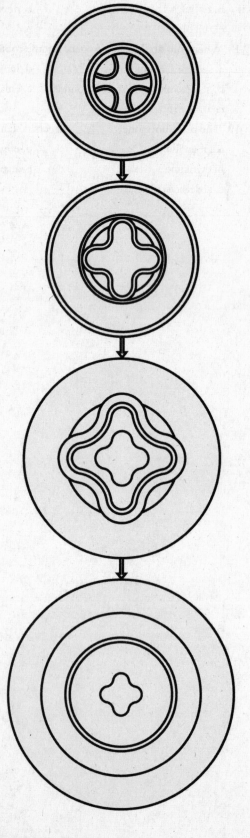

CHAPTER 37

Reproduction in Flowering Plants

This chapter, the fifth in a series of six addressing the integration of plant structure and function, discusses reproduction in flowering plants. As you recall from the first chapter in the series, flowering plants are the largest and most successful group of plants. All flowering plants reproduce sexually; some also reproduce asexually. Sexual reproduction involves flower formation, pollination, fertilization within the flower ovary, and seed and fruit formation. The offspring resulting from sexual reproduction exhibit a great deal of individual variation, due to gene recombination and the union of dissimilar gametes. Although sexual reproduction has some disadvantages, it offers the advantage of new combinations of genes that might make an individual plant better suited to its environment. Asexual reproduction involves only one parent; the offspring are genetically identical to the parent and each other. Asexual reproduction, therefore, is advantageous when the parent is well adapted to its environment. The stems, leaves, and roots of many flowering plants are modified for asexual reproduction. Also, in some plants, seeds and fruits are produced asexually without meiosis or the fusion of gametes.

REVIEWING CONCEPTS

Fill in the blanks.

INTRODUCTION

1. The union of gametes is called (a) _____. Sexual reproduction provides an advantage to the offspring because they are genetically (b) _____ from either parent.

2. Many flowering plants reproduce (a) _____. Genetically these offspring are (b) _____ to their parents.

THE FLOWERING PLANT LIFE CYCLE

3. Plants that spend a portion of their life cycle in a multicellular haploid stage and part in a multicellular diploid stage as said to undergo a reproductive cycle called the

   (a) _____. The haploid part is the

   (b) _____ and the diploid portion is called the

   (c) _____.

Flowers develop at apical meristems

4. The Flowering Locus C gene codes for a transcription factor that represses

   _____.

Each part of a flower has a specific function

5. Flowers are reproductive shoots, usually consisting of four kinds of organs,

   (a) _____, (b) _____,

   (c) _____, and (d) _____.

6. The collective term for all the sepals of a flower is (a) _____. The collective term for all the petals of a flower is (b) _____.

7. Carpels bear _____.

8. Pollen sacs within the anther contain numerous diploid cells called
(a) _____, each of which undergoes meiosis to produce
four haploid cells called (b)_____, each of which divides
mitotically to produce an immature male gametophyte called a
(c)_____. The pollen grain becomes mature when its
generative cell divides to form two nonmotile (d) _____.

POLLINATION

9. The transfer of pollen grains from (a)_____ to
(b)_____ is known as pollination.

Many plants have mechanisms to prevent self-pollination

10. The mating of genetically similar individuals is known as
(a) _____, and the mating of dissimilar individuals is
called (b) _____.

11. A genetic condition in which the pollen is ineffective in fertilizing the same
flower or other flowers on the same plant is known as _____.

Flowering plants and their animal pollinators have coevolved

12. Two things flowers have evolved to attract pollinators are
(a) _____, and (b) _____.

13. Plants pollinated by insects often have (a)_____ (color?)
petals, but usually not (b)_____ petals.

Some flowering plants depend on wind to disperse pollen

14. Bees see ultraviolet light as a color referred to as _____.

15. Flowers pollinated by birds are usually colored
(a) _____, (b) _____, or (c) _____
because birds see well in this range of (d) _____.

16. Flowers pollinated by bats bloom (a) _____ and have
(b) _____ petals. They usually smell like
(c) _____.

17. Some plants have evolved flowers resembling specific shapes. The flower of
one species of orchid actually resembles an insect, a _____.

18. Scientists think humming-bird pollinated flowers arose from
_____ pollinated flowers.

19. Wind-pollinated plants produce many small, inconspicuous _____.

FERTILIZATION AND SEED/FRUIT DEVELOPMENT

20. Once pollen grains have been transferred from anther to stigma, the tube cell
grows a pollen tube down through the (a)_____ and into an
(b)_____ in the ovary.

A unique double fertilization process occurs in flowering plants

21. The tissue with nutritive and hormonal functions that surrounds the
developing embryonic plant in the seed is called _____.

Embryonic development in seeds is orderly and predictable

22. A seed is contained in an (a) _____ and stores (b) _____.

23. The tissue that anchors the developing embryo and aids in nutrient uptake is called the _____.

The mature seed contains an embryonic plant and storage materials

24. The mature embryo within the seed consists of a short embryonic root, or (a)_____; an embryonic shoot; and one or two seed leaves, or (b)_____.

25. The shoot apex or terminal bud is called the _____.

Fruits are mature, ripened ovaries

26. The four basic types of fruits are (a) _____, (b) _____, (c) _____, (d) _____.

27. One type of fruit, the (a)_____ fruit, develops from a single pistil. It may be fleshy or dry. Fruits that have soft tissues throughout, like those in tomatoes and grapes, are called (b)_____, while fleshy fruits that have a hard, stony pit surrounding a single seed are known as (c)_____.

28. Fruits that split along one suture, like the milkweed are an example of a (a) _____. Those that split along just two sutures are an example of a (b) _____, and those that split along two or more (multiple) sutures are an example of a (c)_____.

29. (a)_____ fruits result from the fusion of several developing ovaries in a single flower. The raspberry is an example. Similarly, (b)_____ fruits form from the fusion of several ovaries of many flowers that grow in proximity on a common floral stalk. The pineapple is an example.

30. (a)_____ fruits contain plant tissue in addition to ovary tissue. When one eats a strawberry, for example, he or she is eating the fleshy (b)_____, and when eating apples and pears, one is consuming the (c)_____ that surrounds the ovary.

Seed dispersal is highly varied

31. Flowering plant seeds and fruits are adapted for various means of dispersal, among which four common means are (a) _____, (b) _____, (c) _____, and (d) _____.

GERMINATION AND EARLY GROWTH

32. Dry seeds take up water by a process called _____.

Some seeds do not germinate immediately

33. A process of scratching or scarring the seed coat before sowing it is called _____.

Eudicots and monocots exhibit characteristic patterns of early growth

34. Corn and grasses have a special sheath of cells called a
_____ that surrounds and protects the young shoot.

ASEXUAL REPRODUCTION IN FLOWERING PLANTS

35. Various vegetative structures may be involved in asexual reproduction. Some
of them, for example, are (a)_____, like those in bamboo and
grasses that have horizontal underground stems; (b)_____,
underground stems greatly enlarged for food storage, like those in potatoes;
(c)_____, short underground stems with fleshy storage leaves, as in
onions and tulips; (d)_____, thick underground stems covered with
papery scales; and (e)_____ or horizontal, aboveground stems with
long internodes, like those in strawberries.

36. Some plants reproduce asexually by producing _____,
aboveground shoots that develop from adventitious buds on roots.

Apomixis is the production of seeds without the sexual process

37. The advantage of apomixis over other methods of asexual reproduction is that
the seeds and fruits produced can be _____ by methods
associated with sexual reproduction.

A COMPARISON OF SEXUAL AND ASEXUAL REPRODUCTION

Sexual reproduction has some disadvantages

26. The many adaptations of flowers for different modes of pollination represent
one cost of _____.

BUILDING WORDS

Use combinations of prefixes and suffixes to build words for the definitions that follow.

Prefixes	The Meaning	Suffixes	The Meaning
co-	with, together, in association	-ancy	state of
endo-	within	-ate	to make
dorm-	to sleep	-cotyl	cup
germin-	to sprout	-ome	mass
hypo-	under, below	-sperm	seed
plum-	feather	-ule	small
rhiz-	root		

Prefix	Suffix	Definition
_____	_____	1. Nutritive tissue within seeds.
_____	_____	2. The part of a plant embryo or seedling below the point of attachment of the cotyledons.
_____	-evolution	3. Evolutionary changes that result from close interaction and reciprocal adaptations between two species.
_____	_____	4. The shoot above the cotyledon.
_____	_____	5. A temporary state of arrested physiological activity.
_____	_____	6. When a seed beings to develop and the embryo resumes growth.
_____	_____	7. The sheath of cells that protects and surrounds a young shoot such as in grasses.
_____	_____	8. A horizontal underground stem.

MATCHING

Terms:

a. Aggregate fruit
b. Apomixis
c. Carpel
d. Corolla
e. Cotyledon
f. Drupe
g. Fruit
h. Ovary
i. Ovule
j. Pollen
k. Seed
l. Sepal
m. Stolon

For each of these definitions, select the correct matching term from the list above.

_____ 1. A collective term for the petals of a flower

_____ 2. A ripened ovary.

_____ 3. The outermost parts of a flower, usually leaf-like in appearance, that protect the flower as a bud.

_____ 4. A type of reproduction in which fruits and seeds are formed asexually.

_____ 5. A fruit that develops from a single flower with many separate carpels, such as a raspberry.

_____ 6. A plant reproductive body composed of a young embryo and nutritive tissue.

_____ 7. The male gametophyte in plants.

_____ 8. An above-ground, horizontal stem with long internodes.

_____ 9. The female reproductive unit of a flower.

_____10. A simple fleshy fruit that contains a hard and stony pit surrounding a single seed; such as a plum or peach.

_____11. A structure with the potential to develop into a seed.

_____12. The fist leaf to be produced by seeds, monocots have produce a single one and eudicots produce two.

MAKING COMPARISONS

Fill in the blanks.

Examples of Fruit	Main Type of Fruit	Subtype	Description
Bean	Simple	Dry: opens to release seeds; legume	Splits along the 2 sutures
Acorn	Simple	#1	#2
Corn	Simple	Dry: does not open; grain	#3
Tomato	#4	#5	Soft and fleshy throughout, usually with few to many seeds
Strawberry	#6	None	Other plant tissues and ovary tissue comprise the fruit
Pineapple	#7	None	#8
Raspberry	#9	None	#10
Peach	#11	#12	#13

MAKING CHOICES

Place your answer(s) in the space provided. Some questions may have more than one correct answer.

_____ 1. A modified underground bud in which fleshy storage leaves are attached to a stem is a
   a. corm.   d. fruit.
   b. stolon.   e. capsule.
   c. bulb.

_____ 2. A simple, dry fruit that develops from two or more fused carpels is a
   a. corm.   d. fruit.
   b. stolon.   e. capsule.
   c. bulb.

_____ 3. A short, erect underground stem is a
   a. corm.   d. fruit.
   b. stolon.   e. capsule.
   c. bulb.

_____ 4. A horizontal, above ground stem with long internodes is a
    a. corm.
    b. stolon.
    c. bulb.
    d. fruit.
    e. capsule.

_____ 5. Which of the following is/are somehow related to simple dry fruits?
    a. drupe
    b. legume
    c. grain
    d. single carpel
    e. green bean

_____ 6. The process by which an embryo develops from a diploid cell in the ovary without fusion of haploid gametes is called
    a. lateral bud generation.
    b. dehiscence.
    c. asexual reproduction by means of suckers.
    d. apomixis.
    e. spontaneous generation.

_____ 7. The tomato is actually a
    a. drupe.
    b. berry.
    c. fleshy lateral bud.
    d. fruit.
    e. mature ovary.

_____ 8. The "eyes" of a potato are
    a. diploid gametes.
    b. parts of a stem.
    c. capable of producing complete plants.
    d. axillary buds.
    e. rhizomes.

_____ 9. Horizontal, asexually reproducing stems that run above ground are
    a. diploid gametes.
    b. rhizomes.
    c. tubers.
    d. corms.
    e. stolons.

_____10. Raspberries and blackberries are examples of
    a. follicles.
    b. achenes.
    c. drupes.
    d. aggregate fruits.
    e. multiple fruits.

_____11. A fruit that forms from many separate carpels in a single flower is a/an
    a. follicle.
    b. achene.
    c. drupe.
    d. aggregate fruit.
    e. accessory fruit.

_____12. A fruit composed of ovary tissue and other plant parts is a
    a. follicle.
    b. achene.
    c. drupe.
    d. aggregate fruit.
    e. accessory fruit.

_____13. In flowering plants, the fusion of two gametes
    a. produces the gametophyte generation.
    b. produces the sporophyte generation.
    c. produces a haploid generation.
    d. produces a diploid generation.
    e. is called fertilization.

_____14. In flowering plants, haploid spores are produced by
    a. the sporophyte generation.
    b. the gametophyte generation.
    c. meiosis.
    d. gametes.
    e. self fertilization.

_____15. The outermost and lowest whorl on a floral shoot is made up of

    a. anthers.   d. petals.

    b. carpels.   e. peduncels.

    c.. sepals.

_____16. Terms associated with anthers include

    a. pollen grains.   d. generative cell.

    b. pollen tube.   e. tube cell.

    c. carpel.

_____17. Pollen sacs in an anther

    a. form from pistols.   d. are the site of male gamete formation.

    b. are collectively call the calyx.   e. contain microsporocytes.

    c. contain diploid cells.

_____18. Visible wavelengths in which insect see well include

    a. blue.   d. red.

    b. indigo.   e. violet.

    c. yellow.

_____19. Common flower colors visited by birds include

    a. pink.   d. red.

    b. white.   e. yellow.

    c. purple.

_____20. Features that plants have evolved to attract pollinators include

    a. shape.   d. scent.

    b. nectar.   e. petals.

    c. coloration.

VISUAL FOUNDATIONS

Color the parts of the illustration below as indicated.

RED ☐ ovule

GREEN ☐ sepal

YELLOW ☐ ovary (and derived from ovary)

BLUE ☐ stamen

ORANGE ☐ style

BROWN ☐ stigma

TAN ☐ seed

PINK ☐ floral tube (and derived from floral tube)

VIOLET ☐ petal

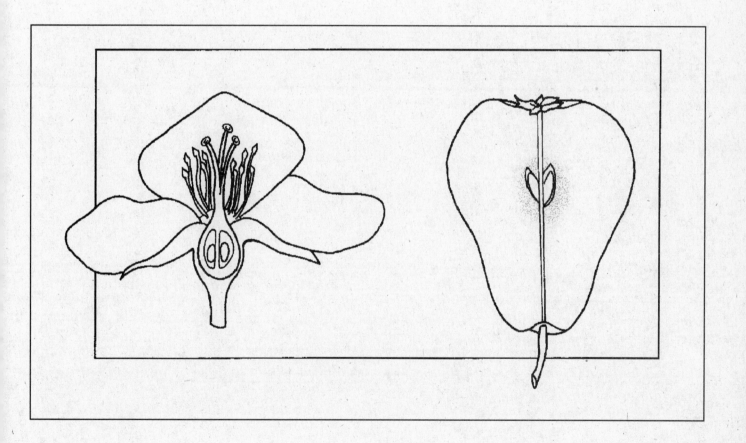

CHAPTER 38

□

Plant Developmental Responses to External and Internal Signals

This is the last chapter in a series of six addressing the integration of plant structure and function. How a particular gene is expressed is determined by a variety of factors, including signals from other genes and from the environment. Environmental cues, such as temperature, light, growth, and touch, influence plant growth and development. All aspects of plant growth and development are affected by hormones, organic compounds present in very low concentrations that act as highly specific chemical signals between cells. Hormones interact in complex ways with one another to produce a variety of responses in plants. Auxins are involved in cell elongation, phototropism, gravitropism, apical dominance, and fruit development. Giibberellins are involved in stem elongation, flowering, and seed germination. Cytokinins promote cell division and differentiation, delay senescence, and interact with auxins in apical dominance. Ethylene has a role in the ripening of fruits, apical dominance, leaf abscission, and wound response. Abscisic acid is involved in stomatal closure due to water stress, and bud and seed dormancy. Brasinosteroids are plant steroid hormones and they are involved in several aspects of plant growth and development. Other chemicals have been implicated in plant growth and development, and are the subject of ongoing research.

REVIEWING CONCEPTS
Fill in the blanks.

INTRODUCTION
1. Growth and development of a plant are controlled by its _____ organic compounds present in low concentration in the plant's tissues that act as chemical signals between cells.

TROPISMS
2. A tropism is a _____.
3. Growth or movement initiated by light is called a (a) _____ response and it is triggered by (b) _____ light.
4. For an organism to have a biological response to light, it must contain a light-sensitive substance, called a _____, to absorb the light.
5. Tropisms are categorized according to the stimulus that causes them to occur. A response to gravity is (b)_____, and movement caused by a mechanical stimulus is (c)_____.

PLANT HORMONES AND DEVELOPMENT
6. The six major classes of hormones that regulate responses in plants are
(a) _____, (b)_____,
(c) _____, (d) _____,
(e) _____, and (f) _____.

347

© 2011 Cengage Learning. All Rights Reserved. May not be scanned, copied or duplicated, or posted to a publicly accessible website, in whole or in part.

Plant hormones act by signal transduction

7. Many plant hormones bind to (a) _____,
   located in the plasma membrane where they trigger (b) _____
   reactions.

Auxins promote cell elongation

8. Bending toward light occurs below the tip of the _____.

9. _____ is the most common and physiologically
   important auxin.

10. The movement of auxins is (a) _____ because it always move in one
    direction, specifically from the (b)_____
    toward the (c) _____.

11. Plants that grow almost entirely at the apical meristem rather than from axillary
    buds are exhibiting _____.

12. Auxins produced by developing seeds stimulated the development of the
    _____.

Gibberellins promote stem elongation

13. A disease seen in rice seedlings causes them to grow extremely tall, fall over,
    and die. It is caused by a _____ that produces gibberellin.

14. In addition to influencing stem elongation, gibberellins are also involved in
    (a) _____, and (b) _____.

Cytokinins promote cell division

15. Cytokinins mainly promote (a) _____and
    (b) _____.

16. They also delay _____ (aging).

Ethylene promotes abscission and fruit ripening

17. Ethylene regulates developmental response to a mechanical stimulus. The
    plant response to a mechanical stimulus is known as
    _____.

Abscisic acid promotes seed dormancy

18. _____ is the temporary state of reduced
    physiological activity in flowering plants.

19. Abscisic acid is an _____ hormone.

20. In seeds, the level of abscisic acid decreases during the winter, and the level of
    _____ increases.

Brassinosteroids are plant steroid hormones

21. *Arabidopsis* mutants that cannot synthesize brassinosteroids develop as
    _____ plants.

Identification of a universal flower-promoting signal remains elusive

22. _____ has been identified as a flower-promoting
    substance.

LIGHT SIGNALS AND PLANT DEVELOPMENT

23. _____ is any response of a plant to the relative lengths of daylight and darkness.

24. Short-day plants are also called _____.

25. (a)_____ plants flower when the night length is equal to or greater than some critical period; (b)_____ plants flower when the night length is equal to or less than some critical period; (c)_____ plants do not flower when night length is either too long or too short; (d)_____ plants do not initiate flowering in response to seasonal changes in the period of daylight and darkness but instead respond to some other type of stimulus, external or internal.

Phytochrome detects day length

26. Phytochrome is the main photoreceptor for _____ and many other light-initiated plant responses.

Competition for sunlight among shade-avoiding plants involves phytochrome

27. Plants tend to grow taller when closely surrounded by other plants, a phenomenon known as _____.

Phytochrome is involved in other responses to light, including germination

28. Seeds with a light requirement must be exposed to light containing (a)_____ wavelengths to convert (b)_____ to (c)_____ and germination occurs.

Phytochrome acts by signal transduction

29. Red light causes (a)_____ channels to open in all cell membranes involved in plant movement. Far-red light causes these channel to close. These ion movements cause a change in turgor pressure producing the observed movement.

Light influences circadian rhythms

30. Internal cycles, known as circadian rhythms, help organisms detect _____.

RESPONSES TO HERBIVOREES AND PATHOGENS

Jasmonic acid activates several plant defenses

31. Jasmonic acid is structurally similar to (a)_____ in animals. Both are (b)_____.

Methyl salicylate may induce systemic acquired resistance

32. Salicyclic acid was first extracted from (a)_____. It is chemically related to (b)_____.

BUILDING WORDS

Use combinations of prefixes and suffixes to build words for the definitions that follow.

Prefixes	The Meaning		Suffixes	The Meaning
senes(c)-	to grow old		-ence	the condition of
photo-	light		-chrom(e)	color
phyto-	plant		-tropism	turn, turning
promo-	to move forward			

Prefix	Suffix	Definition
_____	_____	1. The growth response of an organism to light; usually the turning toward or away from the light source.
gravi-	_____	2. The growth response of an organism to gravity; usually the turning toward or away from the direction of gravity.
_____	_____	3. A blue-green, proteinaceous pigment involved in photoperiodism and a number of other light-initiated physiological responses of plants.
_____	periodism	4. The physiological response of organisms to variations of light and darkness.
_____	_____	5. The natural ageing process that occurs in most cells.
_____	ter	6. A sequence of DNA that is a binding site for RNA polymerase.

MATCHING

Terms:

a. Abscisic acid
b. Apical dominance
c. Auxin
d. Coleoptile
e. Cytokinin
f. Ethylene

g. Florigen
h. Gibberelin
i. Imbibition
j. Nastic movements
k. Necrotic
l. Proteostome

m. Senescence
m. Statolith
o. Thigmotropism
p. Tropism

For each of these definitions, select the correct matching term from the list above.

_____ 1. The inhibition of axillary bud growth by the apical meristem.

_____ 2. A plant hormone involved in dormancy and responses to stress.

_____ 3. The aging process in plants.

_____ 4. A protective sheath that encloses the stem in certain monocots.

_____ 5. A plant hormone involved in apical dominance and cell elongation.

_____ 6. A plant hormone that promotes fruit ripening.

_____ 7. Plant growth in response to contact with mechanical stimuli, such as a solid object.

_____ 8. A plant hormone that promotes rapid cell division and is involved in other aspects of plant growth and development.

_____ 9. A change in the position of a plant root or stem.

_____ 10. A hormone that promotes stem elongation.

____11. Thought to be a flower-promoting substance.

____12. Describes dead areas in plant tissue.

MAKING COMPARISONS

Fill in the blanks.

Principal Action	Hormone(s)
Regulates growth by promoting cell elongation	Auxin and gibberellin
Promotes apical dominance, stem elongation, root initiation, fruit development	#1
Delays leaf senescence, inhibition of apical dominance, embryo development	#2
Promotes seed germination, stem elongation, flowering, fruit development	#3
Fruit ripening, seed germination, root initiation, abscission	#4
Promotes cell division	#5
Promotes seed dormancy	#6

MAKING CHOICES

Place your answer(s) in the space provided. Some questions may have more than one correct answer.

____ 1. Growth in response to gravity is known as

    a. a tropism.

    b. phototropism.

    c. gravitropism.

    d. thigmotropism.

    e. turgor.

____ 2. The hormone(s) that is/are involved in rapid stem elongation just prior to flowering is/are

    a. auxin.

    b. gibberellin.

    c. cytokinin.

    d. ethylene.

    e. abscisic acid.

____ 3. Which of the following is/are correct about hormones?

    a. They are effective in very small amounts.

    b. They are organic compounds.

    c. For the most part, their effects occur near the area where they are produced.

    d. The effects of different hormones overlap.

    e. Each plant hormone has multiple effects.

____ 4. The hormone(s) principally responsible for cell division and differentiation is/are

    a. auxin.

    b. gibberellin.

    c. cytokinin.

    d. ethylene.

    e. abscisic acid.

_____ 5. The hormone(s) principally responsible for the growth of a coleoptile toward light is/are

a. auxin.   d. ethylene.

b. gibberellin.   e. abscisic acid.

c. cytokinin.

_____ 6. If a plant that touches your house continues to grow toward and attach itself to the house, the plant is exhibiting

a. tropism.   d. thigmotropism.

b. phototropism.   e. turgor movement.

c. gravitropism.

_____ 7. The plant hormone(s) about which Charles Darwin gathered information is/are

a. auxins.   d. ethylene.

b. gibberellin.   e. abscisic acid.

c. cytokinin.

_____ 8. The rapid elongation of a floral stalk during the initiation of flowering is know as

a. acceleration.   d. floral enhancement.

b. bolting.   e. internodal elongation.

c. blooming.

_____ 9. The aging process is called

a. dormancy.   d. systemination.

b. bolting.   e. senescence.

c. abscissionation.

_____10. Environmental signals that trigger a response in plants include

a. temperature.   d. light.

b. touch.   e. hours of daylight.

c. hours of darkness.

_____11. The site of gravity perception in root is the

a. meristem.   d. root cap.

b. F-box.   e. amyloplast.

c. thigmotrope.

_____12. Plant hormones may bind to

a. interstitial receptors.   d. intranuclear receptors.

b. enzyme-linked receptors.   e. T1R1 receptors.

c. cytosolic receptors.

_____13. The hormone name derived from the Greek word meaning to enlarge or increase is

a. cytokinin.   d. gibberellins.

b. abscisic acid.   e. auxin.

c. brassinolide.

_____14. Apical dominance

a. promotes growth at the apical meristem.   d. is driven by the levels of indolacetic acid.

b. involves ubiquinylated proteins.   e. leads to phototropism.

c. inhibits lateral bud growth.

_____15. The auxin signaling pathway shares similarities with the

    a. abscisic acid pathway.   d. gibberellin pathway.

    b. brassinosteroid pathway.   e. cytokinin pathway.

    c. ethylene pathway.

_____16. Plants that flower when the night length is equal to or greater than some critical period are

    a. intermediate-day plants.   d. day-neutral plants.

    b. short-day plants.   e. long-day plants.

    c. long-night plants.

_____17. Circadian rhythms

    a. affect sleep movements in plants.   d. affect gene expression in plants.

    b. the rate of photosynthesis.   e. seasonal reproduction in plants.

    c. stomata opening and closing.

VISUAL FOUNDATIONS

Color the parts of the illustration below as indicated.

RED ☐ auxin receptor

VIOLET ☐ repressor protein

TAN ☐ auxin response element

ORANGE ☐ auxin response gene

GREY ☐ enzyme complex

YELLOW ☐ transcription activator

BLUE ☐ DNA

BROWN ☐ newly formed protein

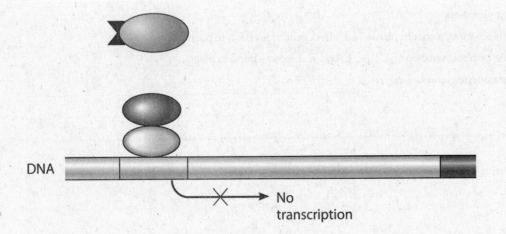

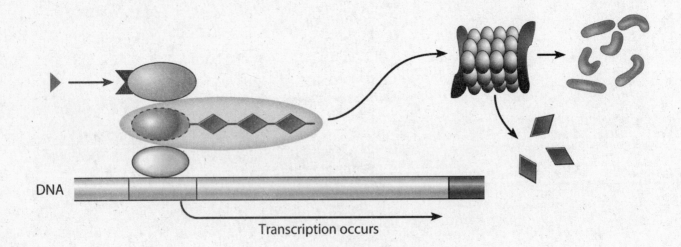

Animal Structure and Function: An Introduction

This is the first in a series of chapters that examine the structural, functional, and behavioral adaptations that help animals meet environmental challenges. This chapter examines the architecture of the animal body. In most animals, cells are organized into tissues, tissues into organs, and organs into organ systems. The principal animal tissues are epithelial, connective, muscular, and nervous. Epithelial tissues are characterized by tight-fitting cells and the presence of a basement membrane. Covering body surfaces and lining cavities, they function in protection, absorption, secretion, and sensation. Connective tissue joins together other tissues, supports the body and its organs, and protects underlying organs. There are many different types of connective tissues, consisting of a variety of cell types. Muscle tissue is composed of cells that are specialized to contract. There are three major types of muscle tissue — cardiac, smooth, and skeletal. Nervous tissue is composed of cells that are specialized for conducting impulses and those that support and nourish the conducting cells. Organs are comprised of two or more kinds of tissues. Complex animals have many organs and organ systems. The organ systems work together to maintain the body's homeostasis.

REVIEWING CONCEPTS

Fill in the blanks.

INTRODUCTION

1. A cell's size is limited by the
   _____.

2. New cells formed by cell division remain associated in multicellular animals. The size of an animal is determined by the (a) _____ of cells that make up its body, not their (b) _____.

3. Bacteria and protists can be small because they rely on diffusion and do not require _____.

TISSUES, ORGANS, AND ORGAN SYSTEMS

Epithelial tissues cover the body and line its cavities

4. A group of cells that carry out a specific function is called a (a) _____, and these groups associate to form (b) _____, which in turn are grouped into the (c) _____ of the body.

5. Epithelial cells form a (a) _____ layer of (b) _____ cells that are attached on one surface to a noncellular (c) _____.

6. Four of the major functions of epithelial cells include
   (a)_____, (b) _____,
   (c) _____, and (d) _____.

7. Epithelial cells can be distinguished on the basis of their shape. (a)_____ cells are flat, (b)_____ cells resemble dice, and (c)_____ cells are tall, slender cells shaped like cylinders.

8. Epithelial tissues also vary in the number of cell layers composed of each shape. For example, (a) _____ epithelium is made up of only one layer of cells and (b) _____ epithelium has two or more cell layers.

9. An epithelial tissue in which the cells appear to be layered but actually form a single layer is said to be _____.

10. A gland is one location where secretory epithelial cells may be located. The two types of glands are (a) _____, for example sweat glands, and (b) _____ which produce hormones.

Connective tissues support other body structures

11. Characteristically, connective tissues contain very few cells, an

(a) _____ in which cells and

(b) _____ are embedded, and a

(c) _____ that is secreted by the cells.

12. The _____ associated with connective tissue is largely responsible for the nature and function of each kind of connective tissue.

13. There are three types of fibers in connective tissues. Collagen fibers are numerous, strong fibers composed of the protein (a) _____. Elastic fibers are composed of (b) _____ and can stretch. Reticular fibers, composed of (c) _____, form delicate networks of connective tissue.

14. Fibroblast cells produce (a) _____, (b) _____, and (c) _____ present in the matrix of connective tissue.

15. _____ function as the body's scavenger cells.

16. The most widely distributed connective tissue in the vertebrate body is

(a) _____ which allows the body parts it connects to (b) _____.

17. Dense connective tissues are predominantly composed of

(a) _____ fibers.

18. (a)_____ are the cords that connect muscles to bones. (b)_____ are the cables that connect bones to one another.

19. All vertebrates have an internal supporting structure called the _____ that is composed of cartilage and/or bone.

20. The (a)_____ that provides support in the vertebrate embryo is largely replaced by (b)_____ in most vertebrate adults.

21. Cartilage cells are called (a) _____ secrete (b)_____ that strengthen the extracellular matrix. As the secreted matrix forms, the cells become isolated in holes called (c) _____.

22. Like cartilage, bone cells are called (a) _____ become isolated in
    (b) _____. Unlike cartilage however, bone is highly
    (c) _____, a term referring to their abundant supply of blood
    vessels.

23. Osteocytes communicate with one another through small channels called
    _____.

24. Compact bone consists of spindle-shaped units called (a) _____.
    Osteocytes are arranged in concentric layers called (b) _____.
    The lamellae surround central microscopic channels known as
    (c) _____, through which blood vessels and
    nerves pass.

25. The noncellular component of blood is the _____.

26. Red blood cells function to _____.

27. White blood cells function to _____.

28. Platelets are small cell fragments that originate in _____.

Muscle tissue is specialized to contract

29. Each muscle cell is called a _____.

30. Contractile proteins called (a) _____, and (b) _____ are
    contained within the elongated (c) _____ of muscle cells.

31. Vertebrates have three kinds of muscles: (a)_____ muscle that
    attaches to bones and causes body movements; (b)_____
    muscle occurring in the walls of the digestive tract, uterus, blood vessels, and
    many other internal organs; and heart or (c)_____ muscle.

Nervous tissue controls muscles and glands

32. Nervous tissue is composed of cells that conduct impulses, (a) _____,
    and support cells which are called (b) _____ cells.

33. Neurons communicate with one another at cellular junctions called
    _____.

34. A _____ is a collection of neurons bound together by connective
    tissue.

35. Neurons generally contain three functionally and anatomically distinct regions:
    the (a) _____, which contains the nucleus; the
    (b)_____, which receive incoming impulses; and the
    (c)_____, which carry impulses away from the cell body.

Tissues and organs make up the organ systems of the body

36. Mammals have 11 organ systems. They are
    (a) _____, (b) _____
    (c) _____, (d) _____
    (e) _____, (f) _____
    (g) _____, (h) _____
    (i) _____, (j) _____,
    and (k) _____.

REGULATING THE INTERNAL ENVIRONMENT

37. A balanced internal environment is referred to as _____.

38. _____ are changes in either the internal or external environment that affects normal body conditions.

Negative feedback systems restore homeostasis

39. In a negative feedback system, a (a) _____ detects a change from the normal condition and an (b) _____ activates hoemostatic mechanisms to restore the steady state.

40. The response from the integrator will be _____ from the output of the sensor.

A few positive feedback systems operate in the body

41. Unlike what occurs in a negative feedback system, the response of a positive feedback system _____ the changing condition detected by the sensor.

THERMOREGULATION

Ectotherms absorb heat from their surroundings

42. An ectotherm's metabolic rate tends to change with the weather; therefore, they have a much _____ daily energy expenditure than endotherms.

Endotherms derive heat from metabolic processes

43. Endotherms can have a metabolic rate as much as _____ (#?) times as high as that of ectotherms.

Many animals adjust to challenging temperature changes

44. When stressed by cold, many animals sink into (a) _____, a decrease in body temperature below normal levels. (b)_____ is long-term torpor in response to winter cold and scarcity of food. (c)_____ is a state of torpor caused by lack of food or water during periods of high temperature.

BUILDING WORDS

Use combinations of prefixes and suffixes to build words for the definitions that follow.

Prefixes	The Meaning	Suffixes	The Meaning
chondro-	cartilage	-blast	embryo
dendr-	tree	-cyte	cell
fibro-	fiber	- ite	part of
homeo-	similar, "constant"	-phage	eat, devour
inter-	between, among	-sis	process of
macro-	large, long, great, excessive	-stasis	equilibrium
multi-	many		
myo-	muscle		
osteo-	bone		
pseudo-	false		
synap-	a union		

Prefix	Suffix	Definition
_____	-cellular	1. Composed of many cells.
_____	-stratified	2. An arrangement of epithelial cells in which the cells falsely appear to be stratified.
_____	_____	3. A cell, especially active in developing ("embryonic") tissue and healing wounds, that produces connective tissue fibers.
_____	-cellular	4. Situated between or among cells.
_____	_____	5. A large cell, common in connective tissues, that phagocytizes ("eats") foreign matter including bacteria.
_____	_____	6. A cartilage cell.
_____	_____	7. A bone cell.
_____	-fibril	8. A thin longitudinal contractile fiber inside a muscle cell.
_____	_____	9. Maintaining a constant internal environment.
_____	_____	10. An extension of a neuron that transmits information to the cell body.
_____	_____	11. The junction where nerves communicate with other nerves, glands, or muscles.

MATCHING

Terms:

a. Acclimitization
b. Adipose tissue
c. Cardiac muscle
d. Cartilage
e. Collagen
f. Elastic
g. Endocrine gland
h. Exocrine gland
i. Gland
j. Glial cell
k. Homeostasis
l. Matrix
m. Organ
n. Organ system
o. Osteon
p. Recticular
q. Skeletal muscle
r. Stressor
s. Striations

For each of these definitions, select the correct matching term from the list above.

_____ 1. Striated (voluntary) muscle.

_____ 2. Tissue in which fat is stored, or the fat itself.

_____ 3. Spindle-shaped unit of bone composed of concentric layers of osteocytes.

_____ 4. Supporting skeleton in the embryonic stages of all vertebrates.

_____ 5. A specialized structure made up of tissues and adapted to perform a specific function or group of functions.

_____ 6. Glands that secrete products directly into the blood or tissue fluid instead of into ducts.

_____ 7. Tissue characterized by the presence of intercalated discs.

_____ 8. A cell that supports and nourishes neurons.

_____ 9. General term for a body cell or organ specialized for secretion.

_____10. A protein in connective tissue fibers.

_____11. Thin, branched fibers that from delicate networks.

_____12. Microscopically helps to identify skeletal and cardiac muscle from smooth muscle.

_____13. Adjustment by animals to seasonal changes.

MAKING COMPARISONS

Fill in the blanks.

Principle Tissue	Tissue	Location	Function
Epithelial tissue	Stratified squamous epithelium	Skin, mouth lining, vaginal lining	Protection; outer layer continuously sloughed off and replaced from below
Epithelial tissue	#1	#2	Allows for transport of materials, especially by diffusion
#3	Pseudostratified epithelium	#4	#5
#6	Adipose tissue	Subcutaneous layer, pads certain internal organs	#7
Connective tissue	#8	Forms skeletal structure in most vertebrates	#9
#10	Blood	#11	#12
Muscle tissue	Cardiac muscle	#13	Contraction of the heart
Muscle tissue	#14	Attached to bones	Movement of the body
#15	Nervous tissue	Brain, spinal cord, nerves	Respond to stimuli, conduct impulses

MAKING CHOICES

Place your answer(s) in the space provided. Some questions may have more than one correct answer.

_____ 1. The nucleus of a neuron is typically found in the
   a. cell body.
   b. synapse.
   c. dendrite.
   d. axon.
   e. glial body.

_____ 2. The major classes of animal tissues include
   a. epithelial.
   b. nervous.
   c. muscular.
   d. skeletal.
   e. connective.

_____ 3. Myosin and actin are the main components of
   a. cartilage.
   b. blood.
   c. collagen.
   d. muscle.
   e. adipose tissue.

_____ 4. The basement membrane
   a. lies beneath epithelium.
   b. contains polysaccharides.
   c. consists of cells that lie beneath the epithelium.
   d. is synonymous with plasma membrane.
   e. is noncellular.

_____ 5. Epithelium cells that are flattened and thin are called
   a. columnar.
   b. cuboidal.
   c. squamous.
   d. stratified.
   e. endothelium.

_____ 6. Chondrocytes are
   a. part of cartilage.
   b. found in bone.
   c. immature osteocytes.
   d. eventually found in the lacunae of a matrix.
   e. embryonic osteocytes.

_____ 7. Large muscles attached to bones are
   a. skeletal muscles.
   b. smooth muscles.
   c. composed largely of actin and myosin.
   d. nonstriated.
   e. multinucleated.

_____ 8. Platelets are
   a. bone marrow cells.
   b. fragments of large cells.
   c. a type of white blood cells.
   d. a type of RBCs.
   e. derived from bone marrow cells.

_____ 9. Blood is one type of
   a. endothelium.
   b. collagen.
   c. mesenchyme.
   d. connective tissue.
   e. cardiac tissue.

_____10. Epithelium located in the ducts of glands would most likely be
   a. simple columnar.
   b. simple cuboidal.
   c. simple squamous.
   d. stratified squamous.
   e. stratified cuboidal.

_____11. The portion of the neuron that is specialized to receive a nerve impulse is the
   a. cell body.
   b. synapse.
   c. dendrite.
   d. axon.
   e. glial body.

_____12. Collagen is
   a. part of blood.
   b. part of bone.
   c. part of connective tissue.
   d. composed of fibroblasts.
   e. fibrous.

_____13. The connective tissue matrix is
   a. noncellular.
   b. mostly lipids.
   c. a gel.
   d. a polysaccharide.
   e. fibrous.

_____14. Haversian canals
   a. contain nerves.
   b. run through cartilage.
   c. run through bone.
   d. are matrix lacunae.
   e. are synonymous with canaliculi.

_____15. Adipose tissue is a type of
    a. cartilage.
    b. epithelium.
    c. connective tissue.
    d. modified blood tissue.
    e. marrow.

_____16. Cells that nourish and support nerve cells are known as
    a. neurons.
    b. glial cells.
    c. neuroblasts.
    d. synaptic cells.
    e. neurocytes.

_____17. A group of closely associated cells that carries out a specific function is a/an
    a. organ.
    b. system.
    c. organ system.
    d. tissue.
    e. clone.

_____18. Cells lining internal cavities that secrete a lubricating mucus are
    a. goblet-cells.
    b. pseudostratified.
    c. exocrine glands.
    d. epithelial.
    e. part of connective tissue.

_____19. Cells that make up the outer layer of skin are
    a. the columnar layer.
    b. the cuboidal layer.
    c. stratified squamous epithelium.
    d. basement membrane cells.
    e. endothelium.

_____20. You would expect the outer layer of skin on the soles of your feet to be
    a. cuboidal.
    b. simple.
    c. connected to a basement membrane.
    d. stratified.
    e. squamous.

_____21. A serous membrane would be expected to
    a. line a body cavity not open to the outside.
    b. line a body cavity not open to the outside.
    c. secrete fluid.
    d. line the abdominal cavity.
    e. contain epithelial cells.

_____22. Both bone and cartilage
    a. are vascularized.
    b. have canaliculi.
    c. have lamellae.
    d. have cells present in lacunae.
    e. are connective tissues.

_____23. Thermoregulation
    a. occurs in ectotherms.
    b. may involve hormones.
    c. occurs in invertebrates.
    d. occurs in endotherms.
    e. may involved hibernation.

_____24. Torpor is or can be associated with the term(s)
    a. hibernation.
    b. acclimatization.
    c. change in temperature.
    d. estivation.
    e. lack of food.

VISUAL FOUNDATIONS

Color the parts of the illustration below as indicated. Also label compact bone and spongy bone.

RED ☐ blood vessel

GREEN ☐ lacuna

YELLOW ☐ osteon

BLUE ☐ osteocyte

ORANGE ☐ Haversian canal

VIOLET ☐ cytoplasmic extensions

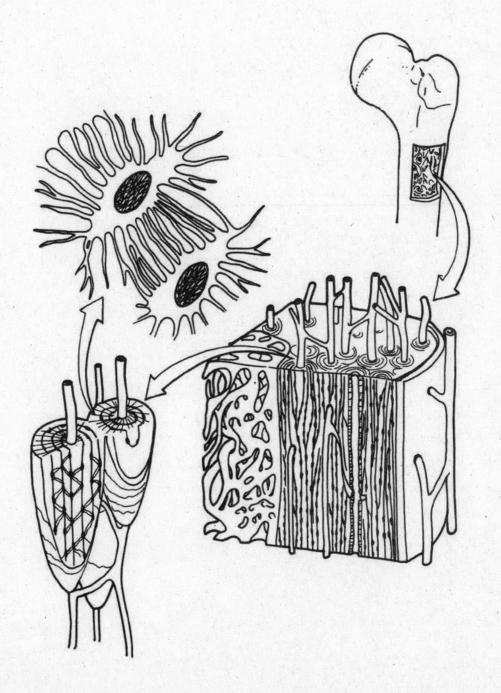

CHAPTER 40

❑

Protection, Support, and Movement

The next several chapters discuss how animals carry out life processes and how the organ systems of a complex animal work together to maintain homeostasis of the organism as a whole. This chapter focuses on epithelial coverings, skeleton, and muscle – systems that are closely interrelated in function and significance. Epithelium covers all external and internal body surfaces. In invertebrates, the external epithelium may contain secretory cells that produce a protective cuticle, secrete lubricants or adhesives, produce odorous or poisonous substances, or produce threads for nests or webs. In vertebrates, specifically humans, the external epithelium (skin) includes nails, hair, sweat glands, oil glands, and sensory receptors. In other vertebrates, it may include feathers, scales, mucous, and pigmentation. The skeleton supports and protects the body and transmits mechanical forces generated by contractile cells. Among the invertebrates are found hydrostatic skeletons and exoskeletons. All exoskeletons are composed of nonliving material above the epidermis. In many invertebrates it prevents growth and requires periodic molting. In invertebrates such as bivalves, clams and oysters, the exoskeleton is added to by secretions from the underlying epithelium as the animal grows and it is not shed. Endoskeletons on the other hand, extensive in echinoderms and chordates, are composed of living tissue that can grow. All animals have the ability to move. Muscle tissue is found in most invertebrates and all vertebrates.

REVIEWING CONCEPTS

Fill in the blanks.

INTRODUCTION

1. The epithelium, skin, of *Rana ridibanda* produces secretions that kill the antibiotic resistant strain of *S. aureus* known as _____.

2. Both the epithelium and the skeletal system provide an animal with

   _____.

3. Muscle can contract. When it is anchored to the _____, muscle contraction causes movement (locomotion).

EPITHELIAL COVERINGS

4. The structure and functions of the external epithelium are adapted to the animal's

   (a) _____ and (b) _____.

Invertebrate epithelium may secrete a cuticle

5. In insects, the outer epithelium secretes a protective and supportive outer covering of nonliving material called the _____.

6. In many species of animals, the epithelium contains cells that secrete

   (a) _____ and (b) _____.

Vertebrate skin functions in protection and temperature regulation

7. The integumentary system of vertebrates includes the (a) _____
   and (b) _____.

8. Structures derived from the epithelium can provide insulation. Examples are the (a) _____ of birds and the (b) _____ of mammals.

9. Examples of structures derived from the epithelium in mammals are
   (a) _____, (b) _____
   (c) _____, (d) _____,
   and (e) _____.

10. The outer layer of skin is called the (a)_____. It consists of several sublayers, or (b)_____.

11. Cells in the _____ continuously divide. As they are pushed upward, these cells mature, produce keratin, and eventually die.

12. Epithelial cells produce an insoluble, elaborately coiled protein called
    (a)_____, which functions in the skin for (b) _____,
    (c) _____, and (d) _____.

13. The dermis consists of dense, fibrous (a) _____ composed mainly of (b) _____ fibers and rests on a layer of (c) _____ tissue composed largely of fat.

SKELETAL SYSTEMS

14. A skeletal system transmits and transforms _____ generated by muscle contractions.

In hydrostatic skeletons, body fluids transmit force

15. Hydrostatic skeletons are made of _____.

16. The hydrostatic skeleton transmits forces generated by _____.

17. Many invertebrates (e.g., hydra) have a (a) _____ in which fluid is used to transmit forces generated by contractile cells or muscle. In these animals, contractile cells are arranged in two layers, an outer
    (b) _____ oriented layer and an inner
    (c) _____ arranged layer. When the (d) _____ (outer or inner?) layer contracts, the animal becomes shorter and thicker, and when the
    (e) _____ (outer or inner?) contracts, the animal becomes longer and thinner.

18. Annelid worms have sophisticated hydrostatic skeletons. The body cavity is divided by transverse partitions called _____, creating isolated, fluid-filled segments that can operate independently.

Mollusks and arthropods have nonliving exoskeletons

19. The support system in some organisms is an _____ consisting of a non-living substance overlying the epidermis.

20. Arthropod exoskeletons, composed mainly of the polysaccharide
    (a) _____. This nonliving skeleton prevents growth, necessitating periodi (b) _____.

Internal skeletons are capable of growth

21. Internal skeletons are also called (a) _____. They are extensively developed only in the echinoderms and the (b) _____.

22. The endoskeletons of echinoderms is composed of spines and plates composed of _____.

23. The vertebrate skeleton provides support, protection, and it
_____.

The vertebrate skeleton has two main divisions

24. The two main divisions of the vertebrate endoskeleton are the

    (a)_____ skeleton along the long axis of the body and the

    (b)_____ skeleton which includes the bones of the limbs and girdles.

25. Components of the axial skeleton include the (a) _____, which consists of cranial and facial bones; the (b)_____, made up of a series of vertebrae; and the (c)_____, consisting of the sternum and ribs.

26. Components of the appendicular skeleton include the (a) _____, consisting of clavicles and scapulas; the (b)_____, which consists of large, fused hipbones; and the (c)_____, each of which terminates in digits.

A typical long bone amplifies the motion generated by muscles

27. The long bones are covered by a connective tissue layer called the

    (a) _____ to which (b)_____ and

    (c) _____ attach.

28. The ends of long bones are called the (b) _____, and the shaft is the (c)_____. A cartilaginous "growth center" in children called the (d)_____ becomes an (e)_____ in adults. (f)_____, made up of osteons, forms a dense, and hard outer shell.

Bones are remodeled throughout life

29. Long bones develop from cartilage templates in a process called

    (a)_____ bone development. Other bones develop from noncartilage connective tissue templates in a process called (b)_____ bone development.

Joints are junctions between bones

30. Junctions between bones are called (a)_____. They are classified according to the degree of their movement: (b)_____, such as sutures, are tightly bound by fibers; (c)_____, like those between vertebrae; and the most common type of joint, the (d)_____.

MUSCLE CONTRACTION

31. All eukaryotic cells contain the contractile protein (a) _____. In many cells, it functions in association with the contractile protein (b) _____.

Invertebrate muscle varies among groups

32. Bivalves mollusks have (a) _____ muscle which they use to shut the shell quickly and (b) _____ muscle which they use to hold the shell tightly closed for long periods of time.

Insect flight muscles are adapted for rapid contraction

33. _____ contractions are muscle contractions that are not synchronized with signals from motor neurons.

Vertebrate skeletal muscles act antagonistically to one another

34. Skeletal muscles pull on cords of connective tissue called _____, which are attached to bones and pull on them.

35. The movement on one muscle can be reversed by the movement of another. This means that muscles are acting _____.

36. The muscle that contracts to produce a particular movement is called the (a) _____ and the muscle that produces the opposite movement is called the (b) _____.

A vertebrate muscle may consist of thousands of muscle fibers

37. A muscle fiber is actually a _____.

38. Within the sacroplasm of each muscle fiber are large, threadlike structures called (a) _____ that run lengthwise through the fiber. These larger structures are made up of smaller structures called (b) _____.

39. Actin is the structural unit of (a) _____ and myosin is the structural unit of (b) _____.

40. The basic unit of skeletal muscle contraction is called a _____.

Contraction occurs when actin and myosin filaments slide past one another

41. A motor neuron releases (a) _____ into the synaptic cleft between the motor neuron and each muscle fiber, where it binds with receptors on the surface of the muscle fiber, depolarizing the sarcolemma and initiating an (b) _____.

42. Depolarization of the T tubules stimulates calcium release from the (a) _____. Calcium then binds to (b) _____ which undergoes a conformational change exposing active sites on the (c) _____ filaments.

43. During the power stroke of muscle contraction the actin filament is pulled toward the _____ of the sarcomere.

ATP powers muscle contraction

44. _____ is the immediate energy source for muscle contraction.

45. _____ is the backup energy storage compound in muscle cells.

46. _____ is the chemical energy stored in muscle fibers.

The type of muscle fibers determines strength and endurance

47. (a) _____ fibers are well-adapted for endurance activities such as swimming and running. They derive most of their energy from (b) _____. These fibers are rich in (c)_____ which enhances oxygen movement from blood into muscle.

48. (a) _____ fibers generate power and carry out rapoid movements by they can only sustain this activity for a short period of time. These fibers obtain most of their energy from (b) _____.

49. (a) _____ fibers contract rapidly and have an intermediate level of fatigue.

Several factor influence the strength of muscle contraction

50. A motor neuron and the givers to which it is functionally connected is called a (a) _____.

51. The single, quick contraction of a skeletal muscle is a _____.

52. Even when not moving skeletal muscles remain in a partially contracted state called _____.

Smooth muscle and cardiac muscle are involuntary

53. (a)_____ muscle is capable of slow, sustained contractions and (b)_____ muscle contracts rhythmically.

BUILDING WORDS

Use combinations of prefixes and suffixes to build words for the definitions that follow.

Prefixes	The Meaning		Suffixes	The Meaning
ax-	center line		-blast	embryo, "formative cell"
chit-	tunic		-chondr(al)	cartilage
endo-	within		-dermis	skin
epi-	upon, over, on		-ial	pertaining to
kerat-	horn		-in	chemical substance
myo-	muscle		-lemma	sheath
osteo-	bone		-mere	a part of
peri-	about, around, beyond		-oste(um)	bone
sarc(o)-	muscle			

Prefix		Suffix	Definition
_____ _____			1. The outermost layer of skin, resting on the dermis.
_____ _____			2. Connective tissue membrane on the surface of bone that is capable of forming bone.
_____ _____			3. Pertains to the occurrence or formation of "something" within cartilage.
_____ _____			4. A bone-forming cell.
_____		-clast	5. Cells that break down bone.
_____		-filament	6. Subunit of myofibril consisting of either actin or myosin.
_____ _____			7. Serves as a diffusion barrier in the skin.

_____ _____ 8. A polysaccharide contained in exoskeletons.

_____ _____ 9. The central skeleton of the body consisting of the skull, vertebral column, ribs, and sternum.

_____ _____ 10. The contractile unit of skeletal muscles.

_____ _____ 11. The plasma membrane of a skeletal muscle fiber.

MATCHING

Terms:

a. Actin
b. Axial skeleton
c. Endoskeleton
d. Endosteum
e. Exoskeleton
f. Keratin
g. Ligament

h. Lacunae
i. Motor unit
j. Oxygen debt
k. Periosteum
l. Sarcomere
m. Sebum

n. Skeletal muscle
o. Stratum basale
p. Tendon
q. T-tubule
r. Vertebral column

For each of these definitions, select the correct matching term from the list above.

_____ 1. A motor neuron and the muscle fibers it stimulates.

_____ 2. A water-insoluble, elaborately coiled protein manufactured by epidermal cells that gives skin mechanical strength and flexibility.

_____ 3. A segment of a striated muscle cell that serves as the basic unit of muscle contraction.

_____ 4. The deepest layer of the epidermis, consisting of cells that continuously divide.

_____ 5. A band of connective tissue that connects bones and limits movement at the joints.

_____ 6. The skull, vertebral column, sternum, and ribs.

_____ 7. The hard exterior covering of certain invertebrates.

_____ 8. Tough cords of connective tissue that anchor muscles to bone.

_____ 9. The oxygen necessary to metabolize the lactic acid produced during strenuous exercise.

_____ 10. The protein composing one type of myofilament.

_____ 11. A mixture of fats and waxes.

_____ 12. The connective tissue layer covering the outer surface of a bone.

_____ 13. Small cavities in bone containing mature osteocytes.

_____ 14. An inward extension of the sacrolemma.

MAKING COMPARISONS

Fill in the blanks.

Organism	Skeletal Material	Kind of Skeleton	Characteristics
Many invertebrates	Fluid	Hydrostatic	Contractile tissue generates force that moves the body
Arthropods	#1	#2	#3
#4	#5	Endoskeleton	Internal shell, some have spines that project to the outer surface
Chordates	#6	#7	#8

Organism	Skeletal Material	Kind of Skeleton	Characteristics	
Characteristics	Slow-Oxidative fibers	Fast-Glycolytic fibers	Fast-Oxidative Fibers	
Contraction speed	Slow	#9	#10	
Rate of fatigue	#11	#12	Intermediate	
Major pathway for ATP synthesis	#13	Glycolysis	#14	
Intensity of contraction	#15	#16	Intermediate	

MAKING CHOICES

Place your answer(s) in the space provided. Some questions may have more than one correct answer.

_____ 1. A major component of the arthropod skeleton is the polysaccharide

    a. keratin.   d. actin.

    b. melanin.   e. myosin.

    c. chitin.

_____ 2. Actin filaments contain

    a. actin.   d. myosin.

    b. tropomyosin.   e. keratin.

    c. troponin complex.

_____ 3. The H zone in a sarcomere consists of

    a. actin.   d. myosin.

    b. tropomyosin.   e. keratin.

    c. troponin complex.

_____ 4. The human skull is part of the

    a. cervical complex.   d. appendicular skeleton.

    b. girdle.   e. atlas.

    c. axial skeleton.

_____ 5. Most of the mechanical strength of bone is due to

    a. spongy bone.   d. endochondral bone.

    b. osteocytes.   e. marrow.

    c. intramembranous bone.

_____ 6. You perceive touch, pain, and temperature through sense organs in your

    a. epidermis.   d. epithelium.

    b. stratum basale.   e. dermis.

    c. stratum corneum.

_____ 7. Myofilaments are composed of

    a. myofibrils.   d. myosin.

    b. fibers.   e. sarcoplasmic reticulum.

    c. actin.

_____ 8. The outer layer of a vertebrate's skin is the
    a. epidermis.   d. epithelium.
    b. stratum basale.   e. dermis.
    c. stratum corneum.

_____ 9. The tissue around bones that lays down new layers of bone is the
    a. metaphysis.   d. periosteum.
    b. epiphysis.   e. endosteum.
    c. marrow.

_____10. The human axial skeleton includes the
    a. ulna.   d. femur.
    b. shoulder blades.   e. breastbone.
    c. skull.

_____11. Cartilaginous growth centers in children are called
    a. metaphysis.   d. periosteum.
    b. epiphysis.   e. endosteum.
    c. marrow.

_____12. The human appendicular skeleton includes the
    a. ulna.   d. femur.
    b. shoulder blades.   e. breastbone.
    c. centrum.

_____13. Vertebrate appendages are connected to
    a. the cervical complex.   d. the appendicular skeleton.
    b. girdles.   e. the atlas.
    c. the axial skeleton.

_____14. Skeletons that operate entirely by hydrostatic pressure are found in
    a. annelids.   d. lobsters and crayfish.
    b. echinoderms.   e. vertebrates.
    c. hydra.

_____15. An opposable digit is found in
    a. lobsters.   d. some insects.
    b. great apes.   e. humans.
    c. crayfish.

_____16. Internal skeletons are found in
    a. annelids.   d. lobsters and crayfish.
    b. echinoderms.   e. chordates.
    c. hydra.

_____17. External skeletons are found in
    a. annelids.   d. insects.
    b. echinoderms.   e. chordates.
    c. hydra.

_____18. The "space" between the terminal axon of one neuron and a dendrite of the next neuron is called the
    a. synaptic cleft.   d. Z line.
    b. chemotrasmitter zone.   e. binding site.
    c. cross bridge.

_____ 19. Melanocytes

    a. are located in the dermis.   d. are continuously replaced.

    b. produce keratin.   e. are located in the stratum basale.

    c. produce melanin.

_____ 20. Epithelial derivatives include

    a. antlers.   d. sebum.

    b. claws.   e. melanin.

    c. glands.

_____ 21. Terms directly associated with the dermis include

    a. blood vessels.   d. cuticle.

    b. collagen.   e. dense connective tissue.

    c. sensory receptors.

_____ 22. Exoskeletons

    a. are rigid and inflexible.   d. transmit forces.

    b. is a one-piece sheath.   e. must be shed in animals that have them.

    c. are found only in invertebrates.

_____ 23. The vertebrate vertebral column has

    a. 7 cervical vertebrae.   d. 14 thoracic vertebrae.

    b. 6 sacral vertebrae.   e. 6 fused vertebrae making up the coccygeal region.

    c. 5 lumbar vertebrae.

_____ 24. When a sarcomere contracts

    a. the myofilaments shorten.   d. the Z-lines get closer together.

    b. the I-band shortens.   e. the H-zone shortens.

    c. the A-band gets longer.

_____ 25. Skeletal muscles use stored _____ as a backup energy supply even though it is present in very limited quantities.

    a. myoglobin.   d. glycogen.

    b. ATP.   e. glucose.

    c. creatine phosphate.

VISUAL FOUNDATIONS

Color the parts of the illustration below as indicated. Also label H zone, A band, I band, sarcomere, and muscle fiber.

RED	☐	Z line
GREEN	☐	mitochondria
YELLOW	☐	sarcoplasmic reticulum
BLUE	☐	T tubule
ORANGE	☐	myofibrils
TAN	☐	sarcomere
PINK	☐	sarcolemma
VIOLET	☐	nucleus

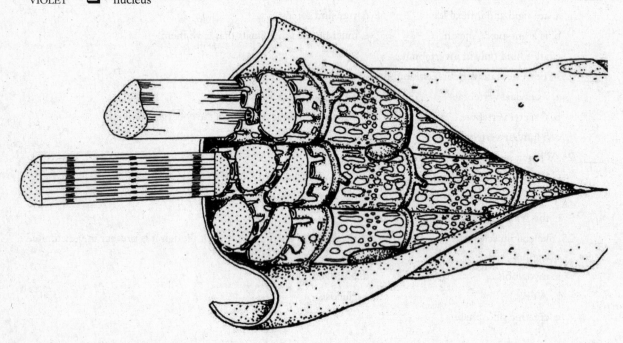

❏

CHAPTER 41

Neural Signaling

Behavior and physiological processes in animals are regulated by the endocrine and nervous systems. Endocrine regulation is generally slow and long-lasting, whereas nervous regulation is typically rapid and brief. Changes within an animal or in the outside world are detected by receptors and transmitted as electrical signals to the central nervous system where the information is sorted and interpreted. An appropriate response is then sent to muscles or glands where the response occurs. The nervous system is composed mainly of two types of specialized cells, glial cells and neurons. Glial cells help support, maintain, and regulate the nervous system and neural function. The neuron is the structural and functional unit that carries electrical impulses. This chapter focuses on the role of the neuron in nervous regulation and the importance of glial cells. The typical neuron consists of a cell body, dendrites, and an axon. The cell body houses the nucleus and has branched cytoplasmic projections called dendrites that receive stimuli. The axon is a single, long structure that transmits impulses to a neuron, muscle, or gland. Transmission involves changes in ion distribution between the inside and outside of the neuron. Transmission from one neuron to the next generally involves the release of chemicals into the space between neurons, the synapse. These chemicals diffuse across the space and bring about a response in the adjacent neuron, muscle, or gland. Hundreds of messages may arrive at a single neuron at the same time and must be integrated prior to a given response. Integration may occur at the level of the neuronal cell body, in the spinal cord, or in the brain. Following integration, the neuron may or may not initiate an impulse along its axon.

REVIEWING CONCEPTS

Fill in the blanks

INTRODUCTION

1. The ability of an organism to survive is largely dependent on its ability to detect and respond appropriately to changes in either the internal or external environments referred to as _____.

2. In animals, two systems are responsible for the regulation of physiological processes and behavior. Regulation by the (a) _____ system is slow and long-lasting, whereas regulation by the (b)_____ system is rapid.

NERUAL SIGNALING: AN OVERVIEW

3. Information flow through the nervous system begins with the detection of a stimulus by a "sensory device," a process called (a) _____. The "sensory device" is associated with (b) _____ neurons (sensory neurons) that often transmit information to (c)_____ (association neurons) where integration of the information occurs.

4. Signals are carried from the central nervous system (CNS) by (a) _____ to (b) _____ which may be a muscle or a gland.

5. _____ neurons carry impulses to skeletal muscle.

NEURONS AND GLIAL CELLS

Neurons receive stimuli and transmit neural signals

6. Neurons are highly specialized cells with a distinctive structure consisting of a cell body and two types of cytoplasmic extensions. Numerous, short (a) _____ receive impulses and conduct them to the (b) _____, which integrates impulses. Once integrated, the impulse is conducted away by the (c) _____, a long process that terminates at another neuron or effector.

7. Branches at the ends of axons, called (a) _____, end in tiny (b) _____ that release a transmitter chemical called a (c) _____.

8. Many axons outside of the vertebrate CNS are surrounded by a series of (a) _____ that form an insulating covering called the (b) _____. Gaps in this insulating covering, called (c) _____, occur between successive Schwann cells.

Certain regions of the CNS produce new neurons

9. _____ is the production of new neurons.

Axons aggregate to form nerves and tracts

10. A (a) _____ is a bundle of axons outside the CNS, whereas a (b) _____ is a bundle of axons within the CNS.

11. Masses of cell bodies outside the CNS are called (a) _____, whereas masses of cell bodies within the CNS are called (b)_____.

Glial cells play critical roles in neural function

12. Collectively, glial cells make up the _____.

13. There are several types of glial cells. (a) _____ are star-shaped glial cells that provide physical support for neurons. (b) _____ form the myelin sheath in the CNS. (c) _____ are ciliated and line internal cavities of the CNS, and (d) _____ mediate responses to injury or disease in the CNS.

TRANSMITING INFORMATION ALONG THE NEURON

14. The voltage measured across the plasma membrane of a cell is called the

_____.

Ion channels and pumps maintain the resting potential of the neuron

15. The membrane potential in a resting neuron or muscle cell is its (a) _____. A typical value for this membrane potential is (b) _____.

16. Two main factors that determine the magnitude of the membrane potential are (a) _____ and (b) _____.

17. When a neuron is not conducting an electrical impulse, the charge in the extracellular fluid relative to the intracellular fluid is (a) _____ (negative, positive) and the intracellular fluid relative to the extracellular fluid is (b) _____ (negative, positive).

18. In most cells, potassium ion concentration is highest (a) _____ (inside or outside?) the cell and sodium ion concentration is highest) (b)_____ (inside or outside?) the cell.

19. The ionic balance across the plasma membrane of neurons is a result of several factors. The (a) _____ is an active transport system that moves (b) _____sodium ions out of the cell and (c) _____ potassium ions into the cell.

Graded local signals vary in magnitude

20. Depolarization is described as an (a) _____ action as compared to hyperpolarization, which is described as an (b)_____ action.

Axons transmit signals called action potentials

21. If a stimulus is strong enough, it depolarizes the membrane to the (a) _____, it will provoke a response from the neuron called an (b) _____.

22. Action potentials are initiated when (a) _____ in the plasma membrane of the neuron open in response to a critical level of voltage change in the membrane potential allowing (b) _____ to enter the neuron.

This response is the (c) _____ phase of an action potential.

23. After a certain period (a) _____ close and the (b) _____ open which is the (c) _____ phase of the action potential.

24. As the action potential moves down the axon, _____ occurs behind it.

25. The time in which a new action potential cannot be generated is called the _____ period.

The action potential is an all-or-none response

26. An action potential either occurs or it does not. This is called an _____ response.

An action potential is self-propagating

27. In (a) _____ conduction in myelinated neurons, depolarization skips along the axon from one (b) _____ to the next.

TRANSMITTING INFORMATION ACROSS SYNAPSES

28. A junction between two neurons, or between a neuron and an effector, is called a _____.

29. The neuron that terminates at a synapse is known as the

    (a)_____ neuron, and the neuron that begins at that synapse is called the (b)_____ neuron.

Signals across synapses can be electrical or chemical

30. In (a) _____, the pre- and postsynaptic neurons

    occur very close together and form gap junctions. Most synapses are

    (b) _____ that involve

    (c) _____ that cross the (d)_____

    between neurons.

Neurons use neurotransmitters to signal other cells

31. Cells that release (a) _____ are called cholinergic

    neurons. This neurotransmitter is released by some neurons in the

    (b) _____ and (c) _____.

32. _____ release norepinephrine.

33. The three neurotransmitters that are called catecholamines are

    (a) _____,

    (b) _____, and

    (c) _____.

34. The neurotransmitters (a) _____ and

    (b) _____ are called biogenic amines.

35. The neuropeptides (a) _____ and

    (b) _____ function mainly as neuromodulators, molecules

    that stimulate long-term changes.

36. The body's endogenous opioids are called _____.

Neurotransmitters bind with receptors on postsynaptic cells

37. Neurotransmitters are stored in the synaptic terminals within small membrane bounded sacs called _____.

Activated receptors can send excitatory or inhibitory signals

38. A depolarization of the postsynaptic membrane that brings the neuron closer to firing is called an (a)_____. A

    hyperpolarization of the postsynaptic membrane that reduces the probability that the neuron will fire is called an (b)_____.

NEURAL INTEGRATION

Postsynaptic potentials are summed over time and space

39. Local responses in the postsynaptic membrane that vary in magnitude, fade over distance, and can be summated are called _____.

    EPSPs and IPSPs are examples.

40. Graded potentials can be added together in a process called (a)_____. When a second EPSP occurs before the depolarization caused by the first EPSP has decayed, (b)_____ occurs. In (c)_____, several synapses in the same area generate EPSPs simultaneously. Adding EPSPs together can bring the neuron to threshold.

NEURAL CIRCUITS: COMPLEX INFORMATION SIGNALING

41. In a (a) _____ circuit a single neuron is controlled by synapses with two or more presynaptic neurons. In a (b) _____ circuit a single presynaptic neuron stimulates may postsynaptic neurons

BUILDING WORDS

Use combinations of prefixes and suffixes to build words for the definitions that follow.

Prefixes	The Meaning	Suffixes	The Meaning
hyper-	above	-glia	glue
inter-	between, among	-neuro(n)	nerve
multi-	many, much, multiple		
neuro-	nerve		
post-	behind, after		
pre-	before, prior to, in advance of		

Prefix	Suffix	Definition
_____	_____	1. A nerve cell that carries impulses from one nerve cell to another, and is between a sense receptor and an effector.
_____	_____	2. Cells providing support and protection for neurons.
_____	-transmitter	3. Substance used by neurons to transmit impulses across a synapse.
_____	-polar	4. Pertains to a neuron with more than two (often many) processes or projections.
_____	-synaptic	5. Pertains to a neuron that begins after a specific synapse.
_____	-synaptic	6. Pertains to a neuron that ends before a specific synapse.
_____	-polarized	7. The state of the membrane as a result of too many potassium ions entering the cell at the end of repolarization.

MATCHING

Terms:

a. Action potential
b. Axon
c. Convergence
d. Dendrite
e. Facilitation
f. Ganglion
g. Integration
h. Myelin sheath
i. Nerve
j. Neuron
k. Refractory period
l. Schwann cell
m. Summation
n. Synapse
o. Threshold level

For each of these definitions, select the correct matching term from the list above.

_____ 1. One of many projections from a nerve cell that conducts a nerve impulse toward the cell body.

_____ 2. A nerve cell; a conducting cell of the nervous system that typically consists of a cell body, dendrites, and an axon.

_____ 3. A mass of neuron cell bodies located outside the central nervous system.

_____ 4. The long extension of the neuron that transmits nerve impulses away from the cell body.

_____ 5. A large bundle of axons wrapped together in connective tissue that are located outside the CNS.

_____ 6. The brief period of time that must elapse after the response of a neuron or muscle cell, during which it cannot respond to another stimulus.

_____ 7. The electrical activity developed in a muscle or nerve cell during activity.

_____ 8. An insulating covering around axons of certain neurons.

_____ 9. The junction between two neurons or between a neuron and an effector.

_____10. The sorting and interpreting of incoming sensory information and determining a response.

_____11. A combination of EPSPs and IPSPs.

_____12. The necessary critical value before an action potential is initiated.

MAKING COMPARISONS

Fill in the blanks.

Kind of Potential	mV	Membrane Response	Ion Channel Activity
Membrane potential	-70mV	Stable	Sodium-potassium pump active, passive sodium and potassium channels open
Resting potential	#1	#2	#3
#4	From -50 to -55mV	Depolarization	#5
Action potential	#6	#7	Voltage-activated sodium ion channels and potassium ion channels open
EPSP	Increasingly positive	#8	#9
IPSP	Increasingly negative	#10	Neurotransmitter-receptor combination may open potassium ion channels or chloride ion channels

MAKING CHOICES

Place your answer(s) in the space provided. Some questions may have more than one correct answer.

_____ 1. The inner surface of a resting neuron is generally _____ compared with the outside.
    a. positively charged   d. 70 mV
    b. negatively charged   e. −70 mV
    c. polarized

_____ 2. The part of the neuron that transmits an impulse from the cell body to an effector cell is the
    a. dendrite.   d. collateral.
    b. axon.   e. hillock.
    c. Schwann body.

_____ 3. Saltatory conduction
    a. occurs between nodes of Ranvier.   d. requires less energy than continuous conduction.
    b. occurs only in the CNS.   e. involves depolarization at nodes of Ranvier.
    c. is more rapid than the continuous type.

_____ 4. The nodes of Ranvier are
    a. on the cell body.   d. gaps between adjacent Schwann cells.
    b. on the dendrites.   e. insulated with myelin.
    c. on the axon.

_____ 5. A neuron that begins at a synapse is called a
    a. presynaptic neuron.   d. neurotransmitter.
    b. postsynaptic neuron.   e. acetylcholine releaser.
    c. synapsing neuron.

_____ 6. Which of the following correctly expresses the movement of ions by the sodium-potassium pump?
    a. sodium out, potassium in   d. more sodium out than potassium in
    b. sodium in, potassium out   e. less sodium out than potassium in
    c. sodium and potassium both in and out, but in different amounts

_____ 7. A nerve pathway or tract in the CNS is
    a. one neuron.   d. a ganglion.
    b. a group of nerves.   e. a bundle of cell bodies.
    c. a bundle of axons.

_____ 8. An axon cannot transmit an action potential no matter how great a stimulus is applied when it is
    a. hyperpolarized.   d. in the relative refractory period.
    b. depolarized.   e. in the resting state.
    c. in the absolute refractory period.

_____ 9. If a neurotransmitter hyperpolarizes a postsynaptic membrane, the change in potential is referred to as
    a. spatial summation.   d. IPSP.
    b. temporal summation.   e. EPSP.
    c. a threshold level impulse.

_____ 10. Aside from the sodium-potassium pump, the resting potential is due mainly to
    a. inward diffusion of chloride ions.   d. open sodium channels in the membrane.
    b. inward diffusion of sodium ions.   e. large protein anions inside the cell.
    c. outward diffusion of potassium ions.

_____11. Branchlike extensions of the cell body involved in receiving stimuli are

a. dendrites.   d. collaterals.

b. axons.   e. hillocks.

c. Schwann bodies.

_____ 12. The first thing that needs to happen in the nerve signaling pathway for transmission of a nerve impulse to occur is

a. transmission of the signal.   d. integration.

b. perception.   e. reception.

c. depolarization.

_____ 13. _____ are phagocytic glial cells.

a. Oligodendrocytes   d. Micoglia

b. Ependymal cells   e. Astrocytes

c. Neuroglia

_____ 14. The word nucleus is used to refer to

a. the part of the eukaryotic cell containing DNA.   d. a region of the CNS containing nerve cell bodies.

b. masses of nerve cell bodies outside the CNS.   e. the region of a nerve where dendrites attach.

c. the site where nerve impulses are integrated.

_____ 15. In neurons, potassium ions

a. move by passive diffusion.   d. are pumped into the cell.

b. move through voltage-gated channels.   e. contribute to the equilibrium potential.

c. are in greater concentration outside the cell.

_____ 16. A graded potential

a. is transmitted for short distances.   d. is determined by the resting potential.

b. varies with stimulus strength.   e. results in hyperpolarization of a neuron.

c. is a localized response to a stimulus.

_____ 17. During depolarization of a neuron

a. voltage-gated sodium channels close.   d. potassium channels open.

b. the neuron is in a relative refractory period.   e. the transmembrane potential becomes more positive.

c. the voltage-gated potassium channels are closed.

_____ 18. _____ are cells that myelinate axons.

a. Node cells   d. Schwann cells

b. Oligodendrocytes   e. Astrocytes

c. Ependymal cells

_____ 19. Chemical synapses

a. are found at gap junctions.   d. occur at clefts about 20 nm wide.

b. are the majority of synapses.   e. are used by many interneurons to communicate.

c. are involved in the escape response of animals.

_____ 20. Adrenergic neurons

    a. bind acetylcholine.   d. form electrical synapses.

    b. release norepinephrine.   e. bind norepinephrine.

    c. release acetylcholine.

_____ 21. Neuropeptides that act as neurotransmitters include

    a. substance P   d. serotonin.

    b. gamma-aminobutyric acid (GABA).   e. dopamine.

    c. norepinephrine.

_____ 22. Neuropeptides

    a. act as neuromodulators.   d. are excitatory molecules on skeletal muscles.

    b. inhibit neurons in the brain.   e. enhance the action of GABA.

    c. inhibit neurons in the spinal cord.

_____ 23. Postsynaptic potentials

    a. may be summed temporally.   d. may be inhibitory.

    b. may be summed spatially.   e. may occur at a subthreshold level.

    c. may involve convergent circuits.

VISUAL FOUNDATIONS

Color the parts of the illustration below as indicated. Also label nodes of Ranvier and Schwann cell.

RED   ☐   synaptic terminals

GREEN   ☐   terminal branches

YELLOW   ☐   myelin sheath

BLUE   ☐   axon

ORANGE   ☐   cellular sheath

PINK   ☐   cell body and dendrite

VIOLET   ☐   nucleus of neuron

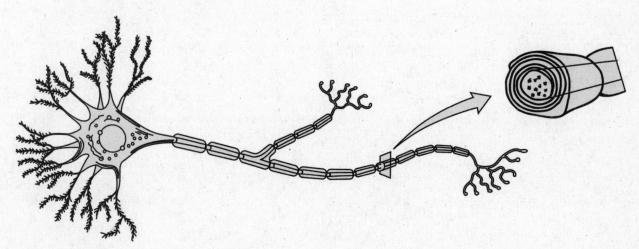

Color the parts of the illustration below as indicated. Also label extracellular fluid and cytoplasm.

RED   ☐ sodium ion   BROWN   ☐ potassium channel

GREEN   ☐ potassium ion   TAN   ☐ large anions

YELLOW   ☐ plasma membrane   PINK   ☐ sodium-potassium pump

BLUE   ☐ arrows indicating diffusion into the cell   VIOLET   ☐ arrows for diffusion out of the cell

ORANGE   ☐ sodium channel

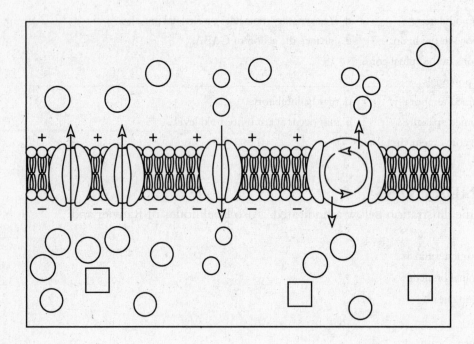

CHAPTER 42

❑

Neural Regulation

This chapter compares several animal nervous systems, and then examines the vertebrate nervous system, with emphasis on the human brain. Also examined are the cellular mechanisms of memory and learning. The simplest organized nervous system is the nerve net found in cnidarians. A nerve net consists of nerve cells scattered throughout the body; there is no central control organ and no definite nervous pathways. The nervous system of echinoderms is more complex, with a nerve ring and nerves that extend into various parts of the body. Bilaterally symmetric animals have more complex nervous systems. The vertebrate nervous system has two main divisions: a central nervous system (CNS) consisting of a complex tubular brain that is continuous with a tubular nerve cord, and a peripheral nervous system (PNS). The CNS provides centralized control, integrating incoming information and determining appropriate responses. The PNS consists of sensory receptors and cranial and spinal nerves. It functions to regulate the organism's internal environment and to help it adjust to its external environment. Learning is a function of the brain, involving the storage of information and its retrieval. Many drugs affect the nervous system, some by changing the levels of neurotransmitters within the brain.

REVIEWING CONCEPTS

Fill in the blanks.

INTRODUCTION

INVERTEBRATE NERVOUS SYSTEMS: TRENDS IN EVOLUTION

1. In general, the lifestyle of an animal is closely related to the design and complexity of its nervous system. For example, the simplest nervous system, called a (a) _____, is found in (b) _____, which includes sessile animals such as Hydra.

2. Echinoderms have a modified nerve net. In this system, a _____ surrounds the mouth, and large, radial nerves extend into each arm. Branches from these nerves coordinate the animal's movements.

3. Planarian flatworms have a (a) _____ nervous system with concentrations of nerve cells in the head called (b) _____; these serve as a primitive brain.

4. Annelids have a ventral nerve cord with the cell bodies of many of the neurons massed into _____.

THE VERTEBRATE NERVOUS SYSTEM: STRUCTURE AND FUNCTION

5. The human brain contains about (a) _____ (#?) neurons. How these neurons are organized is critical in determining an individual's
   (b) _____.

6. The vertebrate nervous system is divided into a (a) _____ system consisting of the brain and spinal cord, and a (b) _____ system consisting of sensory receptors and nerves.

7. The PNS system is divided into the (a) _____division, which is concerned with changes in the external environment, and the (b)_____ division, which regulates the internal environment. The autonomic division has two efferent pathways made up of (c) _____ and (d) _____ nerves.

EVOLUTION OF THE VERTEBRATE BRAIN

8. The (a) _____ of the vertebrate embryo differentiates anteriorly into the (b) _____ and posteriorly into the (c) _____.

The hindbrain develops into the medulla, pons, and cerebellum

9. The (a) _____, which coordinates muscle activity, and the (b) _____, which connects the spinal cord and medulla with upper parts of the brain, are formed from the (c) _____. The (d) _____, made up largely of nerve tracks, is formed from the (e) _____.

10. The brainstem is made up of the (a) _____, (b) _____, and (c) _____.

11. The cerebellum is responsible for (a) _____, (b) _____, and (c) _____.

The midbrain is prominent in fishes and amphibians

12. The midbrain is the most prominent part of the brain in fish and amphibians. It is their main _____ area, linking sensory input and motor output.

13. In mammals, the midbrain consists of the (a) _____ which regulates visual reflexes, and the (b)_____, which regulate some auditory reflexes. It also contains the (c) _____ that integrates information about muscle tone and posture.

The forebrain gives rise to the thalamus, hypothalamus, and cerebrum

14. The forebrain, or proencephalon, differentiates into the diencephalon and the telencephalon. The diencephalon gives rise to the (a) _____ and (b) _____. The telencephalon gives rise to the (c) _____.

15. The cerebrum is divided into right and left cerebral _____.

16. The cerebrum is made up of an inner layer of (a) _____ matter which is predominantly comprised of (b) _____ axons that connect various parts of the brain. The outer layer of (c) _____matter contains cell bodies and dendrites.

17. Some reptiles and all mammals have a type of cerebral cortex called the _____ which consists primarily of association areas.

18. The surface area of the human cerebral cortex is increased by folds called (a)_____. The furrows between the tissue are called (b) _____ if they are shallow and (c) _____ if they are deep.

THE HUMAN CENTRAL NERVOUS SYSTEM

19. The human central nervous system consists of the brain and spinal cord. Both are protected by bone and three meninges, the innermost (a) _____, the middle (b) _____, and the outermost (c) _____, The CNS is bathed by cerebrospinal fluid produced in a specialized capillary bed called the (d)_____.

The spinal cord transmits impulses to and from the brain

20. The spinal cord extends from the base of the brain to the _____ _____ vertebra.

21. Grey matter in a cross section of the spinal cord is shaped like the letter "H." It is surrounded by (a) _____ that contains nerve pathways, or (b)_____.

The most prominent part of the human brain is the cerebrum

22. The human cerebral cortex consists of three functional areas. (a)_____ areas receive incoming sensory information, (b)_____ areas control voluntary movement, and (c)_____ areas link the other two areas.

23. Each hemisphere of the human cerebrum is divided into lobes. The (a) _____ lobes contain primary motor areas and are separated from the primary sensory areas in the (b) _____ lobes by a groove called the (c) _____.

24. Deep in the white matter of the cerebrum lie the paired _____ that are important in coordinating movement.

25. The neurotransmitter(a) _____ found in the substantia nigra helps balance (b) _____ and (c)_____ of neurons involved in motor functions.

The body follows a circadian cycle of sleep and wakefulness

26. The (a) _____ is an endocrine gland located in the hypothalamus. It produces (b) _____ a hormone that plays a role in regulating the sleep-wake cycle.

27. The reticular activating system (RAS) is responsible for maintaining consciousness. It is located within the _____.

28. Electrical activity in the form of brain waves can be detected by a device that produces brain wave tracings called an (a) _____. The (b) _____ waves are associated with times of relaxation, (c)_____ waves have a fast-frequency and are associated with mental activity. (d) _____ and (e) _____ are slow waves that are associated with sleep.

29. There are two main stages of sleep: (a) _____ sleep, which is associated with dreaming, and (b) _____ sleep, which is characterized by theta and delta waves.

The limbic system affects emotional aspects of behavior

30. The limbic system affects the emotional aspects of behavior, evaluates rewards, and is important in motivation. It consists of parts of the _____ as well as parts of the thalamus, and hypothalamus, several nuclei in the midbrain, and the neural pathways that connect these structures.

Learning and memory involve long-term changes at synapses

31. (a)_____ is the process by which information is encoded, stored, and retrieved. (b)_____ is unconscious memory for perceptual and motor skills and (c)_____ involves factual knowledge of people, places, or objects, and requires conscious recall of the information.

32. The _____ functions in the formation and retrieval of memories.

33. The low-frequency stimulation of neurons results in a long-lasting decrease in the strength of synaptic connections. This decrease is called

_____.

Language involves comprehension and expression

34. Wernicke's area is located in the _____ and is an important center for language comprehension.

35. _____ located in the left frontal lobe controls our ability to speak.

THE PERIPHERAL NERVOUS SYSTEM

The somatic division helps the body adjust to the external environment

36. The somatic nervous system consists includes (a) _____ that detect changes in the external environment. (b)_____ transmit information to the CNS, and (c)_____ that adjust the positions of the skeletal muscles that help maintain the body's posture and balance.

37. Neurons are organized into nerves. The 12 pairs of (a) _____ nerves connect to the brain and are largely involved with sense receptors. Thirty-one pairs of (b) _____ nerves connect to the spinal cord.

The autonomic division regulates the internal environment

38. The efferent component of the autonomic division is divided into two systems. The (a) _____ system frequently operates to stimulate organs and to mobilize energy. The (b) _____ system influences organs to conserve and restore energy.

39. The autonomic system uses two neurons between the CNS and the effector. The first neuron is called the (a)_____ and the second is called the (b)_____.

EFFECTS OF DRUGS ON THE NERVOUS SYSTEM

40. Habitual use of almost any mood-altering drug can result in
    _____, in which the user becomes emotionally dependent on the
    drug.

BUILDING WORDS

Use combinations of prefixes and suffixes to build words for the definitions that follow.

Prefixes	The Meaning	sufixes	The Meaning
hypo-	under, below	-ic	pertaining to
limb-	border	-ory	related to
para-	beside, near		
post-	behind, after		
pre-	before, prior to, in advance of		
sens-	perceive		
som	body		

Prefix	Suffix	Definition
_____	-thalamus	1. Part of the brain located below the thalamus; principal integration center for the regulation of the viscera.
_____	-ganglionic	2. Pertains to a neuron located distal to (after) a ganglion.
_____	-ganglionic	3. Pertains to a neuron located proximal to (before) a ganglion.
	-vertebral	4. Pertains to structures located beside the vertebral column.
_____	_____	5. Nerve that receive impulses about the internal and external environment and transmit them to the CNS.
_____	_____	6. A system in the brain that influences the emotional aspects of behavior.
_____	_____	

MATCHING

Terms:

a. Autonomic nervous system
b. Central nervous system
c. Cerebellum
d. Cerebral cortex
e. Corpus callosum
f. Gyrus

g. Limbic system
h. Meninges
i. Parasympathetic nervous system
j. Peripheral nervous system
k. Pons

l. Sensory areas
m. Somatic nervous system
n. Temporal
o. Thalamus

For each of these definitions, select the correct matching term from the list above.

_____ 1. An action system of the brain that plays a role in emotional responses.

_____ 2. The outer layer of the cerebrum, composed of gray matter and consisting of densely-packed nerve cells.

_____ 3. The receptors and nerves that lie outside of the central nervous system.

_____ 4. The subdivision of the brain that coordinates muscle activity and is responsible for muscle tone, posture, and equilibrium.

_____ 5. The connective tissue that envelops the brain and spinal cord.

_____ 6. The portion of the peripheral nervous system that controls the visceral functions of the body.

_____ 7. A large bundle of nerve fibers interconnecting the two cerebral hemispheres.

_____ 8. The bulge on the anterior surface of the brainstem between the medulla and midbrain; connects various parts of the brain.

_____ 9. The nervous system consisting of the brain and spinal column.

_____10. A division of the autonomic nervous system concerned primarily with storing and restoring energy.

_____11. A relay center in the brain for motor and sensory messages.

_____12. Lobe of the brain located just above the ear.

MAKING COMPARISONS

Fill in the blanks.

Human Brain Structure	Subdivision or Location of Structure	Function of Structure
Medulla	Myelencephalon	Contains vital centers and other reflex centers
#1	Metencephalon	Connects various parts of the brain
#2	Mesencephalon	#3
Thalamus	#4	#5
#6	#7	Controls autonomic functions; links nervous and endocrine systems; controls temperature, appetite, and fluid balance; involved in some emotional and sexual responses
Cerebellum	#8	Responsible for muscle tone, posture, and equilibrium
#9	Telencephalon	Complex association functions
#10	Brain stem and thalamus	Arousal system
#11	Certain structures of the cerebrum, diencephalon	Affect emotional aspects of behavior, motivation, sexual behavior, autonomic responses, and biological rhythms

MAKING CHOICES

Place your answer(s) in the space provided. Some questions may have more than one correct answer.

_____ 1. The largest part of the human brain is the
    a. cerebrum.   d. medulla oblongata.
    b. cerebellum.   e. midbrain.
    c. myelencephalon.

_____ 2. The part of the human brain that is the center for intellect, memory, and language is the
    a. cerebrum.   d. medulla oblongata.
    b. cerebellum.   e. midbrain.
    c. myelencephalon.

_____ 3. Skeletal muscles are controlled by the
    a. parietal lobes.   d. temporal lobes.
    b. frontal lobes.   e. mesolimbic ganglia.
    c. central sulcus.

_____ 4. Cerebrospinal fluid (CSF)
    a. is produced by choroid plexi.   d. is within layers of the meninges.
    b. circulates through ventricles.   e. is in the subarachnoid space.
    c. is between the arachnoid and pia mater.

_____ 5. In vertebrates, the cerebrum is typically
    a. divided into two hemispheres.   d. mostly white matter.
    b. mostly gray matter.   e. mainly axons connecting parts of the brain.
    c. mainly cell bodies and some sensory neurons.

_____ 6. Regulation of body temperature under ordinary circumstances is under the control of the
    a. sympathetic n.s.   d. cranial nerve VI.
    b. autonomic n.s.   e. dorsal root ganglion.
    c. forebrain.

_____ 7. When you are eating one of your favorite foods, the cranial nerve(s) directly involved in your perception of this pleasurable experience is/are
    a. facial.   d. XIX.
    b. VII.   e. XI.
    c. X.

_____ 8. The regulation of body temperature, appetite, and fluid balance is mainly controlled by the
    a. cerebellum.   d. hypothalamus.
    b. cerebrum.   e. red nucleus.
    c. thalamus.

_____ 9. In non-REM sleep, as compared to REM sleep, there is/are
    a. higher-amplitude delta waves.   d. more dream consciousness.
    b. faster breathing.   e. more norepinephrine released.
    c. lower blood pressure.

_____10. The brain and spinal cord are wrapped in connective tissue called
    a. gray matter.   d. sulcus.
    b. meninges.   e. neopallium.
    c. gyri.

_____11. The part of the mammalian brain that integrates information about posture and muscle tone is the
- a. cerebrum.
- b. cerebellum.
- c. myelencephalon.
- d. medulla oblongata.
- e. midbrain.

_____12. The part(s) of the brain that coordinate(s) muscular activity is/are the
- a. cerebrum.
- b. cerebellum.
- c. myelencephalon.
- d. medulla oblongata.
- e. midbrain.

_____13. If you are having strong sexual feelings one minute and enraged over little or nothing the next minute, chances are very good that you have just had a very active
- a. frontal lobe.
- b. limbic system.
- c. thalamus.
- d. reticular activating system (RAS).
- e. cerebellum.

_____14. If only one temporal lobe is damaged, one might expect
- a. blindness in one eye.
- b. total blindness.
- c. decrease in hearing acuity in both ears.
- d. partial loss of hearing in both ears.
- e. inability to detect taste.

_____15. The limbic system
- a. is important in motivation.
- b. is involved in biological rhythms.
- c. is present in all mammals.
- d. affects sexual behavior.
- e. plays a role in autonomic responses.

_____16. Flatworms have
- a. eyespots.
- b. bilateral symmetry.
- c. nerve cords.
- d. cerebral ganglia.
- e. a primitive brain.

_____17. Nerve nets
- a. connect to primitive brains.
- b. connect to ganglia.
- c. are found in cnidarians.
- d. are found in some species of flatworms.
- e. coordinate movements in echinoderms.

_____18. The forebrain gives rise to the
- a. myelencphalon.
- b. diencephalon.
- c. metaencephalon.
- d. mesencephalon.
- e. telecephalon.

_____19. The medulla
- a. contains a space, the third ventricle.
- b. contains centers that regulate respiration.
- c. contains nuclei that rely information between the cerebrum and cerebellum.
- d. consists primarily of nerve tracts.
- e. arises from the metancephalon.

_____20. The hypothalamus
- a. controls body temperature.
- b. contains visual reflex centers.
- c. regulates appetite.
- d. connects to olfactory centers.
- e. contains the neocortex.

_____21. Terms associated with the cerebrum include

    a. corpus callosum.

    b. balance and coordination.

    c. prominent in amphibians.

    d. highly developed association functions.

    e. sulci.

_____22. Visual centers are located in the

    a. insular lobe.

    b. parietal lobe.

    c. frontal lobe.

    d. temporal lobe.

    e. occipital lobe.

_____23. The left and right hemispheres are connected by the

    a. fornix.

    b. corpus callosum.

    c. substantia nigra.

    d. paleocortex.

    e. amygdale.

_____24. Heightened mental activity is associated with an increase in

    a. alpha waves.

    b. delta waves.

    c. low frequency brain waves.

    d. theta waves.

    e. beta waves.

_____25. Implicit memory

    a. is unconscious memory for motor skills.

    b. is another name for short term memory.

    c. uses dopamine as the neurotransmitter.

    d. is another name for declarative memory.

    e. involves factual knowledge.

_____26. Formulating a meaningful word order for a sentence involves

    a. involves the parietal lobe.

    b. activation of Broca's area.

    c. involves the right hemisphere of the brain.

    d. activation of Wernicke's area.

    e. the left hemisphere of the brain.

VISUAL FOUNDATIONS

Color the parts of the illustration below as indicated. Also label sensory neuron, interneuron, and motor neuron.

RED	☐	muscle
YELLOW	☐	receptor
ORANGE	☐	location of cell bodies of sensory neuron
TAN	☐	central nervous system

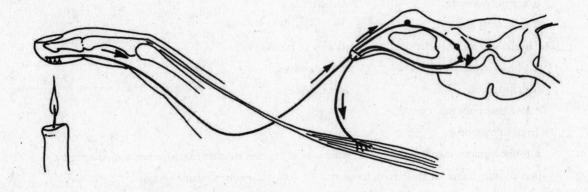

Color the parts of the illustration below as indicated. Also label sulcus. Indicate the vertebrate class represented by each brain illustration.

RED ☐ cerebrum
GREEN ☐ olfactory bulb, tract, and lobe
YELLOW ☐ corpus striatum
BLUE ☐ cerebellum
ORANGE ☐ optic lobe
BROWN ☐ epiphysis
TAN ☐ medulla
VIOLET ☐ diencephalon

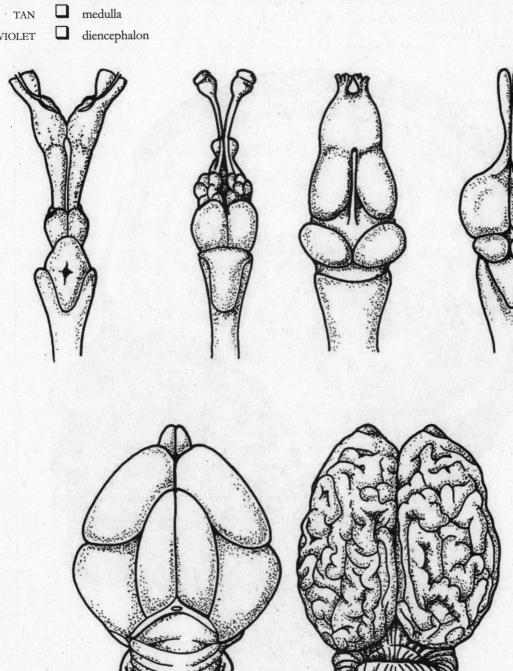

Color the parts of the illustration below as indicated. Also label diencephalon and midbrain.

RED ☐ pituitary
GREEN ☐ thalamus
YELLOW ☐ spinal cord
BLUE ☐ hypothalamus
ORANGE ☐ medulla and pons
BROWN ☐ cerebrum
TAN ☐ cerebellum
PINK ☐ corpus callosum
VIOLET ☐ pineal body

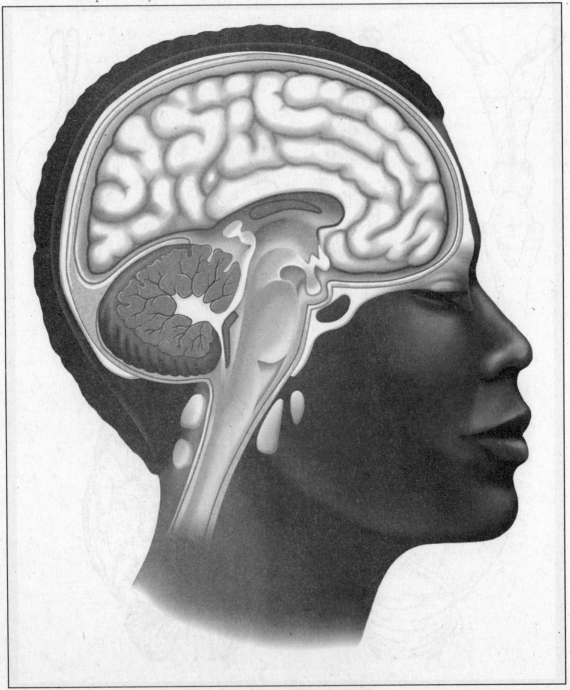

CHAPTER 43

Sensory Systems

Sense organs are specialized structures with receptor cells that detect changes in the internal or external environment and transmit the information to the nervous system. Receptor cells may be neuron endings or specialized cells in contact with neurons. Sensory receptors can be classified according to the location of the stimuli to which they respond or to the type of energy they transduce. They work by absorbing energy and converting it into electrical energy. Depending upon which ion channels are affected, a receptor cell will either depolarize or hyperpolarize. When the state of depolarization reaches a threshold level, an impulse is generated in the axon, thereby transmitting the information to the CNS. Sense organs that respond to heat and cold provide important cues about body temperature, and help some animals locate a warm-blooded host or prey. Those that detect electrical energy are found in some fish where they can detect electric currents in water. Some fish species can generate a shock for defense or to stun prey. Sense organs that detect magnetic fields are used for orientation and migration, and those that detect physical force are responsible for pain reception. Sense organs that respond to mechanical energy include the various tactile receptors in the skin, the lateral line organs in fish, the receptors that respond continuously to tension and movement in muscles and joints, the receptors that respond to body position, and the receptors responsible for hearing. Sense organs that respond to chemical energy are responsible for the sense of taste and smell, and those that respond to light energy are responsible for vision.

REVIEWING CONCEPTS

Fill in the blanks.

INTRODUCTION

1. Dolphins, bats, and a few other vertebrates detect distant objects by _____, sometimes called biosonar.

HOW SENSORY SYSTEMS WORK

Sensory receptors receive information

2. In the process of _____, sensory receptors absorb a small amount of energy from some stimulus.

Sensory receptors transduce energy

3. The energy to which a receptor is responding must be

    (a) _____ into (b) _____ energy which is the form of information used by the nervous system.

4. A change in the transmembrane potential of a receptor cell is called a

    (a) _____ which is a

    (b) _____ because the magnitude of the change depends upon the energy of the stimulus.

5. If a receptor potential (a) _____ the sensory neuron to threshold, it initiates an (b) _____ that travels along the

    (c) _____ neuron to the central nervous system.

Sensory input is integrated at many levels

6. A decrease in the frequency of action potentials in a sensory neuron, even though the stimulus is maintained, is called _____.

7. Sensation is determined by the part of the _____ that receives the message from the sense organ.

8. Impulses from sense organs are encoded into messages based upon the (a)_____ of neurons (or fibers) carrying the message and the (b)_____ of action potentials transmitted by the neuron (fiber).

We can classify sensory receptors based upon location of stimuli or on the type of energy they transduce

9. Sensory perception is the process of (a) _____, (b) _____, and (c) _____ sensory information.

10. Receptors that detect stimuli in the external environment are called (a) _____; and (b)_____ detect changes in such things as pH, osmotic pressure, body temperature, and the chemical composition of the blood.

THERMORECEPTORS

11. Two types of snakes, the (a) _____ and (b) _____, use thermoreceptors to locate prey.

12. In mammals, free nerve endings in the (a) _____ and (b) _____ detect temperature changes in the external environment.

13. Thermoreceptors that detect internal temperature changes in mammals are found in the _____ of the brain.

ELECTRORECEPTORS AND ELECTROMAGNETIC RECEPTORS

14. Some electroreceptors, called _____, are sensitive enough to detect Earth's magnetic field.

NOCICEPTORS

15. Nociceptors transmit signals through sensory neurons to interneurons in the (a) _____. The sensory neurons release the neurotransmitter (b) _____ and several neuropeptides, including (c) _____.

MECHANORECEPTORS

16. Mechanoreceptors are activated when they _____.

Tactile receptors are located in the skin

17. The tactile receptors of many animals are located at the base of either a (a) _____, or a (b) _____.

18. In mammalian skin, _____ forms discs in the deep epidermis that are sensitive to light touch.

19. (a) _____ located in the upper dermis are sensitive to light touch and vibration. (b) _____ located in the dermis detect heavy, continuous pressure and stretching of the skin. (c) _____ located in the deep dermis are sensitive to deep pressure that causes rapid movement of tissues.

Proprioceptors help coordinate muscle movement

20. Vertebrates have three categories of proprioceptors:

(a) _____, which detect muscle movement;

(b)_____, which respond to tension in contracting muscles and in the tendons that attach muscle to bone; and

(c)_____, which detect movement in ligaments.

Many invertebrates have gravity receptors called statocysts

21. Statocysts are one type of receptor found in invertebrates. A statocyst, which is an infolding of the epidermis, is lined with (a) _____ that are stimulated when tiny granules called that (b)_____ are pulled upon by gravity.

Hair cells are characterized by stereocilia

22. A _____ is a true cilium with a 9X2 arrangement of microtubules.

Lateral line organs supplement vision in fishes

23. Lateral line organs detect _____ in the water.

24. Lateral line organs consist of a (a) _____ lined with (b) _____ that runs the length of the animal's lateral surface. Water disturbances move the (c) _____ secreted by the hair cells (receptor cells), generating an electrical response.

The vestibular apparatus maintains equilibrium

25. The (a) _____ and the (b) _____ are gravity detectors in inner ear. They have sensory hair cells covered with a (c) _____ in which are embedded (d)_____ composed of calcium carbonate. This structural arrangement and the detection of gravity are similar to that which occurs in (e)_____.

26. Turning movements, referred to as (a) _____, cause movement of a fluid called (b) _____ in the semicircular canals, which in turn stimulate the hair cells of the (c) _____. There are no (d) _____ associated with these receptors.

Auditory receptors are located in the cochlea

27. Auditory receptors in an inner ear structure of birds and mammals, called the (a)_____, contain (b) _____ that detect pressure waves.

28. In terrestrial vertebrates, sound waves first initiate vibrations in the eardrum, or (a) _____. Three tiny ear bones, the (b) _____, (c) _____, and (d) _____ transmit the vibration to fluids in the inner ear through a membrane covered opening called the (e) _____.

29. The _____ helps maintain equal pressure between the middle ear and the atmosphere.

30. Fluids in the cochlea, in response to vibrations from the oval window, initiate vibrations in the (a) _____, which in turn cause stimulation of hair cells in the (b) _____. The hair cells synapse with (c) _____ which when depolarized initiate impulses along the (d) _____.

31. When detecting sound, pitch is dependent upon the (a) _____ of sound waves and loudness is dependent upon the (b) _____ of sound waves.

CHEMORECEPTORS

32. Throughout the animal kingdom, two special senses (a) _____ (b) _____ allow animals to detect chemical substances in food, water, and air.

Taste receptors detect dissolved food molecules

33. Traditionally four basic tastes have generally recognized. They are (a) _____, (b) _____, (c) _____, and (d) _____. A fifth taste known as (e) _____ has now been identified and another basic taste (f)_____ has been proposed.

The olfactory epithelium is responsible for the sense of smell

34. Most invertebrates depend on _____, or the detection of odors, as their main sensory modality.

35. In terrestrial vertebrates, the sense of smell occurs in the _____.

36. The olfactory cortex is part of the _____ in the brain.

Many animals communicate with pheromones

37. Pheromones are _____ that are secreted into the environment.

PHOTORECEPTORS

38. Cephalopod mollusks, arthropods, and vertebrates all have photosensitive pigments called _____ in their eyes.

Invertebrate photoreceptors include eyespots, simple eyes, and compound eyes

39. Some flatworms have _____, which are photoreceptive organs capable of differentiating the intensity of light.

40. The compound eye in insects and crustaceans consists of _____, which collectively form a mosaic image.

Vertebrate eyes form sharp images

41. The tough outer coat of the mammalian eye, called the (a) _____, helps maintain the (b)_____ of the eyeball.

42. The anterior, transparent part of this coat, which allows the entry of light, is called the _____.

43. The viscous fluid located behind the lens of the eye is called _____.

44. The ciliary processes secrete _____ between the cornea and the lens.

45. The amount of light entering the eye is regulated by the _____.

The retina contains light-sensitive rods and cones

46. Rods and cones function as _____.

47. The (a) _____, a specific region of the retina, has the highest density of (b) _____ and therefore is the region of (c)_____.

Light activates rhodopsin

48. Chemically, rhodposin consists of (a) _____ and (b) _____.

49. _____ occurs when enzymes restore rhodopsin to a form in which it can respond to light after a person has been out in bright light.

Color vision depends on three types of cones

50. The three types of cones found in primates are (a) _____, (b) _____, and (c) _____. Each cone is so named because that is the wavelength that is most strongly absorbed. All cones respond to (d) _____.

Integration of visual information begins in the retina

51. Optic nerves cross in the floor of the (a) _____ forming an X-shaped structure called the (b) _____.

BUILDING WORDS

Use combinations of prefixes and suffixes to build words for the definitions that follow.

Prefixes	The Meaning	Suffixes	The Meaning
chemo-	chemical	-cyst	cell
endo-	within	-(t)ory	place for
inter(o)-	between, among	-ile	pertaining to
olfac-	smell	-lith	stone
oto-	ear		
proprio-	one's own		
stat-	standing still		
tact-	to touch		
thermo-	heat, warm		

Prefix	Suffix	Definition
_____	-ceptor	1. Sense organs in muscles, tendons, and joints that enable the animal to perceive the position of its own body parts.
_____	_____	2. Calcium carbonate "stone" in the inner ear of vertebrates.
_____	-lymph	3. Fluid within the semicircular canals of the vertebrate ear.
_____	-receptor	4. A sense organ or sensory cell that responds to chemical stimuli.
_____	-receptor	5. A sensory receptor that provides information about body temperature.
_____	-ceptor	6. A sensory receptor within (among) body organs that helps maintain homeostasis.
_____	_____	7. Receptors that are sensitive to touch.
_____	_____	8. Simplest organs of equilibrium that sense positional changes relative to gravity.
_____	_____	9. Receptor type involved in detecting odorants.

MATCHING

Terms:

a. Chemoreceptor
b. Cochlea
c. Cone
d. Cornea
e. Electroreceptor
f. Exteroceptor
g. Fovea
h. Interoceptor
i. Iris
j. Mechanoreceptor
k. Pacinian corpuscle
l. Rhodopsin
m. Statocyst
n. Thermoreceptor
o. Tympanic membrane

For each of these definitions, select the correct matching term from the list above.

_____ 1. A light-sensitive pigment found in rod cells of the vertebrate eye.

_____ 2. The structure of the inner ear of mammals that contains the auditory receptors.

_____ 3. A sensory receptor that responds to mechanical energy.

_____ 4. The structure of the vertebrate eye that regulates the size of the pupil.

_____ 5. Sensory receptor that provides information about body temperature.

_____ 6. A mechanoreceptor that is sensitive to deep pressure touches.

_____ 7. The "ear drum."

____ 8. The transparent anterior covering of the eye.

____ 9. A sense organ or sensory cell that responds to chemical stimuli.

____10. A conical photoreceptive cell of the retina that is particularly sensitive to bright light, and, by light of various wave lengths, mediates color vision.

____11. Area of the retina with the greatest density of receptor cells.

____12. A gravity receptor found in invertebrates.

MAKING COMPARISONS

Fill in the blanks.

Stimuli	Receptor Classification: Type of Stimuli	Receptor Classification: Location	Example
Light touch	Mechanoreceptor	Exteroceptor	Meissner's corpuscle in skin
Electrical currents in water	#1	#2	Organ in skin of some fish
#3	#4	Exteroceptor	Organ of Corti in birds
Change in internal body temperature	Thermoreceptor	#5	#6
#7	Mechanoreceptor	#8	Muscle spindle in human skeletal muscle
Heat from prey	#9	Exteroceptor	#10
#11	#12	Exteroceptor	Ocelli of flatworms
Food dissolved in saliva	#13	Exteroceptor	#14
Pheromones	#15	#16	Vomeronasal organ in the epithelia of terrestrial vertebrates

MAKING CHOICES

Place your answer(s) in the space provided. Some questions may have more than one correct answer.

____ 1. The process by which sensory receptors change the form of energy of a stimulus to a form of energy that can be transmitted along neurons is called

a. sensory adaptation.   d. receptor potential.

b. conversion.   e. integration.

c. energy transduction.

____ 2. A change in ion distribution that causes a change in the voltage across the membrane of a sensory receptor is a

a. sensory adaptation.   d. receptor potential.

b. conversion.   e. integration.

c. transduction.

_____ 3. If the release of a neurotransmitter from a presynaptic terminal decreases during a constant input of action potentials, _____ is said to have occurred.

    a. sensory adaptation.   d. receptor potential.

    b. conversion.   e. integration.

    c. transduction.

_____ 4. In humans, visual stimuli are interpreted in the

    a. brain.   d. retina.

    b. photoreceptors.   e. optic nerves.

    c. postsynaptic photoganglia.

_____ 5. The state of depolarization or hyperpolarization in a receptor neuron that is caused by a stimulus is the

    a. depolarization potential.   d. action potential.

    b. resting potential.   e. threshold potential.

    c. receptor potential.

_____ 6. Variations in the quality of sound are recognized by the

    a. number of hair cells stimulated.   d. intensity of stimulation.

    b. pattern of hair cells stimulated.   e. amplitude of response.

    c. frequency of nerve impulses.

_____ 7. The membrane at the opening of the inner ear that is in contact with the stapes is in the

    a. tectorial membrane.   d. eardrum.

    b. basilar membrane.   e. round window.

    c. oval window.

_____ 8. Receptors within muscles, tendons, and joints that perceive position and body orientation are

    a. exteroceptors.   d. mechanoreceptors.

    b. proprioceptors.   e. electroreceptors.

    c. interoceptors.

_____ 9. Movement in ligaments is detected by

    a. muscle spindles.   d. proprioceptors.

    b. joint receptors.   e. tonic sense organs.

    c. Golgi tendon organs.

_____ 10. Sense organs that detect changes in pH, osmotic pressure, and temperature within body organs are

    a. exteroceptors.   d. mechanoreceptors.

    b. proprioceptors.   e. electroreceptors.

    c. interoceptors.

_____ 11. Eyespots that detect light but do not form clear images are called

    a. simple eyes.   d. ocelli.

    b. ommatidia.   e. ciliary bodies.

    c. facets.

_____ 12. In the human eye, the "posterior cavity," between the lens and the retina, is filled with

    a. aqueous fluid.   d. the vitreous humor.

    b. aqueous humor.   e. the vitreous body.

    c. rods and cones.

_____ 13. Accommodation involves

    a. changing the shape of the lens.   d. ciliary muscle contractions.

    b. changing the shape of the fovea.   e. redistribution of photoreceptors in the retina.

    c. flattening of the retina.

_____14. Sensory perception

    a. occurs at the level of the receptor. d. occurs when impulses reach the brain.

    b. is influenced by past experiences. e. involves the interpretation of a sensation.

    c. occurs when light hits the retina.

_____15. Thermorecptors are important to species survival in

    a. ticks.   d. snakes.

    b. mosquitoes.   e. fleas.

    c. blood-sucking arthropods.

_____16. Words associated with pain include

    a. nociceptor.   d. free nerve ending.

    b. dopamine.   e. substance P

    c. glutamate.

_____17. _____ are receptors that are sensitive to light touch.

    a. Ruffini endings.   d. Merkel cells.

    b. Pacinian corpuscles.   e. hair cells.

    c. Meissner corpuscles.

_____18. A structural feature common to both a statocyst and a utricle is the presence of

    a. otoliths.   d. calcium carbonate.

    b. hair cells.   e. a gelatinous matrix.

    c. motor nerves.

_____19. Lateral line organs

    a. can be found in frogs.   d. can be found in aquatic organisms.

    b. can be found in snakes.   e. share structural similarities with the cupula of the inner ear.

    c. can be found in lizards.

_____20. The organ of Corti

    a. is located in the labyrinth.   d. rests on the tectorial membrane.

    b. is surrounded by endolymph.   e. is stimulated by mechanical forces.

    c. is stimulated when receptor cells come in contact with the basilar membrane.

_____21. Cranial nerve I

    a. is involved in hearing.   d. connects to the olfactory bulb.

    b. is involved in sight.   e. is also known as the olfactory nerve.

    c. connects to the brain stem.

VISUAL FOUNDATIONS

Color the parts of the illustrations below as indicated. Also label the locations for endolymph, perilymph, tympanic canal, and vestibular canal.

RED ☐ hair cells ORANGE ☐ basilar membrane

GREEN ☐ cilia BROWN ☐ tympanic canal and vestibular canal

YELLOW ☐ cochlear nerve TAN ☐ bone

BLUE ☐ tectorial membrane PINK ☐ cochlear duct

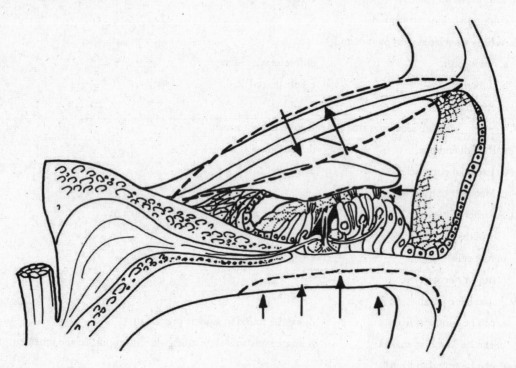

Color the parts of the illustrations below as indicated. Also label the light rays and circle the optic nerve fibers.

RED ☐ rod cell

GREEN ☐ cone cell

ORANGE ☐ bipolar cell

WHITE ☐ ganglion cell

VIOLET ☐ optic nerve fibers

TAN ☐ choroid layer and sclera

PINK ☐ horizontal cell

BLUE ☐ amacrine cell

YELLOW ☐ vitreous body

BROWN ☐ pigmented epithelium

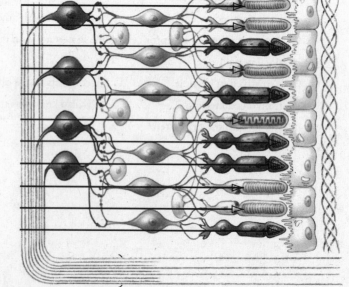

Color the parts of the illustrations below as indicated.

RED	☐	ommatidia
GREEN	☐	facets
YELLOW	☐	optic nerve and ganglion
BLUE	☐	retinular cell
ORANGE	☐	rhabdome
BROWN	☐	iris cell with pigment
TAN	☐	lens
PINK	☐	crystalline cone and stalk

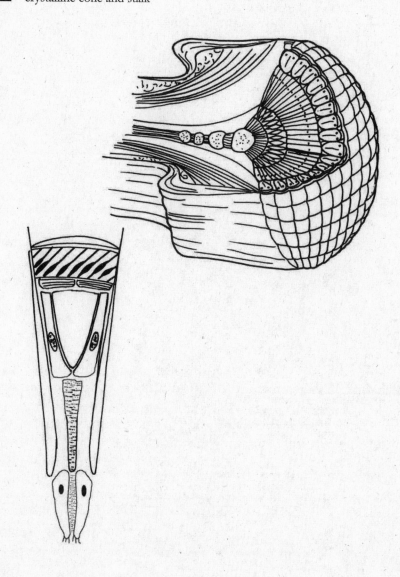

CHAPTER 44

Internal Transport

This chapter discusses how materials get to and from the body cells of animals. Some animals are so small that diffusion alone is effective at transporting materials. Larger animals, however, require a circulatory system. Some invertebrates have an open circulatory system in which blood is pumped from a heart into vessels that have open ends. Blood spills out of the vessels into the body cavity and baths the tissues directly. The blood then passes back into the heart, either directly through openings in the heart (arthropods) or indirectly, passing first through open vessels that lead to the heart (some mollusks). Other animals have a closed circulatory system in which blood flows through a continuous circuit of blood vessels. The walls of the smallest vessels are thin enough to permit exchange of materials between the vessels and the extracellular fluid that baths the tissue cells. A heart can be a muscular organ (vertebrates) or a pulsating vessel (annelids). The muscular heart consists of one or two chambers that receive blood, and one or two that pump blood into the arteries. The lymphatic collects extracellular fluid and returns it to the blood.

REVIEWING CONCEPTS

Fill in the blanks.

INTRODUCTION

1. High-density lipoproteins play a protective role, removing excess _____ from the blood and tissues.

TYPES OF CIRCULATORY SYSTEMS

2. (a) _____, (b) _____,
   (c)_____, (d) _____,
   (e) _____ are animal groups that have no specialized circulatory structures.

3. Some animals have a gastrovascular cavity which serves two different organ functions: It serves as a (a) _____ organ and as a
   (b) _____ organ.

4. The movement and distribution of nutrients and gases or even wastes requires a _____ in which the materials can circulate.

5. Components of a circulatory system include the (a) _____, that is usually pumped by a (b) _____ through a system of (c) _____ or (d) _____.

Many invertebrates have an open circulatory system

6. An open circulatory system consists of a heart connected to (a) _____ vessels from which the blood flows into (b)_____.

7. Blood is called (a) _____ in animals with open circulatory systems because it has been mixed with
   (b)_____ fluid.

8. (a) _____ and (b) _____
_____ have an open circulatory system.

9. Most mollusks have an open circulatory system and a heart that has -
_____(#) of chambers.

10. (a)_____, a blood pigment that imparts a bluish color to
hemolymph in some invertebrates, contains the metal (b) _____.

11. In insects, hemolymph mainly distributes (a) _____ and
(b) _____. Oxygen is supplied to cells through a system of
tubules.

Some invertebrates have a closed circulatory system

12. The three groups of invertebrates having a closed circulatory system are
(a) _____, (b) _____,
and (c) _____.

13. Annelids have a (a)_____ blood vessel that conducts blood anteriorly,
a (b)_____ blood vessel that conducts blood posteriorly, and, in the
anterior part of the worm, (c)_____(#?) pairs of contractile blood vessels
connecting the two.

14. Hemoglobin is found in the _____ of earthworm blood.

Vertebrates have a closed circulatory system

15. All vertebrates have a _____ located heart

16. The vertebrate circulatory system primarily transports
(a)_____ , (b) _____,
(c) _____, and (d) _____.

VERTEBRATE BLOOD

17. Human blood consists of fluid, (a) _____, in which
(b) _____, (c) _____, and
(d) _____ are suspended.

Plasma is the fluid component of blood

18. Plasma is in dynamic equilibrium with (a) _____ fluid, which
bathes cells, and with (b) _____ fluid.

19. The major groups of proteins found in plasma are (a)_____,
a protein involved in blood clotting, (b)_____, some
of which are involved in immunity, and (c)_____.

Red blood cells transport oxygen

20. RBCs are produced in (a) _____, contain the respiratory
pigment (b) _____, and live about (c) _____ days.

21. A deficiency in hemoglobin and/or a decrease in red blood cells are two
situations that can result in a condition known as (a) _____.
Two other causes might be either (b) _____ or
(c) _____.

White blood cells defend the body against disease organisms

22. White blood cells are also known as (a) _____.

23. The two major groups of white blood cells are (a) _____ and (b) _____.

24. The largest white blood cells are called (a) _____. Some of these cells differentiate into (b) _____ which are phagocytic
cells often found in the extracellular space.

24. The three types of granular leukocytes are (a) _____ (b) _____, and (c) _____.

26. Two types of agranular leukocytes are (a) _____ which secrete antibodies, and (b) _____.

Platelets function in blood clotting

27. The blood cells responsible for blood clotting in vertebrates other than mammals are called (a) _____.

28. In mammals, pieces of cells called (b) _____ perform the same function.

VERTEBRATE BLOOD VESSELS

29. (a)_____ carry blood away from the heart; (b)_____ return blood to the heart.

30. The exchange of nutrients and waste products takes place across the thin wall of _____.

31. Blood pressure is affected by constriction and relaxation of smooth muscle associated with the wall of the _____.

EVOLUTION OF THE VERTEBRATE CARDIOVASCULAR SYSTEM

32. Chambers in the heart that pump blood into arteries are called
(a) _____, and chambers that receive blood from veins are called
(b) _____.

33. Fish hearts have (a) _____(#?) atrium(ia) and (b)_____ ventricle(s).

34. The amphibian heart has (a) _____(#?) chambers, (b) _____(#?) atrium(ia) and (c)_____ventricle(s). Oxygen-poor blood returning from the systemic circulation is pumped into the right atrium by the (d) _____.

35. The three groups of animals with a four chambered heart in which the chambers are completely separated are (a) _____,
(b)_____, and (c) _____.

36. Having a heart with four completely separated chambers allows birds and mammals to maintain different blood pressures in the systemic and pulmonary circuits. The pressure in the systemic circulation is _____ (higher/lower) than in the pulmonary circulation.

THE HUMAN HEART

37. The human heart is able to vary the amount of blood it pumps per minute from (a) _____ liters to (b) _____ liters.

38. The human heart is enclosed by a connective tissue sac called the
    (a)_____ which forms a
    (b)_____ sac around the heart that reduces friction when
    the heart beats.

39. The (a)_____ septum separates the two ventricles, and
    the (b)_____ separates the two atria in a four-
    chambered heart.

40. (a)_____(#?) valves control the flow of blood in the human heart. The
    (b) _____ prevents backflow into the
    atria during ventricular contractions. The valve on the right is also known as
    the (c) _____, and the valve on the left is known as
    the (d) _____.
    The (e) _____ "guard" the exits from the heart.

Each heartbeat is initiated by a pacemaker

41. The conduction system of the heart contains a pacemaker called the
    (a) _____, an (b) _____
    which links the atria to the (c) _____,
    which divides, sending a branch into each ventricle.

42. The portion of the cardiac cycle in which contraction occurs is called
    (b) _____, and the portion in which relaxation occurs is called
    (c) _____.

The cardiac cycle consists of alternating periods of contraction and relaxation

43. Of the two main heart sounds, the (a) _____ occurs first and is
    associated with closure of the (b) _____.

44. The (a) _____ sound is produced by the closure of the (b) _____,
    which marks the beginning of ventricular diastole.

The nervous system regulates the heart rate

45. The heart beats independently, but the rate can be regulated by the
    (a) _____ and the (b) _____
    systems.

Stroke volume depends on venous return

46. According to _____, if veins
    deliver more blood to the heart, the heart pumps more blood.

Cardiac output varies with the body's need

47. (a)_____ is the volume of blood pumped by one
    ventricle in one minute. It is determined by multiplying the number of
    ventricular beats per minute times the amount of blood pumped by one
    ventricle per beat, a value referred to as the (b)_____.

48. The cardiac output of a resting adult is about _____.

BLOOD PRESSURE

49. High blood pressure is called (a) _____. It can be
    caused by an (b) _____ (increase or decrease?) in blood volume such
    as frequently occurs with a high dietary intake of (c)_____.

50. The most important factor in determining peripheral resistance to blood flow is the _____.

Blood pressure varies in different blood vessels

51. Flow rate can be maintained in veins at low pressure because they are _____ vessels.

Blood pressure is carefully regulated

52. _____ are sense organs in the walls of some arteries.

53. Sense organs in arteries detect (a) _____ and send the perceived information to (b)_____ in the medulla of the brain.

THE PATTERN OF CIRCULATION

54. A double circulatory system consists of a (a) _____ connecting the heart and lungs, and (b) _____ connecting the heart and body.

The pulmonary circulation oxygenates the blood

55. Pulmonary veins carry oxygen-(a) _____ (rich or poor?) blood to the (b)_____ atrium of the heart.

The systemic circulation delivers blood to the tissues

55. Arteries in the systemic circuit branch from the (a) _____, the largest artery in the body, and serve major body areas. For example, the carotid arteries feed the (b) _____, subclavian arteries supply the (c)_____, and iliac arteries feed the (d)_____.

56. Veins returning blood from the head and neck empty into the large (a) _____ vein, while venous return from the lower body empties into the (b) _____ vein.

THE LYMPHATIC SYSTEM

57. The lymphatic system is considered to be an accessory (a) _____ in vertebrates.

58. The lymphatic system has three main functions. It collects (a) _____ to be returned to the blood, it defends the body by initiating (b) _____ and it surrounds the digestive tract where it (c) _____.

The lymphatic system consists of lymphatic vessels and lymph tissue

59. The fluid contained within lymphatic vessels is called _____.

60. Lymph tissue is organized into masses called _____.

61. The (a) _____, (b) _____, and (c) _____ are organs that consist mainly of lymph tissue and are consider to be part of the lymphatic system.

62. Lymphatic vessels from all over the body join the circulatory system at the base of the (a) _____ by way of ducts: the

(b) _____ duct on the left and the

(c) _____ duct on the right.

The lymphatic system plays an important role in fluid homeostasis

63. The tendency for plasma to leave the blood at the arterial end of a capillary and enter the interstitial fluid is referred to as the

(a) _____pressure. Three forces contribute to this pressure. The main occurs as the result of blood pressure against the capillary wall and is the (b) _____ pressure. This pressure is augmented by the (c) _____ pressure of the interstitial fluid. Principle opposing force is the (d) _____ pressure of the blood.

CARDIOVASCULAR DISEASE

64. (a) _____ is a disease in which the walls of affected arteries are damaged, inflamed, and narrowed as a result of _____ deposits.

65. (a) _____ is said to occur when tissues lack an adequate blood supply as the result of a decrease in the diameter or blockage of a blood vessel. If this situation occurs in the heart chest pains may occur during exercise. This pain is a characteristic of a condition called

(b) _____.

66. A clot that forms in a blood vessel is called a _____.

67. (a) _____ is the insertion of a small balloon into a blocked coronary artery in order to insert a (b) _____ which will be left in the artery to hold it open.

BUILDING WORDS

Use combinations of prefixes and suffixes to build words for the definitions that follow.

Prefixes	The Meaning	Suffixes	The Meaning
baro-	pressure	-ary	pertaining to
capill-	hairy	-cardium	heart
erythro-	red	-coel	cavity
hemo-	blood	-cyte	cell
leuk(o)-	white (without color)	-emia	condition of
neutro-	neutral	-(i)cle	small
peri-	about, around, beyond	-lunar	moon
semi-	half	-phil	loving, friendly, lover
sin-	hollow	-us	thing
vaso-	vessel		
ventr-	belly		

Prefix	Suffix	Definition
_____	_____	1. The blood cavity that comprises the open circulatory system of arthropods and some mollusks.

_____ -cyanin 2. The copper-containing blood pigment in some mollusks and arthropods.

_____ 3. Red blood cell.

_____ _____ 4. General term for all of the body's white blood cells.

_____ _____ 5. The principal phagocytic cell in the blood that has an affinity for neutral dyes.

eosino- _____ 6. WBC with granules that have an affinity for eosin.

baso- _____ 7. WBC with granules that have an affinity for basic dyes.

_____ -emia 8. A form of cancer in which WBCs multiply rapidly within the bone marrow.

_____ -constriction 9. Constriction of a blood vessel.

_____ -dilation 10. Relaxation of a blood vessel.

_____ _____ 11. The tough connective tissue sac around the heart.

_____ _____ 12. Shaped like a half-moon.

_____ -receptor 13. Pressure receptor.

_____ _____ 14. A fluid filled space such as that found in invertebrates which is filled with hemolymph.

_____ _____ 15. The cavity in the mammalian heart from which blood enters the systemic circulation.

_____ _____ 16. The smallest blood vessels in a closed circulatory system.

_____ _____ 17. A cancer caused by the rapid replication of white blood cells in the bone marrow.

MATCHING

Terms:

a. Aorta
b. Arterioles
c. Artery
d. Atrium
e. AV node
f. Blood pressure

g. Interstitial fluid
h. Lymph
i. Lymphatic system
j. Mitral valve
k. Parasympathetic
l. Plasma

m. Pulse
n. SA node
o. Systole
p. Sympathetic
q. Vein

For each of these definitions, select the correct matching term from the list above.

_____ 1. Network of fluid-carrying vessels and associated organs that participate in immunity and in the return of tissue fluid to the main circulation.

_____ 2. The heart valve located between the left atrium and the left ventricle.

_____ 3. The fluid in lymphatic vessels.

_____ 4. The fluid portion of the blood consisting of a pale, yellowish fluid.

_____ 5. A contracting chamber of the heart that forces blood into the ventricle.

_____ 6. The rhythmic expansion and recoil of an artery that may be felt with the finger.

_____ 7. The smallest arteries, which carry blood to the capillary beds.

_____ 8. Contraction of the heart muscle, especially that of the ventricle, during which the heart pumps blood into arteries.

_____ 9. Largest blood vessel in the body through which blood leaves the heart and enters the systemic circulation.

_____10. A blood vessel that carries blood from the tissues toward the heart.

_____11. Nerves that release acetylcholine to slow the heart rate.

_____12. In the electrical conductance system of the heart, it connects to the AV bundle.

MAKING COMPARISONS

Fill in the blanks.

Specific Protein or Cell Type	Blood Component	Function
Fibrinogen	Plasma	Involved in blood clotting
#1	#2	Transport fats, cholesterol
Erythrocytes	#3	#4
Thrombocytes	Cell Component	#5
Monocytes	#6	#7
#8	#9	Help regulate the distribution of fluid between plasma and interstitial fluid
Neutrophils	Cell Component	#10

MAKING CHOICES

Place your answer(s) in the space provided. Some questions may have more than one correct answer.

_____ 1. The pericardium surrounds

    a. atria only.
    b. ventricles only.
    c. blood vessels.
    d. the heart.
    e. clusters of some blood cells.

_____ 2. Very small vessels that deliver oxygenated blood to capillaries are called

    a. arterioles.
    b. venules.
    c. arteries.
    d. veins.
    e. setum vessels.

_____ 3. One would expect an ostrich to have a heart most like a

    a. lizard.
    b. snail.
    c. earthworm.
    d. human.
    e. guppy.

_____ 4. Fibrinogen is
    a. a plasma lipid.
    b. in serum.
    c. a protein.
    d. a gamma globulin.
    e. involved in clotting.

_____ 5. Three chambered hearts generally consist of the following number of atria/ventricles:
    a. 2/1.
    b. 1/2.
    c. 1/1 and an accessory chamber.
    d. 0/3.
    e. 3/0.

_____ 6. Blood enters the right atrium from the
    a. right ventricle.
    b. inferior vena cava.
    c. superior vena cava.
    d. pulmonary artery.
    e. jugular vein.

_____ 7. In general, blood circulates through vessels in the following order:
    a. veins, capillaries, arteries, arterioles.
    b. veins, capillaries, arterioles, arteries.
    c. veins, arteries, arterioles, capillaries.
    d. arteries, arterioles, capillaries, veins.
    e. capillaries, arteries, arterioles, veins.

_____ 8. Prothrombin
    a. is produced in liver.
    b. is a precursor to thrombin.
    c. catalyzes conversion of fibrinogen to fibrin.
    d. requires vitamin K for its production.
    e. is a toxic by-product of metabolism.

_____ 9. Erythrocytes are
    a. also called RBCs.
    b. spherical.
    c. produced in bone marrow.
    d. one kind of leucocyte.
    e. carriers of oxygen.

_____10. Monocytes
    a. are RBCs.
    b. are WBCs.
    c. are produced in the spleen.
    d. are large cells.
    e. can become macrophages.

_____11. The blood-filled cavity of an open circulatory system is called the
    a. lymphocoel.
    b. hemocoel.
    c. atracoel.
    d. sinus.
    e. ostium.

_____12. Pulmonary arteries carry blood that is
    a. low in oxygen.
    b. low in CO_2.
    c. high in oxygen.
    d. high in CO_2.
    e. on its way to the lungs.

_____13. The semilunar valve(s) is/are found between a ventricle and
    a. an atrium.
    b. another ventricle.
    c. the aorta.
    d. the pulmonary vein.
    e. the pulmonary artery.

_____14. Phenomena that tend to be associated with hypertension include
    a. obesity.
    b. increased vascular resistance.
    c. increased workload on the heart.
    d. decrease in ventricular size.
    e. deteriorating heart function.

_____15. A heart murmur may result when
   a. the heart is punctured.
   b. systole is too high.
   c. a semilunar valve is injured.
   d. diastole is too low.
   e. diastole is too low or too high.

_____16. Ventricles receive stimuli for contraction directly from the
   a. SA node.
   b. AV node.
   c. Purkinje fibers.
   d. atria.
   e. intercalated disks.

_____17. If a person's heart rate is 65 and one ventricle pumps 75 ml with each contraction, what is that person's cardiac output in liters?
   a. 3.575
   b. 4.000
   c. 4.525
   d. 4.875
   e. 5.350

_____18. An open circulatory system will
   a. have a pump to move the blood.
   b. have a capillary bed.
   c. be found in cephalopods.
   d. circulate fluid called hemolymph.
   e. have arteries.

_____19. Hemocyanin
   a. contains copper.
   b. is found in earthworms.
   c. contains iron.
   d. circulated oxygen in flatworms.
   e. transports oxygen in arthropods.

_____20. A closed circulatory system is found in
   a. some arthropods.
   b. in annelids.
   c. is some reptiles.
   d. some mollusks.
   e. in echinoderms.

_____21. Functions of a closed circulatory system include
   a. maintaining fluid balance.
   b. transporting carbon dioxide.
   c. transporting oxygen to the gills from the tissues.
   d. transporting hormones.
   e. transporting red blood cells.

_____22. Plasma consists of about _____% water and _____% proteins.
   a. 95/5
   b. 92/7
   c. 97/2
   d. 85/15
   e. 98/1

_____23. Serum lacks
   a. globulins.
   b. hormones.
   c. albumins.
   d. antibodies.
   e. clotting proteins.

_____24. The cellular group granulocytes includes
   a. lymphocytes.
   b. basophils.
   c. neutrophils.
   d. monocytes.
   e. eosinophils.

_____25. Red blood cell(s)

        a. are produced in yellow marrow.   d. are red because they contain hemocyanin.

        b. last about 100 days.   e. formation is regulated by erythropoietin.

        c. are also called leukocytes.

_____26. In a normal individual the greatest number of white blood cells are

        a. lymphocytes.   d. monocytes.

        b. neutrophils.   e. eosinophils.

        c. basophils.

_____27. In a normal individual the white blood cells present in the lowest number are

        a. lymphocytes.   d. monocytes.

        b. neutrophils.   e. eosinophils.

        c. basophils.

_____28. For a normal blood clot to form

        a. vitamin K is required.   d. calcium ions are required.

        b. magnesium ions are required.   e. thrombin must be converted to prothrombin.

        c. fibrinogen is converted to soluble fibrin.

_____29. A closed circulatory system with two completely separated circuits

        a. provides a higher pressure to the pulmonary circuit.   d. is found in some amphibians.

        b. provides more rapid deliver of oxygen to the tissues.   e. is found in annelids.

        c. allows birds to have a high metabolic rate.

_____30. Parasympathetic neurons affect heart rate by

        a. opening calcium channels.   d. releasing norepinephrine.

        b. activating a G protein.   e. opening potassium channels.

        c. by activating a protein kinase.

_____31. An increase in blood pressure

        a. is reversed by the hormone rennin.   d. causes signals to be sent to centers in the pons.

        b. is reversed by angiotensin II.   e. is detected by baroreceptors.

        c. is reversed by inhibitory nerve impulses sent to sympathetic nerves.

_____32. Net filtration will occur in a capillary bed

        a. if blood hydrostatic pressure exceeds blood osmotic pressure.

        b. if interstitial osmotic pressure exceeds blood osmotic pressure.

        c. if blood osmotic pressure exceeds interstitial osmotic pressure.

        d. if blood osmotic pressure exceeds blood hydrostatic pressure.

        e. None of these conditions would result in net filtration in the capillary bed.

VISUAL FOUNDATIONS

Color the parts of the illustration below as indicated. Also circle the capillary bed.

RED	☐	artery and arrows indicating flow of blood away from the heart
PINK	☐	arteriole
VIOLET	☐	capillaries
PURPLE	☐	venuole
BLUE	☐	vein and arrows indicating flow of blood to the heart
GREEN	☐	lymphatic, lymph capillaries
YELLOW	☐	lymph node

Color the parts of the illustration below as indicated. Label artery, vein, and capillary

RED ☐ smooth muscle
GREEN ☐ outer coat (connective tissue)
YELLOW ☐ endothelium

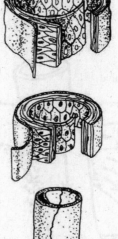

Color the parts of the illustration below as indicated.

RED ☐ aorta and pulmonary artery
GREEN ☐ ventricle
YELLOW ☐ atrium
BLUE ☐ vein from body and pulmonary vein

ORANGE ☐ valve
BROWN ☐ partition
TAN ☐ conus
PINK ☐ sinus venosus

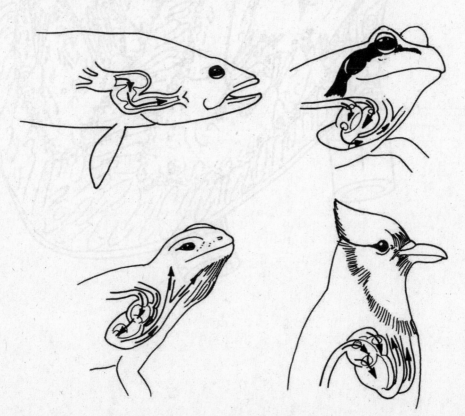

Color the parts of the illustration below as indicated. Also label interventricular septum, superior vena cava, inferior vena cava, pulmonary veins, aorta, and pulmonary artery.

RED ☐ chordae tendineae
GREEN ☐ tricuspid valve
YELLOW ☐ pulmonary semilunar valve
BLUE ☐ mitral valve
ORANGE ☐ aortic semilunar valve
BROWN ☐ papillary muscle
TAN ☐ atrium
PINK ☐ ventricle

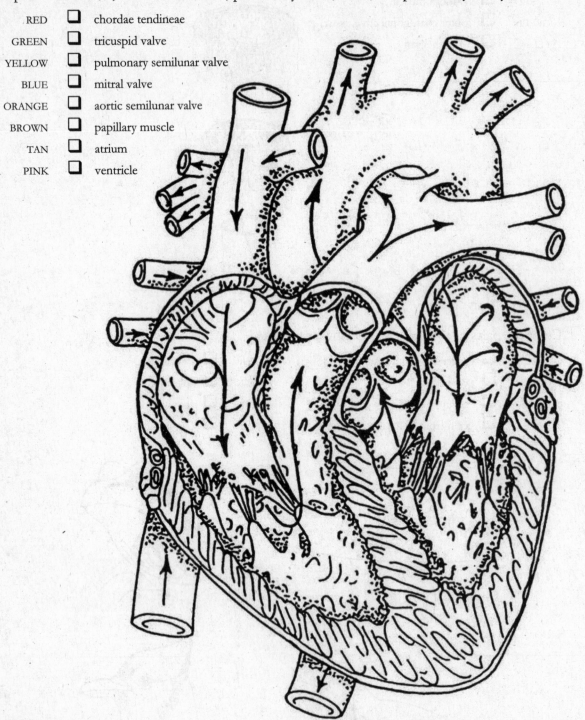

CHAPTER 45

The Immune System: Internal Defense

All animals have the ability to prevent disease-causing microorganisms from entering their body. Because these external defense mechanisms sometimes fail, most animals have developed internal defense mechanisms as well. Before an animal can attack an invader, it must be able to recognize its presence and to distinguish between its own cells and the invader. Whereas most animals are capable of nonspecific responses, such as phagocytosis, it is primarily vertebrates that are capable of specific responses, such as the production of antibodies that target specific pathogens. This type of internal defense is characterized by a more rapid and intense response the second time the organism is exposed to the pathogen. Resistance to many pathogens can be induced artificially by either the injection of the pathogen in a weakened, killed, or otherwise altered state, or the injection of antibodies produced by another person or animal. The immune system is constantly surveying the body for virally-infected cells and tumor cells. It destroys them whenever they arise; failure results in cancer. This same surveillance system is responsible for graft rejection by reading the foreign cells as nonself cells and destroying them. Sometimes the system falsely reads self cells as nonself and destroys them, resulting in an autoimmune disease. Allergies are another example of abnormal immune responses. There are also times that pathogens evade the immune system and cause disease.

REVIEWING CONCEPTS
Fill in the blanks.

INTRODUCTION

1. _____ is the study of internal defense systems.
2. Immune cells and tissues are scattered throughout the body and it is important that they are able to _____ one another.

EVOLUTION OF IMMUNE RESPONSES

3. An immune response is a two-step process that involves
   _____ and reacting to them.
4. There are the two main types of immune responses
   (a) _____,
   and (b) _____.
5. A substance that the immune system specifically recognizes as foreign is called
   an _____.
6. _____ are highly specific proteins that recognize and
   bind to specific antigens.

Invertebrates launch nonspecific immune responses

7. All animals species studied have _____, cells that move
   throughout the body to protect them from foreign matter and bacteria.
8. Annelids are the only invertebrates known to have
   _____ cells, a type of cell that destroys tumor
   cells and virally infected cells.

9. All animals produce an _____
   response when they have an infection.

10. The primary immune response in invertebrates is a _____
   response.

Vertebrates launch nonspecific and specific immune responses

11. The first specific immune response may have evolved in

   _____.

NONSPECIFIC IMMUNE RESPONSES

12. Nonspecific immune responses are also referred to as

   _____.

Physical barriers prevent most pathogens from entering the body

13. An animal's first line of defense is its (a) _____
   which, by being physically intact, blocks the entry of

   (b) _____.

14. _____ are antimicrobial peptides found in saliva.
   They are toxic to bacteria and fungi.

15. _____ is found in tears and it attacks the cell walls of
   gram-positive bacteria.

Pattern recognition receptors activate nonspecific immune responses

16. Bacteria and other microbes have certain proteins and other molecules that are
   not present on the cells of humans and other mammals. These chemical
   compounds that stimulate a nonspecific immune response are called

   _____.

Cells of the nonspecific immune system destroy pathogens and produce chemicals

17. _____ are the main phagocytes in the body.

18. _____ cells are active against tumor cells and
   virally infected cells.

19. (a) _____ cells are found in all epithelial tissues that come in
   contact with the environment. They produce (b) _____
   which are antiviral cytokines.

Cytokines and complement mediate immune responses

20. Two major groups of molecules that are critical in the nonspecific immune
   response are (a) _____ which perform regulatory functions
   and are important signaling molecules and (b) _____
   which consists of over 20 proteins present in plasma and other body fluids.

21. In response to gram-negative bacteria, other pathogens, and some tumor
   cells, _____ stimulates immune cells to
   initiate an inflammatory response.

22. _____ are cytokines that regulate interactions
   between white blood cells such as lymphocytes and macrophages.

23. _____ are cytokines that act as signaling molecules to attract, activate, and direct the movement of cells of the immune system.

Inflammation is a protective response

24. The inflammatory response includes three main processes. These are

   (a) _____,

   (b) _____,

   and (c) _____.

SPECIFIC IMMUNE RESPONSES

Many types of cells are involved in specific immune responses

25. The two main groups of cells that participate in the specific immune response are (a) _____ and (b) _____.

26. The two types of lymphocytes that function in specific immunity are

   (a) _____ which are responsible for cell-mediated immunity and

   (b) _____ which are responsible for antibody-mediated immunity.

27. All lymphocytes develop from stem cells in the _____.

28. Two main types of T cells are: the (a) _____ T cells, or killer T cells, that recognize and destroy cells with foreign antigens on their surfaces, and the (b) _____ T cells that secrete cytokines that activate B cells and macrophages.

29. Within the thymus, T cells become _____, that is capable of immunological response.

30. The three types of antigen-presenting cells are (a) _____,

   (b) _____ and (c) _____.

   When they are involved in an immune response, antigen presenting cells display (d) _____ and

   (e) _____.

The major histocompatibility complex is responsible for recognition of self

31. The MHC genes encode _____, or self antigens, that differ in chemical structure, function, and tissue distribution. These antigens could be considered to be your biochemical

   _____.

32. Class I MHC can be found on the surface of (a) _____ cells and class II MHC are found in the surface of (b) _____.

CELL-MEDIATED IMMUNITY

33. T-cells do not recognize antigens unless they are

   _____.

ANTIBODY-MEDIATED IMMUNITY

34. (a) _____ cells are responsible for antibody-mediated immunity. When these cells are activated and divide by mitosis they give rise to a (b) _____ of cells.

A typical antibody consists of four polypeptide chains

35. Antibodies are also called (a) _____, abbreviated

as (b)_____. These molecules "recognize" specific amino acid sequences on antigens called (c) _____. The antibody molecule is Y-shaped, the two arms functioning as (d) _____.

Antibodies are grouped in five classes

36. Antibodies are grouped into five classes, which are abbreviated as

    (a) _____, (b) _____, (c) _____, (d) _____, and

    (e) _____.

37. In humans, about 75% of the circulating antibodies are (a) _____.

    (b)_____ predominate in secretions, (c) _____ is an important immunoglobulin on the B cell surface, and (d) _____ mediates the release of histamine from mast cells.

The immune system responds to millions of different antigens

38. _____ is the process whereby a lymphocyte responds to a specific antigen by repeatedly dividing, giving rise to a clone of cells with identical receptors.

Monoclonal antibodies are highly specific

39. Monoclonal antibodies are (a) _____ antibodies produced by cells from the clone of (b) _____.

40. Each monoclonal antibody is specific for a single

    _____.

Immunological memory is responsible for long-term immunity

41. Memory B cells have a gene that prevent _____ which is programmed cell death that occurs in plasma cells.

42. The first exposure to an antigen stimulates a (a) _____ response. The principal antibody synthesized in this initial response is (b)_____.

43. A second exposure to an antigen evokes a (a) _____response, which is more rapid and more intense than that in the initial response. The principal antibody synthesized in the second exposure is (b) _____.

44. Active immunity can be either (a) _____ or

    (b) _____ induced.

45. Passive immunity is characterized by the fact that its effects are only

    _____.

46. Passive immunity involves the injection of (a) _____ antibodies. As a result, there will be no (b) _____ cells produced.

47. Babies who are breast-fed receive _____ antibodies in the milk.

Cancer cells evade the immune system

48. Cancer cell antigens are classified according to their

    (a) _____ and (b) _____.

49. _____ inhibitors slow the development of blood vessels which helps to slow the growth of tumors.

Immunodeficiency disease can be inherited or acquired

50. The leading cause of acquired immunodeficiency in children is _____.

HIV is the major cause of acquired immunodeficiency in adults

51. HIV typically enters the body through the epithelial _____.

52. HIV enters the (a) _____ cell and is (b) _____ to form DNA.

In an autoimmune disease, the body attacks its own tissues

53. _____ is the ability to recognize self and is established during lymphocyte development.

54. A lymphocyte that has the potential to be _____ may launch an immune response against self tissues.

Rh incompatibility can result in hypersensitivity

55. Rh incompatibility occurs when individuals produce antibodies against _____.

56. Rh incompatibility can cause _____, which destroys red blood cells in a fetus and may cause death

Allergic reactions are directed against ordinary environmental antigens

57. In an allergic response, an allergen stimulates production of _____ type immunoglobulins.

58. In allergic asthma, (a) _____ cells release mediators that cause the smooth muscle of the airways to (b) _____.

Graft rejection is an immune response against transplanted tissue

59. Graft rejection occurs because the _____ of the host and graft tissue are not compatible.

BUILDING WORDS

Use combinations of prefixes and suffixes to build words for the definitions that follow.

Prefixes	The Meaning	Suffixes	The Meaning
anti-	against, opposite of	-cyte	cell
auto-	self, same	-ment	state of being
comple-	to fill up		
lyso-	loosening, decomposition		
mono-	alone, single, one		
Phag(o)-	to eat		

Prefix	Suffix	Definition
_____	-body	1. A specific protein that acts against pathogens and helps destroy them.
_____	-histamine	2. A drug that acts against (blocks) the effects of histamine.
lympho-	_____	3. A white blood cell strategically positioned in the lymphoid tissue; the main "warrior" in specific immune responses.

_____ -clonal   4. An adjective pertaining to a single clone of cells.

_____ -immune   5. An adjective pertaining to the situation wherein the body reacts immunologically against its own tissues (against "self").

_____ -zyme   6. An enzyme that attacks and degrades the cell walls of gram-positive bacteria.

_____ _____   7. A cell that engulfs foreign antigens.

_____ _____   8. A group of proteins that works with other defensive response.

MATCHING

Terms:

a.	Allergen	g.	Immune response	m.	Memory cell
b.	Antigen	h.	Immunoglobulin	n.	Passive
c.	Complement	i.	Innate	o.	Pathogen
d.	Cytokine	j.	Interferon	p.	Plasma cell
e.	Dendritic	k.	Interleukin	q.	T cell
f.	Histamine	l.	Mast cell	r.	Thymus gland

For each of these definitions, select the correct matching term from the list above.

_____ 1. Any substance capable of stimulating an immune response; usually a protein or large carbohydrate that is foreign to the body.

_____ 2. A gland that functions as part of the lymphatic system.

_____ 3. Substance released from mast cells that is involved in allergic and inflammatory reactions.

_____ 4. A type of white blood cell responsible for cell-mediated immunity.

_____ 5. The process of recognizing foreign and dangerous macromolecules and responding to eliminate them.

_____ 6. A protein produced by animal cells when challenged by a virus.

_____ 7. Type of lymphocyte that secretes antibodies.

_____ 8. An organism capable of producing disease.

_____ 9. A type of cell found in connective tissue; contains histamine and is important in allergic reactions.

_____10. Temporary immunity derived from the immunoglobulins of another organism.

_____11. One type of antigen presenting cells.

_____12. Signaling molecule that may act as either an autocrine or a paracrine agent.

_____13. Nonspecific immune responses.

MAKING COMPARISONS

Fill in the blanks.

Cells	Nonspecific Immune Response	Specific Immune Response: Antibody-Mediated Immunity	Specific Immune Response: Cell-Mediated Immunity
B-lymphocytes	None	Activated by helper T cells and macrophages; multiply into a clone	None
#1	None	Secrete specific antibodies	None
#2	None	Long-term immunity	None
#3	None	None	Activated by helper T cells and macrophages; multiply into a clone
#4	None	Involved in B cell activation	Involved in T cell activation
#5	None	None	Chemically destroy cancer cells, foreign tissue grafts, and cells infected with viruses
#6	None	None	Long-term immunity
#7	Kill virus-infected cells & tumor cells	Kill virus-infected cells and tumor cells	None
#8	Phagocytic, destroy bacteria	Antigen-presenting cells, activate helper T cells	Antigen-presenting cells; stimulates cloning
#9	Phagocytic, destroy bacteria	None	None

MAKING CHOICES

Place your answer(s) in the space provided. Some questions may have more than one correct answer.

____ 1. T-cell receptors
   a. bind antigens.
   b. have no known function.
   c. are identical on all T cells.
   d. are found on killer T cells.
   e. stimulate antibody production.

____ 2. Complement
   a. is a system of several proteins.
   b. is highly antigen-specific.
   c. is stimulated into action by an antibody-antigen complex.
   d. helps destroy pathogens.
   e. is an antibody.

____ 3. The first cells to be involved in a second response to the same antigen are
   a. killer T cells.
   b. memory cells.
   c. plasma cells.
   d. macrophages.
   e. helper T cells.

____ 4. Active immunity can be artificially induced by

    a. transfusions.

    b. injecting vaccines.

    c. passing maternal antibodies to a fetus.

    d. injecting gamma globulin.

    e. stimulating macrophage growth.

____ 5. B cells

    a. are granular.

    b. are lymphocytes.

    c. clone after contacting a targeted antigen.

    d. are derived from plasma cells.

    e. include many antigen-binding forms.

____ 6. Which of the following is true of AIDS?

    a. HIV infects helper T cells.

    b. AIDS is not spread by casual contact.

    c. Both heterosexuals and homosexuals are at risk.

    d. HIV can be transmitted by sharing needles.

    e. There is currently no cure for AIDS.

____ 7. T cells

    a. are lymphocytes.

    b. are called LGLs.

    c. are also called T lymphocytes.

    d. are platelets

    e. are involved in cell-mediated immunity.

____ 8. An allergic reaction involves

    a. killer T cells.

    b. mast cells.

    c. interaction between an allergen and mast cells.

    d. production of IgE.

    e. histocompatibility antigens.

____ 9. Which of the following is/are true of Ig?

    a. They are antibodies.

    b. They contain a C region.

    c. They are produced in response to specific antigens.

    d. They contain an antigenic determinant.

    e. They are also called immunoglobulins.

____ 10. Nonspecific immune responses in vertebrates include

    a. skin.

    b. acid secretions.

    c. inflammation.

    d. phagocytes.

    e. antibody-mediated immunity.

____ 11. The histocompatibility complex is

    a. found in cell nuclei.

    b. called HLA in humans.

    c. different in each individual.

    d. the same in individuals comprising a species.

    e. a group of closely linked genes.

____ 12. Pathogens include

    a. viruses.

    b.bacteria.

    c. fungi.

    d. worms.

    e. protests.

____ 13. The most powerful antigens are

    a. lipids.

    b. polysaccharides.

    c. proteins.

    d. pathogens.

    e. bacteria.

____ 14. Animals that have nonspecific imuunity include

    a. annelids.

    b. humans.

    c. cnidarians.

    d. grasshoppers.

    e. mollusks.

_____15. The most ancient animals to have differentiated white blood cells are

      a. echinoderms.   d. jawed vertebrates.

      b. annelids.   e. tunicates.

      c. arthropods.

_____16. A specific immune response is thought to have evolved first in

      a. jawed vertebrates.   d. primates.

      b. annelids.   e. nematodes.

      c. mollusks.

_____17. Macrophages develop from

      a. neutrophils.   d. basophils.

      b. eosinophils.   e. dendritic cells.

      c. monocytes.

_____18. NK cells

      a. have their origin in bone marrow.   d. are agranular lymphocytes.

      b. will destroy abnormal body cells.   e. use nonspecific processes to destroy target cells.

      c. use specific processes to destroy target cells.

_____19. Dendritic cells can be found in

      a. skin.   d. subcutaneous tissue.

      b. urinary epithelium.   e. the epithelium lining the airways.

      c. digestive epithelium.

_____20. Cytokines and complement

      a. stimulate cell growth and repair.   d. stimulate white blood cells.

      b. lyse bacteria and viruses.   e. are involved in nonspecific immune responses.

      c. have been associated with the development of atherosclerosis.

_____21. Cell-mediated immunity involves

      a. helper T cells.   d. B cells.

      b. cytokines.   e. the antigen-MHC class I complex.

      c. the antigen-MHC class II complex.

_____22. Helper T cells

      a. express the CD_8 molecule on their surface.   d. can be antigen presenting cells.

      b. are also known as PAMP cells.   e. activate T_c cells.

      c. secrete cytokines.

_____23. Antibodies

      a. inactivate antigens.   d. may stimulate phagocytosis.

      b. consist of four polypeptide chains.   e. serve as the B cell receptor for antigen.

      c. are present in tears.

_____24. The Rh system

        a. was first identified in monkeys.

        b. results in individuals naturally producing antibodies against the foreign blood type.

        c. identifies Rh negative individuals as being homozygous recessive.

        d. includes antigen D on the surface of red blood cells.

        e. includes over 40 antigens.

VISUAL FOUNDATIONS

Color the parts of the illustration below as indicated. Label variable region and constant region, and circle the antigen-antibody complex.

RED ☐ antigenic determinants

GREEN ☐ antigen

YELLOW ☐ antibody-heavy chain

BLUE ☐ binding sites

ORANGE ☐ antibody-light chain

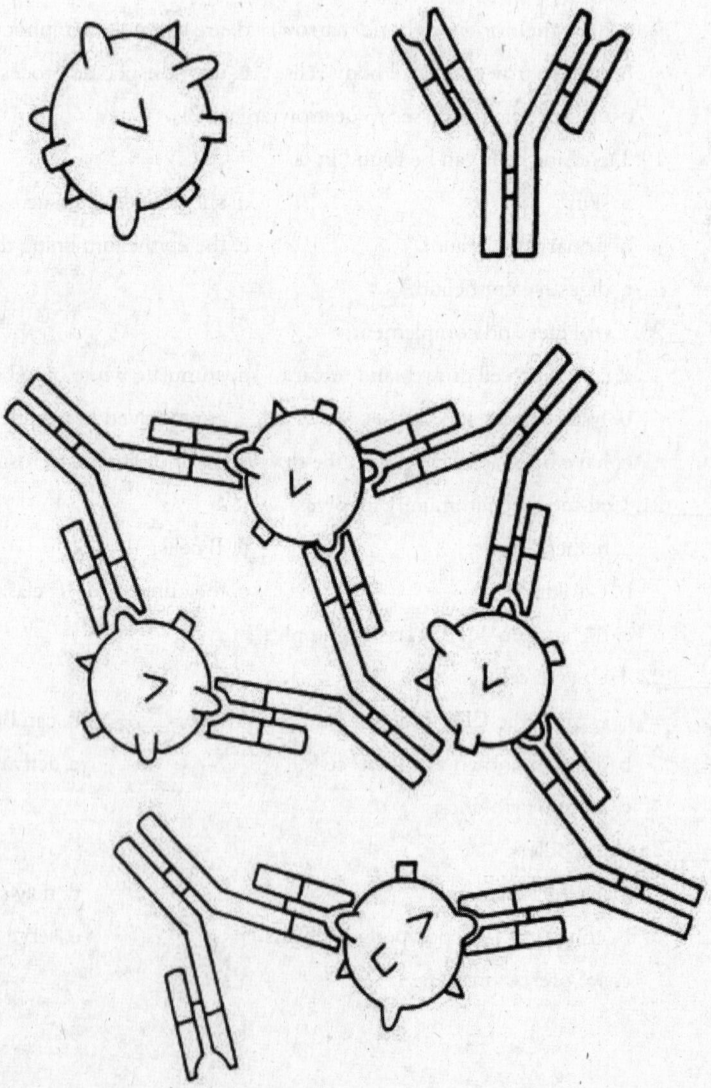

Color the parts of the illustration below as indicated. Label cell-mediated immunity and antibody-mediated immunity. Also label antigen presentation, cooperation, and migration to lymph nodes.

RED ☐ bone marrow
BROWN ☐ thymus
YELLOW ☐ T cell
ORANGE ☐ helper T cell
GREY ☐ cytotoxic T cell
BLUE ☐ memory cell
VIOLET ☐ B cell
PINK ☐ plasma cell
BROWN ☐ natural killer cell
TAN ☐ monocyte, macrophage, and dendritic cell

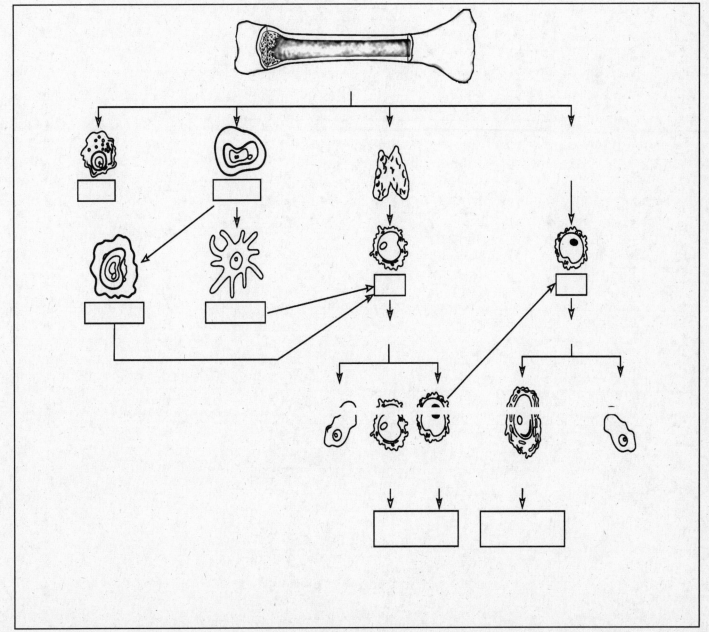

Gas Exchange

In an earlier chapter you studied cellular respiration, the intracellular process that uses oxygen as the final electron acceptor in the mitochondrial electron transport chain and produces carbon dioxide as a waste product. This chapter discusses organismic respiration, that is, how animal cells take up oxygen from the environment in order for it to be made available to the mitochondria, and how the carbon dioxide generated in the mitochondria gets out of the animal body. Oxygen exchange with air is more efficient than with water. This is because air contains a higher molecular concentration of oxygen than water, and oxygen diffuses more rapidly in air than in water. In small, aquatic organisms, gases diffuse directly between the environment and all body cells. In larger, more complex organisms, specialized respiratory structures, such as tracheal tubes, gills, and lungs increase the surface area for gas exchange to occur. Respiratory pigments, such as hemoglobin and hemocyanin, greatly increase respiratory efficiency by increasing the capacity of blood to transport oxygen.

REVIEWING CONCEPTS

Fill in the blanks.

INTRODUCTION

1. The exchange of gases between an organism and the medium in which it lives is called _____.

ADAPTATIONS FOR GAS EXCHANGE IN AIR OR WATER

2. Air contains (a) _____ (more or less?) oxygen than water, and oxygen diffuses (b) _____ (faster or slower?) in air than in water.

3. A mammal uses almost _____ (%) less energy to breathe on land than a fish does to breathe in water.

TYPES OF RESPIRATORY SURFACES

4. All respiratory surfaces need to be _____.

5. The four main types of respiratory surfaces are (a) _____, (b) _____, (c) _____, and (d) _____.

6. Most respiratory systems actively move air or water over their respiratory surfaces, a process known as _____.

The body surface may be adapted for gas exchange

7. In aquatic animals, the body surface is kept moist by the (a) _____ and in terrestrial organisms the respiratory surface is kept moist by (b)_____.

Tracheal tube systems deliver air directly to the cells

8. Arthropods have a respiratory system consisting of a network of

(a) _____; air enters the system through small openings called (b) _____.

9. The respiratory network of an arthropod terminates in small fluid-filled _____ where gas exchange takes place.

Gills are respiratory surfaces in many aquatic animals

10. Sea stars have (a) _____ gills. Gas exchange occurs between the (b) _____ and the (c) _____.

11. In sea stars and bivalves water is moved over the gills by the movement of the _____.

12. The gills of bony fish have many _____ that extend out into the water, within which are numerous capillaries.

13. In bony fish, blood flows through respiratory structures in a direction opposite to water movement, an arrangement called the (a)_____. This system maximizes gas exchange because blood is always coming in contact with water that has a (b) _____ amount of oxygen.

Terrestrial vertebrates exchange gases through lungs

14. Lungs are respiratory structures that develop as ingrowths of the (a)_____ or from the wall of a (b)_____ such as the pharynx.

15. The _____ of spiders are within an inpocketing of the abdominal wall.

16. Modern bony fishes almost all have a homologous structure, a _____ that allows them to control their buoyancy.

17. Birds have the most efficient _____ of any living vertebrate group.

18. In birds, air flows in (a) _____ and is renewed during (b) _____ cycles of ventilation. The air flow first to the (c) _____ and from there is flows into (d) _____. From here the air flows into the (e) _____ and from there out of the body.

THE MAMMALIAN RESPIRATORY SYSTEM

The airway conducts air into the lungs

19. The nasal cavities are lined with a _____ epithelium that is rich in blood vessels.

20. Air flows from the nasal cavity into the (a) _____. From here air passes into the (b) _____ and then the (c) _____.

21. Food and water are prevented from entering the larynx by the _____.

Gas exchange occurs in the alveoli of the lungs

22. The lungs are located in the (a) _____ cavity. The right lung has (b)____(#?) lobes, and the left lung has (c)____(#?) lobes.

23. Each lung is covered by a _____.

24. Each lung is enclosed in a fluid-filled _____.

Ventilation is accomplished by breathing

25. During inspiration, the volume of the thoracic cavity (a) _____ as a result of the diaphragm (b) _____. During exhalation, the volume (c) _____ as a result of the diaphragm (d) _____.

26. Forced inhalation also involves contraction of the
    (a) _____ muscles which has the effect of
    (b) _____ the volume of the thoracic cavity.

27. The work of breathing is reduced by _____ secreted by specialized epithelial cells lining the surface of the alveoli.

The quantity of respired air can be measured

28. (a)_____ is the volume of air that is inhaled and exhaled with each normal resting breath. It averages about (b) _____ (volume?). The volume of air remaining in the lungs after maximal expiration is the (c) _____ and the maximum amount of air that can be inhaled following a maximum exhalation is the (d)_____.

Gas exchange takes place in the alveoli

29. The factor that determines the direction and rate of diffusion of a gas across a respiratory surface is the _____ of that gas.

30. _____ states that the total pressure in a mixture of gases is the sum of the pressures of the individual gases.

31. Fick's law of diffusion says the amount of oxygen or carbon dioxide that diffuses across the alveolar membrane depends on the difference in
    (a) _____ of the gases on the two sides of the membrane and on the amount of (b) _____ of the membrane.

Gas exchange takes place in the tissues

32. The partial pressure of oxygen in arterial blood is (a) _____, while that in venous blood is (b)_____.

Respiratory pigments increase capacity for oxygen transport

33. The most common respiratory pigments found in animals are
    (a) _____ and (b) _____.

34. Hemoglobin is a general term for a group of compounds which consist of an
    (a) _____ or heme which is bound to a protein called (b) _____.

35. Oxygen combines with the element (a) _____ in the (b)_____ group of hemoglobin. This "association" can be illustrated as Hb + O2 → HbO_2 known as (c)_____. The result is a compound that carries oxygen and releases it where it is in lower concentrations.

36. HbO_2 is prone to dissociate in a (a)_____(acidic or basic) environment. A change in the normal HbO_2 dissociation curve caused by a change in pH is known as the (b) _____.

Carbon dioxide is transported mainly as bicarbonate ions

37. Carbon dioxide is transported in the blood in three forms. 10% is transported as (a) _____, 30% is (b) _____ and 60% is transported as (c) _____, the formation of which is catalyzed by an enzyme in RBCs called (d) _____.

38. Any condition that interferes with the removal of carbon dioxide by the lungs can lead to _____ because carbon dioxide is allowed to build up in the blood.

Breathing is regulated by respiratory centers in the brain

39. Respiratory centers are groups of neurons in the _____ that regulate the rhythm of ventilation.

40. Changes in arterial carbon dioxide levels are sensed by chemoreceptors in the (a) _____, (b) _____, and (c) _____.

41. A decrease in pH leads to _____ (faster or slower) breathing.

42. Changes in _____ (carbon dioxide, pH, or oxygen) have the least affect on breathing.

Hyperventilation reduces carbon dioxide concentration

43. A certain concentration of carbon dioxide is needed in the blood to maintain normal _____.

High flying or deep diving can disrupt homeostasis

44. _____ is a deficiency of oxygen that may cause drowsiness, mental fatigue, and headaches.

45. A sudden decrease in environmental pressure may cause the release of dissolved gases in the blood in the form of bubbles that block capillaries, causing a very painful syndrome which, in the vernacular, is called "the bends," and in the scientific community is called

_____.

Some mammals are adapted for diving

46. The diving reflex includes three physiological mechanisms; these are:

(a) _____

(b) _____

(c) _____.

BREATHING POLLUTED AIR

47. When dirty air is breathed into the lungs,

_____

occurs, manifested by the narrowing of the bronchi.

BUILDING WORDS

Use combinations of prefixes and suffixes to build words for the definitions that follow.

Prefixes	The Meaning		Suffixes	The Meaning
alveol-	small cavity			
derm-	skin		-al	pertaining to
diffuse-	to pour out		-ion	the process of
hyper-	over		-ox(ia)	containing oxygen
hyp(o)-	under		-(a)tion	the process of
ox(y)-	containing oxygen		-us	thing
respir-	to breathe			
ventila-	to fan			

Prefix	Suffix	Definition
_____	-ventilation	1. Excessive rapid and deep breathing; a series of deep inhalations and exhalations.
_____	_____	2. Oxygen deficiency.
_____	_____	3. The process whereby animals actively move air or water over their respiratory surfaces.
_____	-hemoglobin	4. A complex of oxygen and hemoglobin.
_____	_____	5. The combined process of inhalation and exhalation.
_____	_____	6. The movement of gas from an area of higher to lower concentration.
_____	_____	7. The type of gills found in echinoderms.
_____	_____	8. An individual air sac in the lung.

MATCHING

Terms:

a. Alveolus
b. Bronchioles
c. Bronchus
d. Diaphragm
e. Epiglottis
f. Gill
g. Larynx
h. Lung
i. Operculum
j. Pharynx
k. Pleural cavity
l. Pleural membrane
m. Respiratory center
n. Trachea

For each of these definitions, select the correct matching term from the list above.

_____ 1. "Windpipe."
_____ 2. An air sac of the lung through which gas exchange with the blood takes place.
_____ 3. A type of respiratory organ of aquatic animals.
_____ 4. The membrane that lines the thoracic cavity and envelopes the lungs.
_____ 5. One of the branches of the trachea and its immediate branches within the lung.
_____ 6. The throat region in humans.
_____ 7. Tiny air ducts of the lung that branch to form the alveoli.
_____ 8. The dome-shaped muscle that forms the floor of the thoracic cavity.
_____ 9. The bony plate covering the gills in fish.
_____ 10. The organ at the upper end of the trachea that contains the vocal cords.
_____ 11. Closes to prevent food and water from entering the larynx.
_____ 12. Located in the medulla.

MAKING COMPARISONS

Fill in the blanks.

Organism	Type of Gas Exchange Surface
Amphibians	Body surface and lungs
Fish	#1
Spiders	#2
Reptiles	#3
Birds	#4
Nudibranch mollusks	#5
Insects	#6
Sea stars	#7
Clams	#8
Mammals	#9

MAKING CHOICES

Place your answer(s) in the space provided. Some questions may have more than one correct answer.

_____ 1. A cockroach obtains oxygen for tissues located deep in its body by means of
    a. book gills.
    b. pseudolungs.
    c. circulation of hemolymph.
    d. a countercurrent exchange system.
    e. branching tracheal tubes.

_____ 2. Air passes through structures of a mammal's respiratory system in the following order:
    a. trachea, pharynx, bronchi, bronchioles.
    b. larynx, trachea, bronchioles, bronchi.
    c. pharynx, larynx, bronchi, bronchioles.
    d. larynx, trachea, bronchi, bronchioles.
    e. trachea, larynx, bronchi, bronchioles.

_____ 3. The ability of oxygen to be released from oxyhemoglobin is affected by
    a. temperature.
    b. pH.
    c. concentration of carbon dioxide.
    d. oxygen concentration in tissues.
    e. ambient oxygen concentration.

_____ 4. If a patient is found to have inelastic air sacs, trouble expiring air, an enlarged right ventricle, obstructed air flow, and large alveoli, he/she probably
    a. has lung cancer.
    b. has emphysema.
    c. has chronic obstructive pulmonary disease.
    d. lives at a high altitude.
    e. experiences chronic bronchitis.

_____ 5. If you are SCUBA diving for a lengthy time at 200 feet, you might get the bends if

    a. you surface quickly.

    b. you are using a helium mixture.

    c. nitrogen in your blood is rapidly absorbed by your tissues.

    d. you come to the surface very slowly.

    e. you stay at 200 feet even longer.

_____ 6. The diaphragm and rib or intercostal muscles alternately contract and relax, these actions causing respectively

    a. inspiration and expiration.

    b. inhalation and exhalation.

    c. expiration and inspiration.

    d. exhalation and inhalation.

    e. oxygen intake and carbon dioxide output.

_____ 7. Characteristics that all respiratory surfaces share in common include

    a. moist surfaces.

    b. thin walls.

    c. specialized structures such as tracheal tubes, gills, or lungs.

    d. large surface to volume ratio.

    e. alveoli or spiracles.

_____ 8. If the PO_2 in the tissue of your pet dog is 10, atmospheric PO_2 is 150 mm Hg, and arterial PO_2 is 110 mm Hg, you might expect the dog to

    a. function normally.

    b. accumulate carbon dioxide.

    c. have a serious, but not lethal, oxygen deficit.

    d. die.

    e. become dizzy from too much oxygen.

_____ 9. Reasonable rates of expiration and inspiration fall in the range of

    a. 25/min.

    b. 14/min.

    c. 72/hr.

    d. 0.2/sec.

    e. 840/hr.

_____ 10. Most of the mucus produced by epithelial cells in the nasal cavities is disposed of by means of

    a. a large handkerchief.

    b. nose picking.

    c. evaporation.

    d. absorption in surrounding lymph ducts.

    e. swallowing.

_____ 11. Gas exchange in air

    a. requires lungs.

    b. requires a moist surface.

    c. is easier because there is a higher molecular concentration of oxygen.

    d. uses more energy than in water.

    e. can lead to desiccation.

_____ 12. Ventilation

    a. is an active process.

    b. occurs only in animals with lungs.

    c. is the exchange of gases between the blood and the cells.

    d. in sponges utilizes flagella.

    e. occurs only in animals that live on land.

_____ 13. Words or phrases that can be associated with tracheal tubes include

    a. tracheole.

    b. respiratory system.

    c. spider.

    d. grasshopper.

    e. spiracle

_____ 14. Gills can be found in animals that are

    a. mollusks.

    b. chordates.

    c. crustaceans.

    d. vertebrates.

    e. annelids.

_____15. A countercurrent exchange system
    a. allows blood high in oxygen to come in contact with water high in oxygen.
    b. allows blood high in oxygen to come in contact with water that is low in oxygen.
    c. allows blood low in oxygen to come in contact with water that is high in oxygen.
    d. allows blood low in oxygen to come in contact with water that is low in oxygen.
    e. occurs in birds.

_____16. In amphibians
    a. most gas exchange occurs across the body surface.
    b. the lungs are long, simple sacs.
    c. the lungs are better developed than in reptiles.
    d. the lungs have a spongy texture.
    e. the lungs have filaments.

_____17. The bird respiratory system
    a. connects to air spaces in some bones.
    b. is less efficient than reptiles.
    c. is similar to the human system.
    d. has a crosscurrent arrangement.
    e. ends in tiny, thin-walled parabronchi open at both ends.

_____18. In the human respiratory system
    a. inhaled air contains about 15% oxygen.
    b. exhaled air contains about 50 times more carbon dioxide than inhaled air.
    c. the concentration of carbon dioxide is lower in the cells than the blood.
    d. pressure of a gas determines its movement.
    e. the majority of carbon dioxide is transported bound to hemoglobin.

_____19. An increase in ventilation rate
    a. would lead to respiratory alkalosis.
    b. would be triggered by receptors in the jugular vein.
    c. would decrease the amount of oxygen bound to hemoglobin.
    d. would lead to respiratory acidosis.
    e. would increase mitochondrial activity.

_____20. If a person become submerged in very cold water
    a. metabolic rate increases.
    b. oxygen is shunted to internal organs.
    c. oxygen demand goes up.
    d. blood pressure increases.
    e. heart rate increases.

VISUAL FOUNDATIONS

Color the parts of the illustration below as indicated.

RED ☐ lungs

GREEN ☐ gills

YELLOW ☐ body surface used for gas exchange

BLUE ☐ tracheal tube

ORANGE ☐ book lung

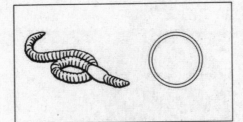 

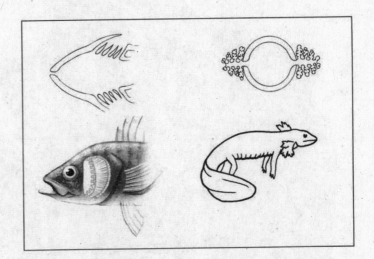

Color the parts of the illustration below as indicated.

RED ☐ red blood cell

BROWN ☐ bronchiole

YELLOW ☐ epithelial cell of alveolus

BLUE ☐ macrophage

PINK ☐ capillary

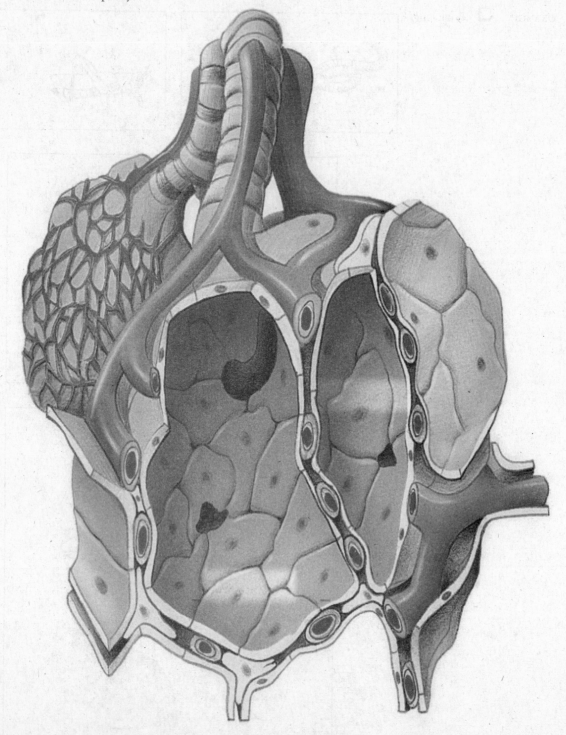

CHAPTER 47

Processing Food and Nutrition

The processing of food involves several steps — taking food into the body, breaking it down into its constituent nutrients, absorbing the nutrients, and eliminating the material that is not broken down and absorbed. Animals eat either plants, animals, or both. Some animals have no digestive systems, with digestion occurring intracellularly within food vacuoles. Others have incomplete digestive systems with only a single opening for both food to enter and wastes to exit. Still other animals have complete digestive systems, in which the digestive tract is a complete tube with two openings, a mouth where food enters and an anus where waste is expelled. The human digestive system has highly specialized structures for processing food. All animals require the same basic nutrients — carbohydrates, lipids, proteins, vitamins, and minerals. Carbohydrates are used by the body as fuel. Lipids are also used as fuel, and additionally, as components of cell membranes, and as substrates for the synthesis of steroid hormones and other lipid substances. Proteins serve as enzymes and as structural components of cells. Vitamins and minerals are needed for many biochemical processes. Serious nutritional problems result from eating too much food, eating too little food, or not eating a balanced diet.

REVIEWING CONCEPTS

Fill in the blanks.

INTRODUCTION

1. Organisms that obtain their main source of energy from organic molecules synthesized by other organisms are known as _____.

NUTRITIONAL STYLES AND ADAPTATIONS

2. Feeding is the (a) _____, (b) _____.
   and (c) _____ of food.

3. Undigested material is removed from the digestive tracts of simpler animals by a process known as _____.

Animals are adapted to their mode of nutrition

4. Animals that feed directly on producers are called (a) _____
   or (b) _____.

5. Animals cannot digest the (a) _____ of plant cell walls. Therefore many herbivores have developed a (b) _____ relationship with microorganisms.

6. Cattle, sheep, and deer are an example of a group of animals called
   _____.

7. Carnivores are (a) _____ or higher-level consumers and are therefore (b) _____. They have well-developed
   (c)_____ teeth for stabbing prey.

8. Consumers of plant and animal material are called _____.

9. Animals that obtain food as suspended particles in water are called
   _____.

Some invertebrates have a digestive cavity with a single opening

10. Examples of animals in which a gastrovascular cavity can be found are

    (a) _____ and (b) _____.

Most animal digestive systems have two openings

11. The vertebrate digestive system is a complete tube extending from the

    (a) _____ to the (b) _____.

12. _____ is responsible for the movement of ingested food along the digestive tract.

THE VERTEBRATE DIGESTIVE SYSTEM

13. The specialized portions of the vertebrate digestive tube, in order, are the:

    mouth → (a)_____ → (b)_____ → stomach

    → (c)_____ → large intestine → anus.

14. The wall of the digestive tract has four layers. The innermost is the

    (a) _____ lines the lumen and consists of a layer of epithelial tissue and the underlying connective tissue. Next is a connective tissue layer the

    (b) _____ in which are blood vessels and nerves. It is surrounded by the muscle layer. The fourth layer, the outer connective tissue layer, is the (c) _____.

Food processing begins in the mouth

15. Ingestion and the beginning of mechanical and enzymatic breakdown of food take place in the mouth. The teeth of mammals perform varied functions:

    (a) _____ are designed for biting, (b) _____ for tearing, and (c) _____ crush and grind food. Each tooth is covered by a coating of very hard (d) _____, under which is the main body of the tooth, the (e) _____, which resembles bone. The

    (f)_____ contains blood vessels and nerves.

16. The salivary glands of terrestrial vertebrates moisten food and release _____, an enzyme that initiates carbohydrate (starch) digestion.

The pharynx and esophagus conduct food to the stomach

17. A lump of food passing through the esophagus is called a _____.

18. During swallowing the _____ closes the opening to the airway.

19. Waves of muscular contractions called _____ move a lump of food through the esophagus into the stomach.

Food is mechanically and enzymatically digested in the stomach

20. The stomach is able to increase and decrease in volumes because the lining is folded into pleats called _____.

21. Pits in the stomach contain (a) _____ which are composed of different types of cells. (b)_____ cells secrete hydrochloric acid and (c)_____ cells secrete pepsinogen, an enzyme precursor that converts to (d)_____.

Most enzymatic digestion takes place in the small intestine

22. The three regions of the small intestine are
    (a) _____, (b) _____,
    and (c) _____. Most chemical digestion of food in
    vertebrates takes place in the (d)_____ portion.

23. The surface area of the small intestine is increased by small fingerlike projections
    called (a)_____, and by (b)_____, which are projections
    of the plasma membrane of the simple columnar epithelial cells of the villi.

The liver secretes bile

24. The function of bile which is secreted by the liver is
    _____.

25. Bile is composed of water, bile (a) _____ and (b) _____
    (c) _____, salts and (d) _____.

26. Bile produced by the liver is stored in the (a) _____ where it
    is also (b) _____.

The pancreas secretes digestive enzymes

27. Pancreatic juice contains (a) _____ and
    (b) _____ that digests polypeptides to dipeptides.

28. The enzyme (a)_____ breaks down fats, and
    (b)_____ breaks down most
    carbohydrates.

29. The enzyme maltose produced by cells lining the small intestine digests
    maltose into (a) _____ and dipeptidases digest
    small peptides into (b) _____.

30. Bile (a) _____ fats and enzymes break fats down into
    (b) _____, (c) _____,
    (d) _____, and (d) _____.

Nerves and hormones regulate digestion

31. Hormones such as secretin, gastrin, CCK, and GIP are polypeptides secreted
    by _____ in the mucosa.

Absorption takes place mainly through the villi of the small intestine

32. Most digested nutrients are absorbed through the (a) _____
    of the small intestine.

33. Amino acids and glucose are transported directly to the _____.

34. Triacylglycerols, along with absorbed cholesterol and phospholipids, are
    packaged into protein-covered fat droplets called (a)_____
    which enter the (b) _____ of the villus.

35. Following digestion and absorption, material remaining at the end of the small
    intestine passes into the large intestine through the _____.

The large intestine eliminates waste

36. The large intestine absorbs sodium and water, cultures bacteria, and eliminates wastes. It is made up of seven regions, which beginning at the junction of the small and large intestine are in the following order: the (a) _____, a blind pouch near the junction of the small and large intestines; the (b) _____; the (c)_____; the (d)_____; the (e)_____; the (f)_____, the last portion of the tube; and the (g)_____, the opening at the end of the tube.

37. Intestinal bacteria benefit their host by producing vitamin (a) _____ and some (b) _____ vitamins that can be absorbed and used by the host.

38. (a)_____ is the process of getting rid of metabolic wastes, while (b)_____ is the process of getting rid of digestive wastes that never participated in metabolism.

REQUIRED NUTRIENTS

39. According to the most recent guidelines, (a) _____ to _____% of our calories should come from carbohydrates, (b) _____ to _____% from fat, and (c) _____ to ____% from protein.

40. The amount of energy in food is expressed in terms of (a) _____, which are equivalent to (b)_____.

Carbohydrates provide energy

41. Most carbohydrates are ingested in the form of (a) _____ and (b) _____.

42. Polysaccharides are referred to as _____.

43. When we eat an excess of carbohydrate-rich food, the liver cells become packed with (a) _____ and convert excess glucose to (b) _____ and (c) _____.

44. A rapid rise in blood glucose concentration stimulates the pancreas to release (a) _____ which lowers blood sugar levels.

Lipids provide energy and are used to make biological molecules

45. About 95% of ingested dietary lipids are in the form of _____.

46. The three polyunsaturated fatty acids that are essential fatty acids that must be obtained from food are (a) _____, (b) _____, and (c) _____.

47. The recommended daily dietary amount of cholesterol is _____mg.

48. Macromolecular complexes of cholesterol or triacylglycerols bound to proteins are called _____.

49. Two types of these important complexes are the (a) _____, which apparently decrease the risk of heart disease by transporting excess cholesterol to the liver; and (b)_____, which transport cholesterol to the cells and have been associated with coronary artery disease.

50. In general, animal foods are rich in both (a) _____ and
    (b) _____, whereas most plant foods are rich in
    (c) _____ and no (d) _____.

Proteins serve as enzymes and as structural components of cells

51. There are approximately (a) _____ (#?) amino acids that cannot be
    synthesized in adult humans. These are called (b) _____
    _____ and need to be provided in a person's diet.

52. Liver cells (a) _____ amino acids and in the process
    produce (b) _____.

Vitamins are organic compounds essential for normal metabolism

53. Vitamins are divided into two broad groups: the (a) _____
    vitamins such as A, D, E, and K, and the (b) _____
    vitamins that include (c)_____.

54. Surpluses from overdoses of the _____ vitamins can
    accumulate to harmful levels.

Minerals are inorganic nutrients

55. The essential minerals required by the body in amounts of 100 mg or more are
    (a) _____, (b) _____,
    (c) _____, (d) _____,
    (e) _____, (f) _____,
    (g) _____, and (h) _____.

56. Minerals required in amounts of less than 100 mg per day are known as
    _____.

Antioxidants protect against oxidants

57. Antioxidants destroy _____, or molecules
    with one or more unpaired electrons.

58. Vitamins (a) _____, (b) _____, and (c) _____ have strong antioxidant
    activity.

Phytochemicals play important roles in maintaining health

59. Diets rich in (a) _____ may be more
    important for health than diets with a low intake of (b) _____.

ENERGY METABOLISM

60. (a)_____ is a measure of the rate
    of energy used during resting conditions.
    (b)_____ is the sum of an individuals' BMR
    and the energy needed to carry out daily activities.

Energy metabolism is regulated by complex signaling

61. The _____ regulates energy metabolism and food intake.

62. When the stomach is empty, it secretes the peptide hormone _____ which stimulates appetite.

63. Neurons in the hypothalamus produce the appetite-stimulated neurotransmitter _____.

64. _____ a hormone released by adipose tissue, is important in the regulation of body weight.

Obesity is a serious nutritional problem

65. Body mass index (BMI) is an index of (a) _____ in relation to (b) _____.

Undernutrition can cause serious health problems

66. Individuals suffering from undernutrition generally are consuming a diet that is most often deficient in _____.

67. Severe protein malnutrition results in a condition known as _____.

BUILDING WORDS

Use combinations of prefixes and suffixes to build words for the definitions that follow.

Prefixes	The Meaning		Suffixes	The Meaning
cec-	blind end		-al	pertaining to
epi-	upon, over, on		-amine	of chemical origin
intrins-	internally		-ic pertaining to	
jejun-	empty		-itis	inflammation
micro-	small		-micro(n)	small, "tiny"
omni-	all		-um	structure
pariet-	wall		-vore	eating
sub-	under, below			
vit-	life			

Prefix	Suffix	Definition
herbi-	_____	1. An animal that eats plants.
carni-	_____	2. An organism that eats flesh.
_____	_____	3. An organism that eats both plants and animals.
_____	-mucosa	4. A layer of connective tissue below the mucosa that binds it to the muscle layer beneath.
periton-	_____	5. Inflammation of the peritoneum.
_____	-glottis	6. A flap of tissue over the airway that prevents food and drink from entering the airway when swallowing.
_____	-villi	7. Small projections of the cell membrane that increase the surface area of the cell.

chylo-   _____   8. Tiny droplets of lipid that are absorbed from the intestine into the lymph circulation.

_____   _____   9. A factor needed for adequate absorption of vitamin B_{12}.

_____   _____   10. Cells in the gastric pit that secrete HCl.

_____   _____   11. The region of the small intestine right after the duodenum.

_____   _____   12. A blind pouch formed where the small and large intestines meet.

_____   _____   13. Organic compounds required in small amounts and they are often components of coenzymes.

MATCHING

Terms:

a.   Absorption
b.   Adventitia
c.   Bile
d.   Digestion
e.   Elimination

f.   Gastrin
g.   Mineral
h.   Mucosa
i.   Pancreas
j.   Pepsin

k.   Peristalsis
l.   Rugae
m.   Stomach
n.   Villus
o.   Vitamin

For each of these definitions, select the correct matching term from the list above.

_____ 1. A cell layer that lines the digestive tract and secretes a lubricating layer of mucous.

_____ 2. The ejection of waste products, especially undigested food remnants, from the digestive tract.

_____ 3. The chief enzyme of gastric juice; hydrolyses proteins.

_____ 4. A large digestive gland located in the vertebrate abdominal cavity, having both exocrine and endocrine functions.

_____ 5. An organic compound necessary in small amounts for the normal metabolic functioning of a given organism; usually acts as a coenzyme.

_____ 6. Folds in the stomach wall, giving the inner lining a wrinkled appearance.

_____ 7. The taking up of a substance, as by the lining of the digestive tract.

_____ 8. Digestive juice produced by the liver and stored in the gallbladder.

_____ 9. Powerful, rhythmic waves of muscular contraction and relaxation in the walls of hollow tubular organs that aid in the movement of substances through the tube.

_____10. A minute "finger-like" projection from the surface of a membrane.

_____11. Inorganic nutrient absorbed in the form of salts dissolved in food and water.

_____12. Lines the lumen of the digestive tract.

MAKING COMPARISONS

Fill in the blanks.

Enzyme	Function	Site of Production
Salivary amylase	Enzymatic digestion of starch	Salivary glands
Maltase	#1	#2
#3	Proteins to polypeptides	#4
#5	Polypeptides to dipeptides	#6
Ribonuclease	#7	#8
#9	Degrades fats	Pancreas
#10	Splits small peptides to amino acids	#11
Lactase	#12	Small intestine

MAKING CHOICES

Place your answer(s) in the space provided. Some questions may have more than one correct answer.

_____ 1. Some important functions of minerals include their role in/as
    a. cofactors.
    b. nerve impulses.
    c. neurotransmitters.
    d. maintaining fluid balance.
    e. digestive hormones.

_____ 2. The relative amounts of nutrient types consumed by impoverished societies, as compared to affluent societies, would likely be
    a. more protein, less carbohydrate.
    b. more lipid, less carbohydrate.
    c. more carbohydrate, less protein.
    d. more carbohydrate, less lipid.
    e. more vegetable matter, less animal matter.

_____ 3. Most absorption of macromolecular subunits occurs in the
    a. stomach.
    b. rectum.
    c. large intestine.
    d. small intestine.
    e. colon.

_____ 4. Products of complete fat digestion include
    a. fatty acids.
    b. amino acids.
    c. chylomicrons.
    d. glycerol.
    e. triacylglycerol.

_____ 5. Nerves and blood vessels in teeth are located in the
    a. pulp.
    b. enamel.
    c. dentin.
    d. cementum.
    e. canines only, not other teeth.

_____ 6. A herbivore is a/an
    a. plant consumer.
    b. primary consumer.
    c. carnivore.
    d. omnivore.
    e. mutualistic symbiote.

_____ 7. Which of the following is true concerning human nutrition?

    a. Water is essential.

    b. Excess nutrients are converted to fat.

    c. We cannot synthesize some required fatty acids.

    d. Proteins cannot be used for energy.

    e. The liver synthesizes amino acids.

_____ 8. The principal function(s) of villi is/are to

    a. absorb nutrients.

    b. secrete enzymes.

    c. increase surface area.

    d. stimulate digestion.

    e. secrete hormones.

_____ 9. All animals are

    a. omnivores.

    b. carnivores.

    c. primary consumers.

    d. consumers.

    e. heterotrophs.

_____ 10. Lipids are used to

    a. maintain body temperature.

    b. build membranes.

    c. provide energy.

    d. synthesize steroid hormones.

    e. create vitamins.

_____ 11. Bile is principally involved in the digestion of

    a. carbohydrates.

    b. lipids.

    c. starch.

    d. nucleic acids.

    e. proteins.

_____ 12. The single most crucial intermediate molecule in the metabolism of lipids and most other nutrients is

    a. keto acid.

    b. ATP.

    c. acetyl CoA.

    d. NAPH.

    e. glucose.

_____ 13. A lacteal is a

    a. lymph vessel.

    b. capillary bed.

    c. digestive gland.

    d. milk protein enzyme.

    e. salivary enzyme.

_____ 14. Peristalsis results from activity of tissues in the

    a. submucosa.

    b. mucosa.

    c. adventitia.

    d. muscle layer.

    e. peritoneum.

_____ 15. Which of the following is true regarding cellulose in the human diet?

    a. It's harmful.

    b. It's a major source of protein.

    c. We can't digest it.

    d. It's an important carbohydrate nutrient.

    e. It's an important source of fiber.

_____ 16. Pepsin is principally involved in the digestion of

    a. carbohydrates.

    b. lipids.

    c. starch.

    d. nucleic acids.

    e. proteins.

_____ 17. Important sources of energy in the human diet are

    a. proteins.

    b. lipids.

    c. carbohydrates.

    d. starches and sugars.

    e. meat.

_____ 18. The total metabolic rate

    a. encompasses the basal metabolic rate.

    b. is less than the basal metabolic rate.

    c. refers to metabolic rate after exercise.

    d. is the rate when at rest.

    e. is a nonexistent phrase.

_____19. Which of the following has/have a complete digestive system?

    a. birds.   d. jelly fish.

    b. earthworms.   e. fish.

    c. sponges.

_____20. A herbivore would likely have

    a. well-developed claws.   d. a short gut.

    b. flattened molars.   e. symbiotic microorganisms.

    c. large, sharp canines.

_____21. Animals that are omnivores include

    a.earthworms.   d. some whales.

    b. black bears.   e. fish.

    c. filter feeders.

_____22. Cnidarians and flatworms

    a. digest prey extracellularly.   d. have a mouth and an anus.

    b. have a pharynx.   e. ingest whole prey.

    c. have a gastrodermis that aides in digestion.

_____23. Having a complete digestive tract is more of an advantage than having a gastrovascular cavity because a complete digestive tract

    a. digests a wider variety of foods.   d. has regions with specific activities.

    b. is lined with absorptive cells.   e. provides better mixing of enzymes with the food.

    c. allows the animal to take in more food before all digestive activity is complete.

_____24. A tooth

    a. is arranged enamel, dentin, pulp   d. is arranged enamel, pulp, dentin

    b. is arranged crown, root   e. has an artery, a vein, a nerve, a lymph vessel

    c. is held in the socket by cementum.

_____25. The liver

    a. stores bile.   d. stores iron.

    b. stores glycogen.   e. regulates blood nutrient levels.

    c. is divided into lobes.

_____26. In which of the following are the two terms correctly paired?

    a. amylase : digests maltose

    b. pancreatic amylase : digests cellulose

    c. trypsin: activated by enterokinase

    d. substance P : stimulation of smooth muscle contraction

    e. Pancreas : production of ribonuclease

_____27. Gastrin

    a. is produced by the small intestine.   d. stimulates HCl production and secretion.

    b. stimulates the release of bile.   e. stimulates insulin secretion.

    c. is released when the stomach becomes distended.

_____28. Hormones associated with the duodenal mucosa include

    a.cholecystokinin   d. secretin.

    b. gastric stimulatory peptide.   e. glucose-dependent insulinotropic peptide.

    c. inhibin.

VISUAL FOUNDATIONS

Color the parts of the illustration below as indicated. Label the mucosa, submucosa, muscle layer, visceral peritoneum, intestinal glands, and villi.

RED	☐	artery
GREEN	☐	lymph
YELLOW	☐	nerve fiber
BLUE	☐	vein
VIOLET	☐	goblet cell
PINK	☐	epithelial cell of villus

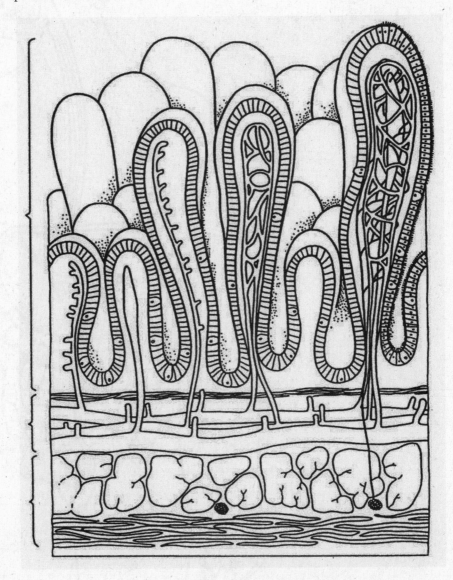

Color the parts of the illustration below as indicated.

RED	☐	liver
GREEN	☐	gall bladder
YELLOW	☐	esophagus
BLUE	☐	pancreas
ORANGE	☐	stomach
BROWN	☐	large intestine
TAN	☐	small intestine
PINK	☐	rectum and anus
VIOLET	☐	vermiform appendix

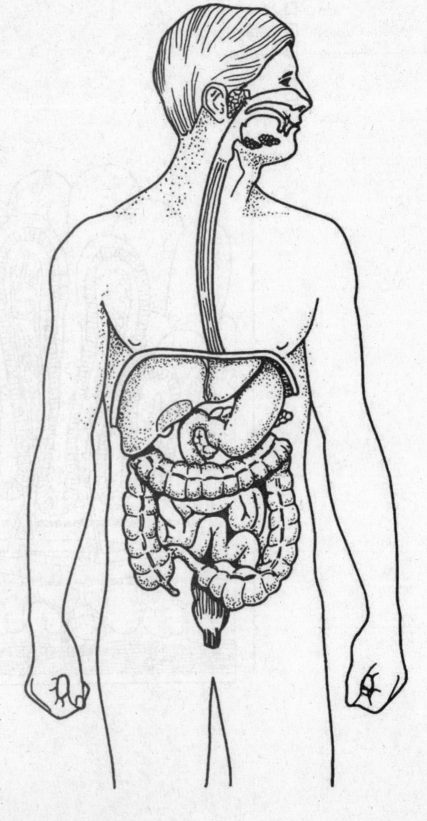

□

CHAPTER 48

Osmoregulation and Disposal of Metabolic Wastes

The water content of the animal body, as well as the concentration and distribution of ions in body fluids is carefully regulated. Most animals have excretory systems that function to rid the body of excess water, ions, and metabolic wastes. The excretory system collects fluid from the blood and interstitial fluid, adjusts its composition by reabsorbing from it the substances the body needs, and expels the adjusted excretory product, e.g., urine in humans. The principal metabolic wastes are water, carbon dioxide, and nitrogenous wastes. Excretory systems among invertebrates are diverse and adapted to the body plan and lifestyle of each species. The kidney, with the nephron as its functional unit, is the primary excretory organ in vertebrates.

REVIEWING CONCEPTS
Fill in the blanks.

INTRODUCTION
1. _____ is the medium in which most metabolic reactions take place.
2. Two processes that maintain homeostasis of fluids in the body are _____ and _____.

MAINTAINING FLUID AND ELECTROLYTE BALANCE
3. Compounds such as inorganic salts, acids, and bases that form ions in solution are called _____.
4. A unit of osmotic pressure equal to the molarity of the solution divided by the number of particles produced when the solute dissolves is an

   _____.
5. _____ controls the concentration of salt and water so that body fluids don't become too dilute or too concentrated.

METABOLIC WASTE PRODUCTS
6. The principal metabolic wastes in most animals are (a) _____,
   (b) _____, and (c) _____.
7. The catabolism of proteins leads to the production of three nitrogen containing waste products. They are (a) _____,
   (b)_____, and (c) _____.
8. The production of (a) _____ is an important water-conserving adaptation in many terrestrial animals, including insects, certain reptiles, and birds. It is produced from (b) _____ and by the
   (c)_____.

9. The major nitrogenous waste product produced by amphibians is

_____.

OSMOREGULATION AND EXCRETION IN INVERTEBRATES

10. The body fluid of most marine invertebrates is (a)_____

with the sea water. As a result, these animals are said to be

(b) _____.

11. Animals living in water environments much less stable than the ocean and

those that live on land maintain an optimal salt concentration independent of

the environmental conditions. These animals are called _____.

Nephridial organs are specialized for osmoregulation and/or excretion

12. The nephridial organs of many invertebrates consist of tubes that open to the

outside of the body through (a) _____. In flatworms,

these excretory organs are called (b) _____, and in

annelids and mollusks they are known as (c) _____.

Malpighian tubules conserve water

13. Malpighian tubules are slender extensions of the (a) _____, and

they have blind ends that lie in the (b)_____ where they are

surrounded by (c) _____.

14. Wastes are transferred from blood to the Malpighian tubules by either

(a) _____ or (b) _____.

OSMOREGULATION AND EXCRETION IN VERTEBRATES

15. The main osmoregulatory and excretory organ in most vertebrates is the

(a) _____, but the (b) _____, (c) _____ or

(d)_____, and the digestive system also help maintain fluid balance and

secrete metabolic wastes.

Freshwater vertebrates must rid themselves of excess water

16. A freshwater fish is _____ to the water around it.

17. In freshwater fish the majority of nitrogenous waste is excreted by the

_____.

18. In freshwater fishes, (a) _____ is the main nitrogenous

waste; about 10% of nitrogenous wastes are excreted as (b) _____.

19. As part of osmoregulation, both freshwater fish and amphibians produce a

large amount of _____.

Marine vertebrates must replace lost fluid

20. Marine bony fishes are _____ to the surrounding water.

21. To compensate for fluid loss, marine bony fishes _____.

22. To help them maintain osmotic balance with the surrounding water, marine

cartilaginous fishes accumulate (a) _____ in their tissues. This makes

these fishes (b) _____ to their environment.

23. Marine mammals need to produce a _____ urine which

helps them excrete the high amounts of ammonia that they produce as a result

of consuming a lot of protein in their diet.

Terrestrial vertebrates must conserve water

24. Birds conserve water by excreting nitrogen as (a) _____;
    mammals excrete (b)_____.

THE URINARY SYSTEM

25. The outer portion of the kidney tissue is the (a) _____
    and the inner portion is the (b) _____.

26. The urinary system is the principal excretory system in human beings and
    other vertebrates. Urine produced in the kidneys is transported through two
    tubes called (a) _____ to the (b) _____
    where it is stored until it passes to the outside of the body through the
    (c)_____.

The nephron is the functional unit of the kidney

27. The functional units of the kidneys are the nephrons. Each nephron consists
    of a cup-shaped (a) _____ and a (b)_____.
    The filtrate passes through structures in the nephron in this sequence:
    (c) _____ → proximal convoluted tubule →
    (d) _____ → distal convoluted tubule →
    (e) _____.

28. There are two types of nephrons in kidneys. Those that have a small
    glomerulus and lie almost entirely within the cortex and outer medulla are
    (a)_____ nephrons. Those that have a large glomerulus and loops of
    Henle that extend deep into the medulla are called
    (b)_____ nephrons.

Urine is produced by filtration, reabsorption, and secretion

29. Urine is produced by a combination of three processes (a) _____,
    (b) _____, and (c) _____.

Filtration is not selective with regard to ions and small molecules

30. Plasma is filtered out of the glomerular capillaries and into
    _____.

31. Three major factors contribute to filtration and the formation of filtrate.
    They are (a) _____, (b) _____,
    and (c) _____.

32. Specialized epithelial cells called (a) _____ surround the
    capillaries of the glomerulus. These cells are separated by narrow gaps called
    (b) _____ which are one part of the
    (c) _____ that holds back blood cells and
    most plasma proteins.

Reabsorption is highly selective

33. The renal tubules reabsorb about (a) ____% of the filtrate into the blood,
    leaving only about (b) _____ L to be excreted as urine in a 24 hour period.

34. About 65% of the filtrate is reabsorbed as it passes through the first section of
    the nephron tubule called the _____.

35. The maximum rate at which a substance can be reabsorbed is its

_____.

Some substances are actively secreted from the blood into the filtrate

36. When the circulating level of potassium ions is too high, the heart rhythm

    (a) _____. The hormone (b) _____

    stimulates secretion of excess potassium by the kidney.

Urine becomes concentrated as it passes through the renal tubule

37. Filtrate is concentrated as it moves downward through the (a) _____

    loop and diluted as it moves upward through the (b)_____

    loop. This movement of filtrate in opposite directions, called the

    (c)_____, helps maintain a

    hypertonic interstitial fluid that draws water out of the collecting ducts.

38. The _____ are capillaries that extend from the efferent

    arterioles of the juxtamedullary nephrons; they collect water from interstitial

    fluid.

Urine consists of water, nitrogenous wastes, and salts

39. _____ is the adjusted filtrate consisting of water, nitrogenous

    wastes, salts, and traces of other substances.

Hormones regulate kidney function

40. Urine volume is regulated by the hormone (a)_____. It

    is released by the (b)_____.

41. The "thirst center" that responds to dehydration is located in the

    _____.

The rennin-angiotensin-aldosterone pathway increases sodium reabsorption

42. The hormone (a) _____ secreted by the adrenal cortex

    stimulates sodium reabsorption in the (b) _____ and the

    (c) _____.

43. When blood pressure falls, cells of the (a) _____ secrete

    the enzyme rennin. Renin converts the plasma protein

    (b) _____ into (c) _____.

    The enzyme ACE then converts (d) _____ into

    (e) _____ which stimulates aldosterone secretion.

Atrial natriuretic peptide inhibits sodium reabsorption

44. Atrial natriurectic peptide (ANP) is a hormone that is produced by the

    (a) _____. ANP increases the excretion of (b) _____

    and decreases (c) _____.

BUILDING WORDS

Use combinations of prefixes and suffixes to build words for the definitions that follow.

Prefixes	The Meaning	Suffixes	The Meaning
glomerul-	little ball	-cyte	cell
juxta-	beside, near	-us	thing
podo-	foot		
proto-	first, earliest form of		

Prefix	Suffix	Definition
_____	-nephridium	1. The flame cell excretory organs of flatworms and nemerteans; the earliest form of specialized excretory organ.
_____	_____	2. A specialized epithelial cell possessing elongated foot processes, which cover the surfaces of most of the glomerular capillaries.
_____	-medullary	3. Pertains to nephrons situated nearest the medulla of the kidney.
_____	_____	4. A cluster of capillaries located inside of Bowman's capsule.

MATCHING

Terms:

a. Aldosterone
b. Antidiuretic hormone
c. Bowman's capsule
d. Osmoconformer
e. Osmoregulator
f. Glomerulus
g. Malpighian tubule
h. Metanephridium
i. Nephron
j. Osmoregulation
k. Reabsorption
l. Urea
m. Ureter
n. Urethra
o. Uric acid

For each of these definitions, select the correct matching term from the list above.

_____ 1. An animal that maintains an optimal salt concentration of its body fluids despite changes in the salinity of its surroundings.

_____ 2. A hormone produced by the vertebrate adrenal cortex that functions to increase sodium reabsorption.

_____ 3. The knot of capillaries at the proximal end of a nephron; enclosed by the Bowman's capsule.

_____ 4. The tube that conducts urine from the bladder to the outside of the body.

_____ 5. The principal nitrogenous excretory product of many terrestrial animals, including insects and certain birds and reptiles.

_____ 6. The active regulation of the osmotic pressure of body fluids.

_____ 7. One of the paired ducts that conducts urine from the kidney to the bladder.

_____ 8. The principal nitrogenous excretory product of mammals; one of the water-soluble end products of protein metabolism.

_____ 9. The functional unit of the vertebrate kidney.

_____10. A hormone secreted by the posterior lobe of the pituitary that controls the rate of water reabsorption by the kidney.

_____11. An animal whose internal salt composition reflects that of its environment.

MAKING COMPARISONS

Fill in the blanks.

Organism	Excretory Mechanism/Structure	
Marine sponges	Diffusion	
Vertebrates	#1	
Insects	#2	
Earthworms	#3	
Flatworms	#4	

MAKING CHOICES

Place your answer(s) in the space provided. Some questions may have more than one correct answer.

_____ 1. The amount of needed substances that can be reabsorbed from renal tubules is a function of

    a. Tm.

    b. blood pH.

    c. tubular transport maximum.

    d. concentration of urea in forming urine.

    e. maximum rate.

_____ 2. The principal functional unit(s) in the vertebrate kidneys is/are

    a. nephridia.

    b. nephrons.

    c. Bowman's capsules.

    d. antennal glands.

    e. Malpighian tubules.

_____ 3. Water is reabsorbed by interstitial fluids when osmotic concentration in interstitial fluid is increased by

    a. salts from filtrate.

    b. urea from filtrate.

    c. free ions.

    d. concentration of filtrate.

    e. deamination in kidney cells.

_____ 4. Urea is a principal nitrogenous waste product produced by

    a. amphibians.

    b. the kidneys.

    c. nephrons.

    d. mammals.

    e. the liver.

_____ 5. Most reabsorption of filtrate in the kidney takes place at the

    a. ureter.

    b. loop of Henle.

    c. proximal convoluted tubule.

    d. urethra.

    e. collecting duct.

_____ 6. Urea is synthesized

    a. from uric acid.

    b. in kidneys.

    c. from ammonia and carbon dioxide.

    d. in the urea cycle.

    e. in aquatic invertebrates.

_____ 7. Relative to sea water, fluids in the bodies of marine organisms

    a. are isotonic.

    b. are hypotonic.

    c. are hypertonic.

    d. lose water.

    e. gain water.

_____ 8. The duct in humans that leads from the urinary bladder to the outside is the

a. ureter.
b. loop of Henle.
c. proximal convoluted tubule.
d. urethra.
e. collecting duct.

_____ 9. Which of the following statements most accurately describes changes in the concentration of filtrate in the two portions of the loop of Henle?

a. decreases in both
b. increases in both
c. increases in descending/decreases in ascending
d. decreases in descending/increases in ascending
e. remains essentially the same in both

_____10. A potato bug would excrete wastes by means of

a. nephridia.
b. nephrons.
c. a pair of kidney-like structures.
d. green glands.
e. Malpighian tubules.

_____11. Relative to fresh water, fluids in the bodies of aquatic organisms

a. are isotonic.
b. are hypotonic.
c. are hypertonic.
d. lose water.
e. gain water.

_____12. Excretion is specifically defined as

a. maintaining water balance.
b. homeostasis.
c. removal of metabolic wastes from the body.
d. the elimination of undigested wastes.
e. the concentration of nitrogenous products.

_____13. The main way(s) that *any* excretory system maintains homeostasis in the body is/are to

a. excrete metabolic wastes.
b. eliminate undigested food.
c. regulate body fluid constituents.
d. regulate salt and water.
e. concentrate urea in urine.

_____14. The group(s) of animals that has/have no specialized excretory systems include

a. insects.
b. cnidarians.
c. annelids.
d. sponges.
e. sharks and rays.

_____15. The principal form(s) of nitrogenous waste products in various animal groups include(s)

a. uric acid.
b. carbon dioxide.
c. amino acids.
d. ammonia.
e. urea.

_____16. The principal nitrogenous waste product(s) in human urine is/are

a. uric acid.
b. carbon dioxide.
c. amino acids.
d. ammonia.
e. urea.

_____17. The first step in the catabolism of amino acids

a. is conversion of ammonia to uric acid.
b. is deamination.
c. produces ammonia.
d. is removal of the amino group.
e. is accomplished with peptidases.

_____18. The type(s) of excretory organ(s) found in animals that collect wastes in flame cells is/are

a. nephridia.
b. nephrons.
c. branching tubes that open to the outside through pores.
d. green glands.
e. Malpighian tubules.

_____19. The nitrogenous waste requiring the largest amount of water for adequate excretion is
    a. urea.
    d. uracil.
    b. uric acid.
    e. ammonia.
    c. an amino acid.

_____20. Urea is synthesized in the liver from
    a. uric acid and ammonia.
    d. ammonia and carbon dioxide.
    b. uric acid and oxygen.
    e. ammonia and oxygen.
    c. ammonia and nitrogen.

_____21. Ammonia is produced in
    a. reptiles.
    d. amphibians.
    b. fish.
    e. mammals.
    c. birds.

_____22. Excretory systems
    a. maintain homeostasis.
    d. selectively adjust the amount of salt and water.
    b. are found in all animals.
    e. evolved first in amphibians.
    c. help to avoid dehydration.

_____23. Most amphibians
    a. produce a dilute urine.
    d. have osmoregulatory systems similar to mammals.
    b. are semi-aquatic.
    e. produce a concentrated urine to conserve water.
    c. produce uric acid.

_____24. Salt glands can be found in some
    a. sharks.
    d. birds.
    b. reptiles.
    e. dolphins.
    c. whales.

_____25. Urine enters the renal pelvis from the
    a. papillae.
    d. collecting duct.
    b. cortex.
    e. renal tubules.
    c. medullary duct.

_____26. The kidneys produce
    a. erythropoietin.
    d. rennin.
    b. vitamin D_3.
    e. factor K.
    c. cortical hormone.

_____27. Under normal circumstances, the total amount of blood passing through a kidney is
    a. 180 L in 24 hours.
    d. 1.2 L per minute.
    b. 5 L per minute.
    e. about ¼ of the cardiac output.
    c. 100 L per hour.

_____28. Each day the kidney tubules reabsorb
    a. 1200 g of salt.
    d. 500 g of chloride.
    b. 20 L of water.
    e. 250 g of glucose.
    c. 180 L of water.

_____29. The loop of Henle
    a. is permeable to water in the ascending limb.
    d. is permeable to salt in the descending limb.
    b. is a countercurrent mechanism.
    e. connects to the collecting duct.
    c. is specialized to maintain a high interstitial sodium chloride concentration.

_____30. Antidiuretic hormone
    a. is released by the hypothalamus.   d. is produced by the pituitary gland.
    b. is produced by the hypothalamus.   e. is released by the pituitary gland.
    c. stimulates water reabsorption.

_____31. Which of the following is correctly paired.
    a. atrial natriuretic peptide : collecting ducts   d. angiotensin II : adrenal glands
    b. aldosterone : afferent arteriole   e. aldosterone: distal tubules
    c. antidiuretic hormone : collecting ducts

VISUAL FOUNDATIONS

Color the parts of the illustration below as indicated.

RED   ☐ abdominal aorta
GREEN   ☐ adrenal gland
YELLOW   ☐ ureter
BLUE   ☐ inferior vena cava
ORANGE   ☐ urethra
BROWN   ☐ kidney
TAN   ☐ urinary bladder
PINK   ☐ renal artery
VIOLET   ☐ renal vein

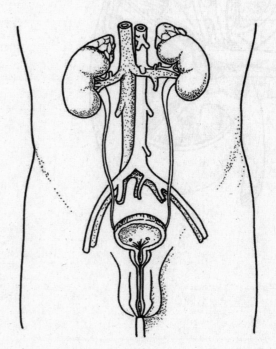

Color the parts of the illustration below as indicated. Label the juxtamedullary nephron and the cortical nephron.

RED	☐ artery	ORANGE	☐	loop of Henle
GREEN	☐ renal pelvis	BROWN	☐	medulla
YELLOW	☐ collecting duct	BLUE	☐	vein
PINK	☐ glomerulus	TAN	☐	cortex
VIOLET	☐ proximal convoluted tubule and distal convoluted tubule			

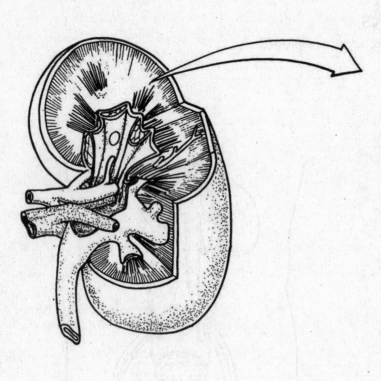
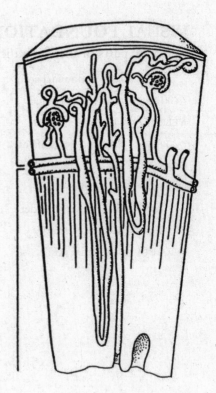

CHAPTER 49

Endocrine Regulation

This chapter discusses the actions of a variety of hormones and examines how overproduction or deficiency of various hormones interferes with normal functioning. The endocrine system is a diverse collection of glands and tissues that secrete hormones, chemical messengers that signal other cells. The majority of hormones are transported by the blood to the target tissue where they stimulate a physiological change. The endocrine system works closely with the nervous system to maintain the steady state of the body. They diffuse from the blood into the interstitial fluid and then combine with receptor molecules on or in the cells of the target tissue. Hormones may be steroids, peptides, proteins, or derivatives of amino acids or fatty acids. Some hormones activate genes that lead to the synthesis of specific proteins. Others activate a second messenger that relays the hormonal message to the appropriate site within the cell. Most invertebrate hormones are secreted by neurons rather than endocrine glands. They regulate growth, metabolism, reproduction, molting, and pigmentation. In vertebrates, hormonal activity is controlled by the hypothalamus, which links the nervous and endocrine systems. Vertebrate hormones regulate growth, reproduction, salt and fluid balance, and many aspects of metabolism. Malfunction of any of the endocrine glands can lead to specific disorders.

REVIEWING CONCEPTS

Fill in the blanks.

INTRODUCTION

1. The endocrine system regulates many physiological processes including

   (a)
   _____,

   (b)
   _____,

   (c)
   _____,

   (d)
   _____,

   and (e)
   _____.

AN OVERVIEW OF ENDOCRINE REGULATION

2. Endocrine glands differ from exocrine glands in that they have no

   (a) _____ and secrete their hormones

   (b) _____.

3. About _____ (#) discrete endocrine glands have been identified.

4. Hormones produce a response only after they

   (a) _____.

The endocrine system and nervous system interact to regulate the body

5. The endocrine system responds more _____ (slowly, quickly) than the nervous system.

6. The response of the endocrine system last a _____ (longer, shorter) time than the nervous system.

Negative feedback systems regulate endocrine activity

7. Most endocrine action is regulated by (a) _____ that act to restore (b) _____ .

Hormones are assigned to four chemical groups

8. The four chemical groups of hormones are

(a) _____, (b) _____,

(c) _____, and (d) _____.

9. Signaling molecules produced by neurons are called _____.

TYPES OF ENDOCRINE SIGNALING

10. Steroid and thyroid hormones are transported

_____.

Neurohormones are transported in the blood

11. Neurohormones are transported down (a) _____ and released into the (b) _____.

Some local regulators are considered hormones

12. There are two types of local regulation. In (a) _____ signaling a hormone acts on the very cell that produces it and in (b)_____ signaling the released hormone works in nearby cells.

13. _____ is a signaling molecule that is stored in mast cells.

14. _____ is a gaseous signaling molecule.

MECHANISMS OF HORMONE ACTION

15. Receptors are continuously synthesized and degraded. Their numbers can be increased or decreased by (a)_____ and (b)_____.

Some hormones enter target cells and activate genes

16. Steroid and thyroid hormones bind receptors that are located

(a) _____ because these hormones are (b) _____.

Many hormones bind to cell-surface receptors

17. Peptide hormones bind to specific receptors on the plasma membrane because _____.

18. With respect to a hormones action, the conversion of an extracellular signal to an intracellular signal is called _____.

19. _____ converts ATP to cAMP.

20. _____ are enzymes that phosphorylate proteins.

21. The products of a reaction catalyzed by the membrane-bound enzyme phospholipase C interacting with its membrane lipid target are
    (a) _____ and (b) _____.

22. Most enzyme-linked receptors are _____ kinases.

NEUROENDOCRINE REGULATION IN INVERTEBRATES

23. Most invertebrate hormones are secreted by _____ rather than by endocrine glands. They help to regulate regeneration, molting, metamorphosis, reproduction, and metabolism.

24. Hormones control growth and development in insects. Neurosecretory cells in the brain of insects produce (a) _____ which is stored in the corpora cardiaca. When released from the corpora cardiaca, this hormone stimulates the (b) _____ glands to produce (c)_____ also called (d) _____. This hormone stimulates growth and molting.

ENDOCRINE REGULATION IN VERTEBRATES

Homeostasis depends on normal concentrations of hormones

25. A target cell is over stimulated if (a) _____ of the hormone occurs and it is under stimulated if (b)_____ occurs.

The hypothalamus regulates the pituitary gland

26. The pituitary gland connects to the hypothalamus by the _____.

27. The pituitary is often referred to as the _____ because of the number of body activities it controls.

The posterior lobe of the pituitary gland releases hormones produced by the hypothalamus

28. The hypothalamus produces vasopressin, also known as _____.

29. The hypothalamus also produces _____, which stimulates milk production in nursing mothers.

The anterior lobe of the pituitary gland regulates growth and other endocrine glands

30. The hypothalamus produces (a) _____ hormones and (b) _____ hormones that regulate the activity of the anterior pituitary.

31. _____ hormones stimulate other endocrine glands.

32. Melanocyte-stimulating hormones (MSH) are secreted by the (a) _____ gland. In humans MSH (b) _____ and is involved in regulation of energy and body weight.

33. The anterior pituitary secretes growth hormone (GH), also called _____ which promotes tissue growth.

34. Hypersecretion of GH during childhood may cause a disorder known as (a) _____. In adulthood, hypersecretion can cause "large extremities," or (b) _____.

Thyroid hormones increase metabolic rate

35. The thyroid secretes (a) _____, also known as T4 because each molecule contains four atoms of (b)_____, and (c) _____, also known as T3 because each molecule contains three atoms of (d) _____

36. Thyroid secretion is regulated by a negative feedback system between the thyroid gland and the (a) _____ gland, which releases less (b)_____ when thyroid hormone blood titers rise above normal.

37. Hyposecretion of thyroid hormones during infancy and childhood may result in retarded development, a condition known as _____.

38. The most common type of hyperthyroidism is _____ which is actually an autoimmune disease.

The parathyroid glands regulate calcium concentration

39. Parathyroid hormone regulates calcium levels in body fluids by stimulating release of calcium from (a)_____ and calcium reabsorption by the (b)_____.

40. A negative feedback system induces the (a) _____ gland to secrete (b)_____, which inhibits parathyroid hormone activity.

The islets of the pancreas regulate glucose concentration

41. Numerous clusters of cells in the pancreas, called (a) _____, secrete hormones that regulate glucose concentration in the blood. When blood glucose levels are high, (b) _____ cells release the hormone (c) _____; when it is low, (d) _____ cells release (e)_____.

42. Insulin lowers the concentration of glucose in the blood by stimulating uptake of glucose by (a) _____, (b) _____, and (c) _____ cells.

43. In the endocrine metabolism disorder known as _____, cells are unable to utilize glucose properly and turn to fat and protein for fuel.

The adrenal glands help the body respond to stress

44. The adrenal medulla and the adrenal cortex secrete hormones that help the body cope with stress. The two secreted by the adrenal medulla are (a) _____ and (b) _____; they increase heart rate, metabolic rate, and strength of muscle contraction, and reroute the blood to organs that require more blood in time of stress.

45. The adrenal cortex produces three types of hormones in appreciable amounts; these are (a)_____,
    (b)_____, and (c)_____.
    The sex hormone precursors are converted to (d)_____,
    the principal male sex hormone, and
    (e)_____, the principal female sex hormone.

46. Kidneys reabsorb more sodium and excrete more potassium in response to the mineralocorticoid hormone _____.

47. The main function of _____ is to enhance gluconeogenesis in the liver.

48. Stress stimulates secretion of CRF, which is the abbreviation for
    (a) _____,
    secreted by the hypothalamus. CRF stimulates the anterior pituitary to release
    (b) _____, which regulates the secretion of glucocorticoids and aldosterone.

Many other hormones are known

49. The (a) _____ gland in the brain produces the hormone
    (b)_____, which influences the onset of sexual maturation.

50. The thymus gland produces the hormone _____ which plays a role in immunity.

51. The heart secretion ANF, which stands for _____, lowers blood pressure.

BUILDING WORDS

Use combinations of prefixes and suffixes to build words for the definitions that follow.

Prefixes	The Meaning		Suffixes	The Meaning
gluc-	sweet		-agon	to fight
hom(eo)-	same			
hyper-	over		-megaly	enlargement
			-oid	resembling
hypo-	under		-stasis	to control the process of
neuro-	nerve			
ster-	hard, solid			

Prefix	Suffix	Definition
_____	-hormone	1. A hormone secreted by certain nerve cells.
_____	-endocrine	2. Refers to a nerve cell that secretes a neurohormone.
_____	-secretion	3. An excessive secretion; over secretion.
_____	-secretion	4. A diminished secretion; under secretion.
_____	-thyroidism	5. A condition resulting from an overactive thyroid gland.
_____	-glycemia	6. An abnormally low level of glucose in the blood.
_____	-glycemia	7. An abnormally high level of glucose in the blood.
acro-	_____	8. An abnormal condition characterized by enlargement of the head, and sometimes other structures.
_____	_____	9. The balanced internal environment of the body.
_____	_____	10. A hormone with 4 carbon rings as a part of their structure. Testosterone for example.
_____	_____	11. A hormone produced by alpha cells that raises blood glucose levels.

MATCHING

Terms:

a. Adrenal gland
b. Aldosterone
c. Beta cells
d. Calcitonin
e. Endocrine gland

f. Glucagon
g. Hormone
h. Insulin
i. Oxytocin
j. Prostaglandin

k. Thymus
l. Thyroid gland
m. Thyroxine
n. Tropic hormone

For each of these definitions, select the correct matching term from the list above.

_____ 1. A hormone produced by the hypothalamus and released by the posterior lobe of the pituitary; causes the uterus to contract and stimulates the release of milk from the mammary glands.

_____ 2. A hormone secreted by the thyroid gland that rapidly lowers the calcium content in the blood.

_____ 3. An endocrine gland that lies anterior to the trachea and releases hormones that regulate the rate of metabolism.

_____ 4. Paired endocrine glands, each located just superior to each kidney.

_____ 5. General term for an organic chemical produced in one part of the body and transported to another part where it affects some aspect of metabolism.

____ 6. Cells of the pancreas that secrete insulin.

____ 7. A gland that secretes products directly into the blood or tissue fluid instead of into ducts.

____ 8. General term for a hormone that helps regulate another endocrine gland.

____ 9. One of the hormones produced by the thyroid gland.

____10. An endocrine gland that produces thymosin; important in the development of the immune response mechanism.

MAKING COMPARISONS

Fill in the blanks.

Hormone	Chemical Group	Function
Testosterone	Steroid	Develops and maintains sex characteristics of males; promotes spermatogenesis
Aldosterone	#1	#2
Thyroxine	#3	#4
#5	Peptide	Stimulates reabsorption of water; conserves water
#6	#7	Helps body adapt to long-term stress; mobilizes fat; raises blood glucose levels
#8	#9	Affects wide range of body processes; may interact with other hormones to regulate metabolic activities
ACTH	Peptide	#10
Epinephrine	#11	#12
#13	#14	Stimulates uterine contraction

MAKING CHOICES

Place your answer(s) in the space provided. Some questions may have more than one correct answer.

____ 1. GH is referred to as an anabolic hormone because it

    a. suppresses appetite .

    b. stimulates other endocrine glands.

    c. promotes tissue growth.

    d. regulates the anterior pituitary gland.

    e. affects neural membrane potentials.

____ 2. The thyroid gland is located

    a. in the neck.

    b. behind the trachea.

    c. in front of the trachea.

    d. above the larynx.

    e. below the larynx.

_____ 3. The hormone and target tissue that are involved in raising glucose concentration in blood by glycogenolysis and gluconeogenesis are

    a. insulin and pancreas.

    b. glucagon and liver.

    c. thyroid-stimulating hormone and thyroid gland.

    d. adrenocorticotropic hormone and adrenal cortex.

    e. thyroxine and various metabolically active cells.

_____ 4. Activity of the anterior lobe of the pituitary gland is controlled by

    a. the hypothalamus.

    b. ADCH.

    c. epinephrine and norepinephrine.

    d. releasing hormones.

    e. inhibiting hormones.

_____ 5. A chemical produced by one cell that has a specific regulatory effect on another cell is the definition of a

    a. neurohormone.

    b. hormone.

    c. exocrine gland secretion.

    d. pheromone.

    e. prohormone.

_____ 6. A three-year old cretin and an adult suffering from myxedema most likely have

    a. a low metabolic rate.

    b. Cushing's disease.

    c. Addison's disease.

    d. a low level of thyroid hormones.

    e. a high level of thyroid hormones.

_____ 7. A person with a fasting level of 750 mg glucose per 100 ml of blood

    a. is hyperglycemic.

    b. is hypoglycemic.

    c. is about normal.

    d. probably has cells that are not using enough glucose.

    e. has too much insulin.

_____ 8. Increased skeletal growth results directly and/or indirectly from activity of

    a. aldosterone.

    b. growth hormone.

    c. hormones from the hypothalamus.

    d. andosterides.

    e. epinephrine and/or norepinephrine.

_____ 9. Diabetes in an adult with ample numbers of adequately functioning beta cells and normal concentrations of insulin

    a. is type I diabetes.

    b. is type II diabetes.

    c. indicates that target cells are not using insulin.

    d. indicates that not enough insulin is being produced.

    e. is unusual since this condition usually occurs in children.

_____10. Small, hydrophobic hormones that form a hormone-receptor complex with intranuclear receptors include

    a. steroids.

    b. cyclic AMP.

    c. chemicals that activate genes.

    d. thyroid hormones.

    e. prostaglandins.

_____11. A possible negative consequence of hyposecretion of insulin is

    a. Addison's disease.

    b. Cushing's disease.

    c. myxedema.

    d. diabetes mellitus.

    e. dwarfism.

_____12. Which of the following is mismatched?

    a. exocrine glands : ductless

    b. negative feedback : calcium regulation

    c. neuropeptide : oxytocin

    d. steroid hormone : G protein

    e. GTP : adenylyl cyclase

_____13. The two hormones involved in the regulation of calcium levels are
   a. thyroid hormone and parathyroid hormone
   b. thyroid hormone and calcitonin
   c. growth hormone and thyroid hormone
   d. parathyroid hormone and calcitonin
   e. a mineralocorticoid and a glucocorticoid

_____14. Which hormone is incorrectly matched with its function?
   a. oxytocin : stimulates smooth muscle contraction.
   characteristics
   b. antidiuretic hormone : stimulates water secretion
   c. melatonin : regulates biological rhythms
   d. progesterone : development of male
   e. epinephrine : stimulates heart rate

_____15. Which gland is incorrectly matched with its hormone?
   a. thyroid gland : parathyroid hormone
   b. anterior pituitary : oxytocin
   c. adrenal cortex : epinephrine
   d. posterior pituitary : antidiuretic hormone
   e. hypothalamus : prolactin

_____16. Melanocyte stimulating hormones
   a. are secreted by the anterior pituitary.
   b.are steroid hormones.
   c. are secreted by neurons in the hypothalamus.
   d. are important in animal camouflage.
   e. regulate secretion of anabolic hormone.

_____17. Goiter
   a. may result from hypersecretion of thyroid hormone.

   b. may occur when there is an iodine deficiency.

   c. may result from hyposecretion of thyroid hormone.

   d. may be associated with TSH receptors and Graves disease.

   e. is an autoimmune disease.

_____18. Diabetes
   a. may cause a disruption in fat metabolism.
   b. may cause a disruption in protein metabolism.
   c. may cause a disruption in carbohydrate metabolism.
   d. involves alpha cells and insulin.
   e. may cause an electrolyte imbalance.

_____19. Which disease and gland is not correctly paired?
   a. Addison's disease : adrenal gland
   b. Cushing's syndrome : adrenal gland
   c. diabetes mellitus : pancreas
   d. myxedema : hypothalamus
   e. credtinism : thyroid gland

VISUAL FOUNDATIONS

Color the parts of the illustration below as indicated. Also label transcription, translation, and the target cell.

RED	☐	receptor molecule	BROWN	☐	endocrine gland cell
YELLOW	☐	hormone molecules	TAN	☐	nucleus
BLUE	☐	DNA	PINK	☐	blood vessel
ORANGE	☐	mRNA	VIOLET	☐	protein molecule

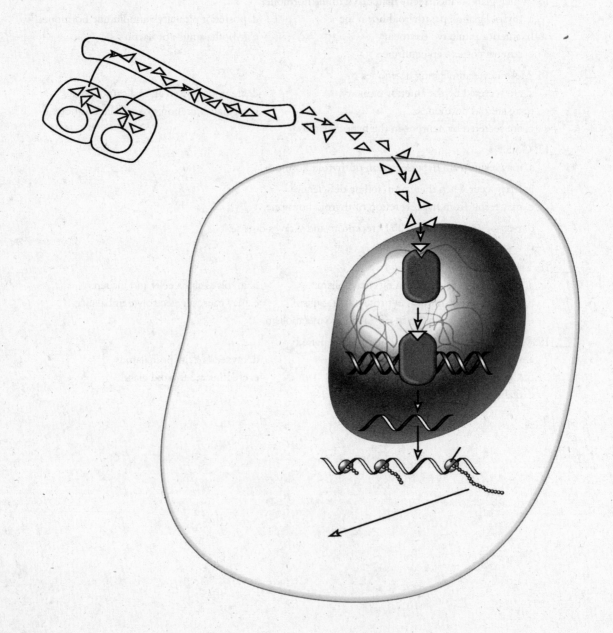

Color the parts of the illustration below as indicated.

RED ☐ receptor
GREEN ☐ hormone
YELLOW ☐ cytosol
BLUE ☐ second messengers
ORANGE ☐ extracellular fluid
BROWN ☐ plasma membrane

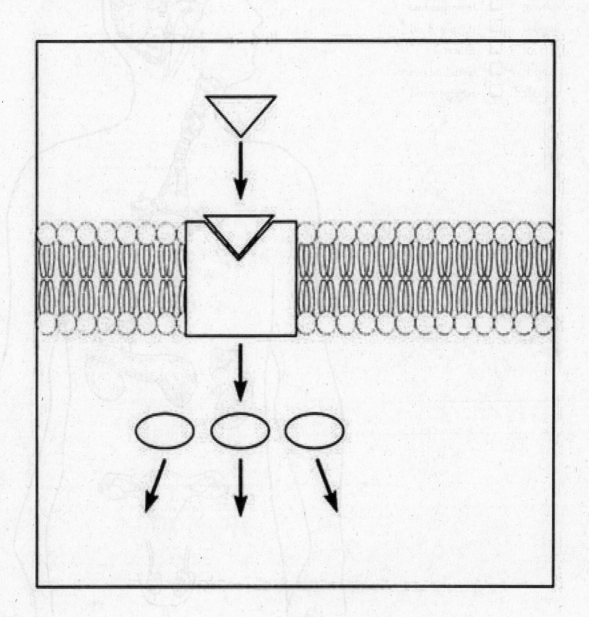

Color the parts of the illustration below as indicated.

RED	☐	pituitary gland
GREEN	☐	pineal gland
YELLOW	☐	thyroid gland
BLUE	☐	parathyroid gland
ORANGE	☐	hypothalamus
BROWN	☐	thymus gland
TAN	☐	pancreas
PINK	☐	testis and ovary
VIOLET	☐	adrenal gland

CHAPTER 50

❑

Reproduction

For a species to survive, it is essential that it replace individuals that die. Animals do this either asexually or sexually. In asexual reproduction, a single parent splits, buds, or fragments, giving rise to two or more offspring that are genetically identical to the parent. In sexual reproduction, sperm contributed by a male parent, and an egg contributed by a female parent, unite, forming a zygote that develops into a new organism. This method of reproduction promotes genetic variety among the offspring, and therefore, gives rise to individuals that may be better able to survive than either parent. Depending on the species, fertilization occurs either outside or inside the body. Some animals have both asexual and sexual stages. In some, the unfertilized egg may develop into an adult. In still others, both male and female reproductive organs may occur in the same individual. More typically, animals have only a sexual stage, fertilization is required for development to occur, and male and female reproductive organs occur only in separate individuals. The complex structural, functional, and behavioral processes involved in reproduction in vertebrates are regulated by hormones secreted by the brain and gonads. Several hormones regulate the birth process in humans. There are a variety of effective methods of birth control. Next to the common cold, sexually transmitted diseases are the most prevalent communicable diseases in the world.

REVIEWING CONCEPTS

Fill in the blanks.

INTRODUCTION

1. Hermaphrodite have (a) _____ and (b) _____ reproductive organs.

ASEXUAL AND SEXUAL REPRODUCTION

Asexual reproduction is an efficient strategy

2. In asexual reproduction, the offspring are _____ to the parent that produces them.

3. Sponges and cnidarians can reproduce by _____, which occurs when a group of expendable cells that separate from the "parent's" body.

4. _____ is a form of asexual reproduction whereby a "parent" breaks into several pieces, each piece giving rise to a new individual.

5. Parthenogenesis refers to the production of a complete organism from a(n) _____.

Most animals reproduce sexually

6. The union of an egg and a sperm cell produces a _____.

7. Many aquatic animals practice _____ in which the gametes meet outside the body.

8. _____ is a form of sexual reproduction in which a single individual produces both eggs and sperm.

Sexual reproduction increases genetic variability

    9. Sexual reproduction is thought to remove (a) _____ from a population and at the same time it maintains (b) _____.

HUMAN REPRODUCTION: THE MALE

The testes produce gametes and hormones

    10. The production of sperm is called (a) _____. The initial development of a sperm begins with an undifferentiated stem cell called a (b)_____, which enlarges to form the (c)_____ that goes through meiosis.

    11. The _____ at the front of a sperm contains enzymes that assist the sperm in penetrating an egg.

    12. _____ cells secrete nutrients for the developing sperm and secrete hormones and signaling molecules.

    13. The testes, housed in an external sac called the (a) _____. Within the body of the testes are the (b)_____ where sperm are produced.

    14. The _____ is the passage way that a testis takes as it descends through the abdominal wall into the scrotum.

A series of ducts store and transport sperm

    15. Sperm complete their maturation and are stored in the (a) _____, from which they are moved up into the body through the (b) _____. The next portion of the tubule, the (c) _____ passes through the prostate gland and then opens into the (d) _____.

The accessory glands produce the fluid portion of semen

    16. Human semen contains an average of about_____(#?) sperm.

    17. Sperm are suspended in fructose rich secretions that are produced by two paired glands called the (a) _____. The (b) _____ secretes an alkaline fluid to neutralize acidic conditions in the female vagina.

    18. Prior to intercourse, the _____ glands release a mucous secretion that lubricates the penis.

The penis transfers sperm to the female

    18. The penis consists of an elongated (a) _____ that terminates in an expanded portion called the (b)_____, which is partially covered by a fold of skin called the (c)_____. An operation that removes this "extra" skin is known as (d)_____.

    19. The penis contains three columns of erectile tissue, two (a) _____ and one (b) _____. When they become engorged with (c)_____, the penis erects.

Testosterone has multiple effects

    20. Testosterone is classified as an (a) _____ and is the principal male sex hormone. In some target tissues, testosterone is converted to other (b) _____. Interestingly, in brain cells testosterone is converted to (c) _____, a female sex hormone.

The hypothalamus, pituitary gland, and testes regulate male reproduction

21. The two gonadotropic hormones, (a) _____ and (b) _____, along with testosterone, stimulate sperm production.

22. LH stimulates secretion of the hormone (a) _____, which is responsible for establishing and maintaining primary and secondary sex characteristics in the male. This male hormone is produced by the (b) _____ cells in the testes.

HUMAN REPRODUCTION: THE FEMALE

The ovaries produce gametes and sex hormones

23. Formation of ova is called (a) _____. In this process, stem cells in the fetus, the (b) _____, enlarge, forming the (c) _____, which begin meiosis before birth.

24. A primary oocyte and the cells immediately around it comprise the (a) _____. At puberty, the hormone (b) _____ stimulates development of these cells, causing some of them to complete meiosis. One of the resulting cells, the (c) _____, continues development to form the ovum.

25. After ovulation, the portion of the follicle that remains in the ovary develops into a temporary endocrine gland called the (a) _____ which secretes (b) _____.

The oviducts transport the secondary oocyte

26. In a tubal pregnancy, the embryo begins to develop in the _____ rather than the uterus.

The uterus incubates the embryo

27. Embryos normally implant in the (a) _____ of the uterus. If there is no embryo, or the embryo fails to implant, this layer is shed during the period of (b) _____.

28. The lower portion of the uterus, the _____, extends slightly into the vagina.

The vagina receives sperm

29. The vagina serves as a _____ for sperm.

The vulva are external genital structures

30. The female external genitalia are collectively known as the _____.

31. The most sensitive part in the vulva is the _____.

The breasts function in lactation

32. The gland cells of the breast are arranged in grapelike clusters called _____.

33. _____ is the production of milk.

34. Milk production begins as a result of stimulation by the hormone _____.

The hypothalamus, pituitary gland, and ovaries regulate female reproduction

35. The first day of menstrual bleeding marks the first day of the menstrual cycle. _____ occurs at about day 14 in a 28-day menstrual cycle.

36. FSH stimulates follicle development. The developing follicles release (a) _____, which stimulates development of the endometrium. The two hormones (b) _____ and (c) _____ stimulate ovulation, and (d)_____ also promotes development of the corpus luteum.

37. The corpus luteum secretes _____, which stimulates final preparation of the uterus for possible pregnancy.

Menstrual cycles stop at menopause

38. A change in the (a) _____ triggers menopause, a period when (b)_____ are no longer produced and a woman becomes (c)_____.

Most mammals have estrous cycles

39. In an estrous cycle, the endometrium is _____ if conception does not occur.

FERTILIZATION, PREGNANCY, AND BIRTH

Fertilization is the fusion of sperm and egg

40. Fertilization and the establishment of pregnancy together are referred to as _____.

41. After being deposited in the female reproductive tract, sperm generally retain the ability to fertilize and ovum for about (a) _____ hours. The ovum remains fertile for up to (b) _____ hours after ovulation.

42. Uterine and oviduct smooth muscle contractions help to move the sperm toward the ovum. These contractions are stimulated by (a) _____ secreted by the female and (b) _____ present in the semen.

Hormones are necessary to maintain pregnancy

43. Membranes that develop around an embryo secrete a peptide hormone called _____that signals the corpus luteum to continue to function.

44. Membranes surrounding the embryo and uterine tissue form the _____, an organ of exchange between the embryo and the mother.

The birth process depends on a positive feedback system

45. The mechanisms that terminate pregnancy and initiate the birth process are called _____.

46. _____ is the hormone that is responsible for initiating uterine contractions as part of labor.

HUMAN SEXUAL RESPONSE

47. Sexual stimulation results in two basic physiological responses:

(a) _____ and

(b) _____.

48. The four phases of the sexual response cycle are

(a) _____

(b) _____

(c) _____ and (d) _____.

BIRTH CONTROL METHODS AND ABORTION

Many birth control methods are available

49. Any deliberate separation of sexual intercourse from reproduction is called
_____.

Most hormone contraceptives prevent ovulation

50. The most common oral contraceptives are combinations of

(a) _____, and (b) _____.

Intrauterine devices are widely used

51. The IUD is used by an estimated _____(#?)
women around the world.

Barrier methods of contraceptive methods include the diaphragm and condom

52. The condom is the only contraceptive device that affords some protection
against (a) _____,

and (b) _____.

Emergency contraception is available

53. The most common types of emergency contraception are
_____, which can decrease the probability of
pregnancy by about 89%.

Sterilization renders an individual incapable of producing offspring

54. Male sterilization is accomplished by performing a

(a)_____ in which the (b) _____ is cut.

55. Female sterilization is accomplished by performing a

(a) _____ in which the (b) _____ is cut.

Future contraceptives may control regulatory peptides

56. One method of contraception being investigated involves the stimulation of
the immune system to produce _____ which would act
against hormones such as hCG.

Abortions can be spontaneous or induced

57. The drug RU-496 is used to interrupt a pregnancy. It works by
_____,

but it does not activate them. This causes the endometrium to break down and
uterine contractions begin.

SEXUALLY TRANSMITTED DISEASES

58. (a)_____ is the most
common STD in the United States. About
(b)_____(#?) people are currently infected. At least
(c)_____(%?) of sexually active males and females will become infected at
sometime during their lives.

BUILDING WORDS

Use combinations of prefixes and suffixes to build words for the definitions that follow.

<u>Prefixes</u>	<u>The Meaning</u>	<u>Suffixes</u>	<u>The Meaning</u>
acro-	extremity, height	-ectomy	excision
circum-	around, about	-gen(esis)	production of
contra-	against, opposite, opposing	-some	body
endo-	within		
intra-	within		
oo-	egg		
partheno-	virgin		
post-	behind, after		
pre-	before, prior to, in advance of, early		
spermato-	seed, "sperm"		
vaso-	vessel		

<u>Prefix</u>	<u>Suffix</u>	<u>Definition</u>
_____	_____	1. The production of an adult organism from an egg in the absence of fertilization; virgin development.
_____	_____	2. The production of sperm.
_____	_____	3. The organelle (body) at the extreme tip of the sperm head that helps the sperm penetrate the egg.
_____	-cision	4. The removal of all or part of the foreskin by cutting all the way around the penis.
_____	_____	5. The production of eggs or ova.
_____	-metrium	6. The lining within the uterus.
_____	-ovulatory	7. Pertains to a period of time after ovulation.
_____	-ovulatory	8. Pertains to a period of time prior to ovulation.
vas-	_____	9. Male sterilization in which the vas deferentia are partially excised.
_____	-congestion	10. The engorgement of erectile tissues with blood.
_____	-uterine	11. Located or occurring within the uterus.
_____	-ception	12. Birth control; techniques or devices opposing conception.

MATCHING

Terms:

a.	Budding	f.	Fragmentation	k.	Prostate gland
b.	Clitoris	g.	Ovary	l.	Scrotum
c.	Corpus luteum	h.	Oviduct	m.	Semen
d.	Ejaculatory duct	i.	Ovulation	n.	Testosterone
e.	Fertilization	j.	Progesterone	o.	Zygote

For each of these definitions, select the correct matching term from the list above.

_____ 1. Fluid composed of sperm suspended in various glandular secretions that is ejaculated from the penis during orgasm.

_____ 2. A hormone produced by the corpus luteum of the ovary and by the placenta; acts with estradiol to regulate menstrual cycles and to maintain pregnancy.

_____ 3. A gland in male mammals that secretes an alkaline fluid that is part of the seminal fluid.

_____ 4. A small, erectile structure at the anterior part of the vulva in female mammals; homologous to the male penis.

_____ 5. One of the paired female gonads; responsible for producing eggs and sex hormones.

_____ 6. Pocket of endocrine tissue in the ovary that is derived from the follicle cells; secretes the hormone progesterone and estrogen.

_____ 7. The fusion of the male and female gametes; results in the formation of a zygote.

_____ 8. A short duct that passes through the prostate gland and connects the vas deferens to the urethra.

_____ 9. The release of a mature "egg" from the ovary.

_____ 10. The external sac of skin found in most male mammals that contains the testes and their accessory organs.

_____ 11. A form of asexual reproduction in which the adult breaks into pieces and new offspring are formed.

_____ 12. This hormone is produced by interstitial cells.

MAKING COMPARISONS

Fill in the blanks.

Endocrine Gland	Hormone	Target Tissue	Action in Male	Action in Female
Hypothalamus	GnRH	Anterior pituitary	Stimulates release of FSH and LH	Stimulates release of FSH and LH
#1	FSH	#2	#3	#4
Anterior pituitary	LH	Gonad	#5	#6
#7	#8	General	Establishes and maintains primary & secondary sex characteristics	Not produced
#9	#10	General	Not produced	Establishes and maintains primary & secondary sex characteristics

MAKING CHOICES

Place your answer(s) in the space provided. Some questions may have more than one correct answer.

_____ 1. The hormone in urine or blood that indicates pregnancy is
    a. GnRH.
    b. FSH.
    c. hCG.
    d. inhibin.
    e. ABP.

_____ 2. The tube(s) that transport(s) gametes away from gonads is/are the
    a. vas deferens.
    b. seminal vesicle.
    c. sperm cord.
    d. vagina.
    e. oviduct.

_____ 3. A human female at birth already has
    a. primary oocytes in the first prophase.
    b. preovulatory Graafian follicles.
    c. all the formative oogonia she will ever have.
    d. some secondary oocytes in meiosis II.
    e. zona pellucida.

_____ 4. The _basic_ physiological responses that result from effective sexual stimulation are
    a. vasocongestion.
    b. erection.
    c. vaginal lubrication.
    d. muscle tension.
    e. orgasm.

_____ 5. Human chorionic gonadotropin (hCG)
    a. is produced by embryonic membranes.
    b. is secreted by the endometrium.
    c. affects functioning of the corpus luteum.
    d. is originally produced in the pituitary.
    e. causes menstrual flow to begin.

_____ 6. The hormone(s) primarily responsible for secondary sexual characteristics is/are
    a. luteinizing hormone.
    b. estrogen.
    c. follicle-stimulating hormone.
    d. testosterone.
    e. human chorionic gonadotropin.

_____ 7. The corpus luteum is directly or indirectly involved in
    a. secreting estrogen.
    b. secreting progesterone.
    c. stimulating glands to produce nutrients for an implanted embryo.
    d. responding to signals from embryonic membrane secretions.
    e. maintaining a newly implanted embryo.

_____ 8. Follicle-stimulating hormone (FSH) stimulates development of
    a. seminiferous tubules.
    b. interstitial cells.
    c. sperm cells.
    d. the corpus luteum.
    e. ovarian follicles.

_____ 9. When used properly, the most effective form of birth control is
    a. condoms.
    b. douching.
    c. oral contraception.
    d. the rhythm method.
    e. spermicidal sponges.

_____10. The first measurable response to effective sexual stimulation
    a. is orgasm.
    b. is ejaculation.
    c. occurs in the excitement phase.
    d. is penile erection.
    e. is vaginal lubrication.

_____11. The form(s) of birth control that can definitely deter contraction of sexually transmitted diseases is/are

   a. condoms.   d. diaphragms.
   b. douching.   e. spermicides.
   c. oral contraception.

_____12. Your authors consider a count of 200 million sperm in a single ejaculation as

   a. clinical sterility.   d. about average.
   b. above average.   e. impossible.
   c. below average.

_____13. A Papanicolaou test (Pap smear)

   a. can detect early cancer.   d. looks at cells from the menstrual flow.
   b. requires an abdominal incision.   e. examines cells scraped from the cervix.
   c. involves examination of cells taken from the lower portion of the uterus.

_____14. Primitive, undifferentiated stem cells that line the seminiferous tubules are called

   a. sperm.   d. secondary spermatocytes.
   b. spermatids.   e. spermatogonia.
   c. primary spermatocytes.

_____15. The external genitalia of the human female are collectively known as the

   a. vagina.   d. major and minor lips.
   b. clitoris.   e. vulva.
   c. mons pubis.

_____16. In gametogenesis in the human male, the cell that develops after the first meiotic division is the

   a. sperm.   d. secondary spermatocyte.
   b. spermatid.   e. spermatogonium.
   c. primary spermatocyte.

_____17. The portion of the ovarian follicle that remains behind after ovulation

   a. develops into the corpus luteum.   d. is surrounded by the zona pellucida.
   b. becomes a gland.   e. transforms into the endometrium.
   c. is called the Graafian follicle.

_____18. The term gonad refers to the

   a. penis.   d. ovaries.
   b. testes.   e. gametes.
   c. vagina.

_____19. Human sperm cell production occurs in the

   a. epididymis.   d. accessory glands.
   b. sperm tube.   e. seminiferous tubules.
   c. testes.

_____20. The normal, periodic flow of blood from the vagina is associated with

   a. ovulation.   d. removal of the inner lining of the vagina.
   b. orgasm.   e. miscarriages.
   c. sloughing of the endometrium.

_____21. Human female breast size is primarily a function of the amount of

   a. alveoli.   d. milk the woman can produce after giving birth.
   b. milk glands.   e. colostrum.
   c. adipose tissue (fat).

_____22. Which of the following is/are true of the human penis?

    a. It has three erectile tissue bodies.   d. It contains a bone.

    b. Circumcision involves removal of the prepuce.   e. Erection results from increased blood pressure.

    c. The glans is an expanded portion of the spongy body.

_____23. The term "virgin development" refers to which method of reproduction?

    a. budding   d. hermaphroditism

    b. fragmentation   e. parthenogenesis

    c. fission

_____24. Testosterone

    a. stimulates descent of the testes into the scrotum.   d. is produced by interstitial cells in the testes.

    b. secretion is under the influence of FSH.   e. is produced by Sertoli cells in the testes.

    c. is down regulated by inhibin.

_____25. Seminal vesicles

    a. secrete fructose.   d. secrete an alkaline fluid.

    b. secrete fibrinogen.   e. secrete clotting enzymes.

    c. secrete prostaglandins.

_____26. Low sperm counts may be the result of

    a. smoking.   d. decreased PSA levels.

    b. alcohol use.   e. high levels of ABP.

    c. exposure to PCBs.

_____27. Which of the following is incorrectly paired?

    a. inhibin : anterior pituitary

    b. androgen binding protein : Sertoli cells

    c. oxytocin : stimulation of uterine contractions

    d. luteinizing hormone : development of the endometrium

    e. follicle stimulating hormone : Sertoli cells

VISUAL FOUNDATIONS

Color the parts of the illustrations below as indicated. Label the first meiotic division and second meiotic division.

RED ☐ sperm

PINK ☐ large chromosome

BLUE ☐ small chromosome

ORANGE ☐ primary spermatocyte

BROWN ☐ secondary spermatocyte

YELLOW ☐ spermatids

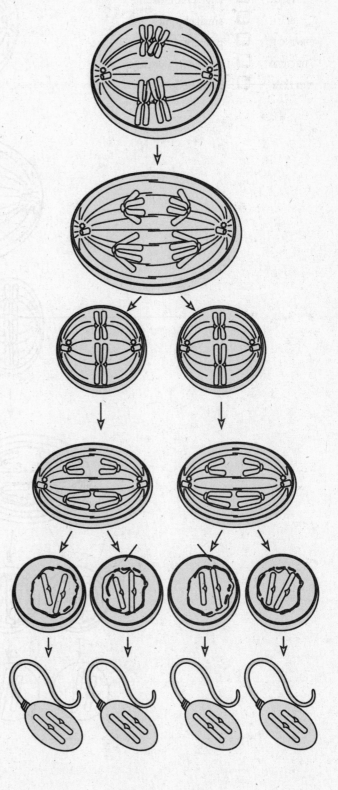

Color the parts of the illustrations below as indicated. Label first and second meiotic divisions.

RED	☐	ovum
PINK	☐	large chromosome
BLUE	☐	small chromosome
ORANGE	☐	primary oocyte
BROWN	☐	secondary oocyte
YELLOW	☐	polar body

Animal Development

Development encompasses all of the changes that take place during the life of an organism from fertilization to death. Guided by instructions encoded in the DNA of the genes, the organism develops from one cell to billions, from a formless mass of cells to an intricate, highly specialized and organized organism. Development is a balanced combination of several interrelated processes — cell division, increase in the number and size of cells, cellular movements that arrange cells into specific structures and appropriate body forms, and biochemical and structural specialization of cells to perform specific tasks. The stages of early development, which are basically similar for all animals, include fertilization, cleavage, gastrulation, and organogenesis. All terrestrial vertebrates have four extraembryonic membranes that function to protect and nourish the embryo. Environmental factors, such as nutrition, vitamin and drug intake, cigarette smoke, and disease-causing organisms, can adversely influence developmental processes. Aging is a developmental process that results in decreased functional capacities of the mature organism.

REVIEWING CONCEPTS

Fill in the blanks.

INTRODUCTION

1. During development, growth occurs primarily by an increase in

   (a) _____ and

   (b) _____.

DEVELOPMENT OF FORM

2. The process by which cells become specialized is called _____.

3. Cells undergo _____, a process during which they become increasingly organized, shaping the intricate pattern of tissues and organs that characterizes a multicellular animal.

FERTILIZATION

4. Fertilization is the result of a sperm cell and an egg cell fusing to form a

   _____.

5. Fertilization restores the (a) _____ number of chromosomes, during which time the (b) _____ of the individual is determined.

The first step in fertilization involves contact and recognition

6. A mammalian egg is enclosed by a thick, noncellular (a) _____ which is surrounded by a layer of (b) _____ cells.

7. _____, a species-specific protein on the acrosome, adheres to receptors on the vitelline envelope.

8. Before a mammalian sperm can participate in fertilization, it must first undergo (a) _____, a maturation process that occurs in the female reproductive tract. During this process, sperm become increasingly motile and capable of undergoing an (b) _____ when they encounter an egg.

Sperm entry is regulated

9. Fertilization of the egg by more than one sperm is _____.

10. Two reactions prevent multiple sperm from entering an egg. During the (a) _____ the egg cell membrane depolarizes as the result of the opening of (b) _____ channels. During the (c) _____, enzymes and other proteins are released into the space between the plasma membrane of the egg and the vitelline envelope.

Fertilization activates the egg

11. During activation, aerobic respiration (a) _____ (increases, decreases), enzymes and proteins become active and a burst of (b)_____ synthesis occurs.

Sperm and egg pronuclei fuse, restoring the diploid state.

12. It is believed that after the sperm enters the egg, it is guided toward the egg nucleus by a system of _____.

CLEAVAGE

13. The inherent potential of the egg to form all of the different cell types of the fully differentiated individual is referred to as _____.

14. The zygote undergoes cleavage, first forming a morula and then a _____.

The pattern of cleavage is affected by yolk

15. The isolecithal eggs of (a) _____ undergo (b) _____ cleavage, that is, the entire egg divides into cells of about the same size. (c)_____ cleavage of isolecithal eggs is common in chordates and echinoderms, while some annelids and mollusks undergo (d)_____ cleavage.

16. The (a) _____ cleavage that occurs in the telolecithal eggs of reptiles and birds is restricted to the (b)_____.

Cleavage may distribute developmental determinants

17. Mosaic development is a consequence of the _____ of important materials in the cytoplasm of the zygote.

18. In some species fertilization triggers a rearrangement of the cytoplasm. The rearranged cytoplasm directly opposite the site of sperm penetration forms a region called the _____ which is thought to contain growth factors.

Cleavage provides building blocks for development

19. _____ are important in helping cells recognize one another and determining which ones adhere to form tissues.

GASTRULATION

20. During gastrulation, the _____ becomes a three-layered embryo.

21. Embryonic layers are collectively called (a) _____. The outermost is the (b), _____ next is the (c) _____, and the innermost is the (d) _____.

The amount of yolk affects the pattern of gastrulation

22. In the amphibian, cells from the animal pole move down over the yolk-rich cells and invaginate, forming the dorsal lip of the _____.

23. In the bird, invagination occurs at the _____ and no archenteron forms.

ORGANOGENESIS

24. The (a) _____ induces the formation of the neural plate, from which cells migrate downward thus forming the (b) _____, flanked on each side by neural folds. When neural folds fuse, the (c) _____ is formed. This structure gives rise to the brain and spinal cord also known as the (d) _____.

25. (a)_____ meet the (b)_____ thereby forming branchial arches that give rise to elements of the face, jaws, and neck.

EXTRAEMBRYONIC MEMBRANES

26. Terrestrial vertebrates have four extraembryonic membranes. The outermost is the (a) _____ In birds and reptiles the primary function of this membrane is (b) _____. The (c)_____ surrounds the fluid-filled space in which the embryo develops. The (d) _____ is an outgrowth of the digestive tract. In birds and reptiles it stores (e) _____. The fourth membrane is the (f) _____.

HUMAN DEVELOPMENT

27. The (a) _____ dissolves when the embryo enters the uterus from the ovarian tube. At this time, the embryo is in the morula stage, but now begins to differentiate into a (b) _____.

28. The outer layer of embryonic cells, or _____, forms the chorion and amnion.

29. While the developing embryo floats in the uterus prior to implantation it is nourished by _____.

30. The trophoblast cells of the blastocyst secrete enzymes that digest an opening into the uterine wall in a process called (a) _____. The process begins at about the (b) _____ day of development.

The placenta is an organ of exchange

31. In placental mammals, the placenta develops from the (a) _____ of the embryo and from (b) _____ tissue of the mother.

32. From the time the embryo begins to implant the trophoblastic cells release
   (a) _____ which signals the
   (b) _____ that the pregnancy has begun.

Organ development begins during the first trimester

33. Gastrulation occurs during the _____ and _____ weeks of human development.

34. At the beginning of the _____ month, the embryo is referred to as a fetus.

35. Sex can be determined by external observation by the
   _____.

Development continues during the second and third trimesters

36. During the (a) _____ month of pregnancy the cerebrum grows rapidly and develops (b)_____.

More than one mechanism can lead to a multiple birth

37. (a) _____ twins develop when two eggs have been fertilized and (b) _____ twins develop when the inner cell mass subdivides into two separate masses.

38. A sonogram is a picture taken by the technique known as _____.

Environmental factors affect the embryo

39. Drugs or other substances that interfere with morphogenesis are called
   _____.

40. Ultrasound imaging techniques (a) _____, the analysis of intrauterine fluids, and (b) _____ are helpful in anticipating potential birth defects.

The neonate must adapt to its new environment

41. A neonate's first breath may be initiated by _____ in the blood after the umbilical cord is cut.

Aging is not a uniform process

42. Women live an average of _____(#?) years longer than men.

Homeostatic response to stress decreases during aging

43. Cell aging appears to be related to the fact that most somatic cells lose the ability to _____.

BUILDING WORDS

Use combinations of prefixes and suffixes to build words for the definitions that follow.

Prefixes	The Meaning		Suffixes	The Meaning
arch-	primitive		-age	collection of
acr(o)	tip		-blast	embryo
blast(o)-	bud, sprout		-chord	cord
cleav-	to divide		-cyst	sac
fet-	pregnant		-enteron	gut
holo-	whole, entire		-mere	part
iso-	equal		-por(e)	opening
mero-	part, partial		-som(e)	body
neo-	new, recent		-us	person
noto-	back			
telo-	end			
tri-	three			
tropho-	nourishment			

Prefix	Suffix	Definition
_____	-lecithal	1. Having an accumulation of yolk at one end (vegetal pole) of the egg.
_____	-lecithal	2. Having a fairly equal distribution of yolk in the egg.
_____	_____	3. Any cell that is a part of the early embryo (specifically, during cleavage).
_____	_____	4. The blastula of the mammalian embryo.
_____	_____	5. The opening into the archenteron of an early embryo.
_____	-blastic	6. A type of cleavage in which the entire egg divides.
_____	-blastic	7. A type of cleavage in which only the blastodisc divides; partial cleavage.
_____	_____	8. The central cavity resulting from gastrulation; the primitive digestive system.
_____	_____	9. The extraembryonic part of the blastocyst that chiefly nourishes the embryo or that develops into fetal membranes with nutritive functions.
_____	-mester	10. A term or period of three months.
_____	-nate	11. A newborn child.
_____	_____	12. The cap at the head of a sperm cell.
_____	_____	13. A rapid series of mitotic divisions with no period of growth during each cell cycle.
_____	_____	14. Supporting rod of mesodermal cartilage-like cells.
_____	_____	15. The time of intrauterine development between week 9 and birth.

MATCHING

Terms:

a. Amnion
b. Animal pole
c. Blastula
d. Chorion
e. Cleavage

f. Ectoderm
g. Endoderm
h. Fertilization
i. Gastrula
j. Germ layer

k. Implantation
l. Morphogenesis
m. Placenta
n. Vegetal pole

For each of these definitions, select the correct matching term from the list above.

_____ 1. A membrane that forms a fluid-filled sac for the protection of the developing embryo.

_____ 2. The first of several cell divisions in early embryonic development that converts the zygote into a multicellular blastula.

_____ 3. Any of the three embryonic tissue layers.

_____ 4. The yolk pole of a vertebrate or echinoderm egg.

_____ 5. Usually a spherical structure produced by cleavage of a fertilized ovum; consists of a single layer of cells surrounding a fluid-filled cavity.

_____ 6. The outer germ layer that gives rise to the skin and nervous system.

_____ 7. The attachment of the developing embryo to the uterus of the mother.

_____ 8. The development of the form and structures of an organism and its parts.

_____ 9. Early stage of embryonic development during which the embryo has three layers and is cup-shaped.

_____10. The fusion of the egg and the sperm, resulting in the formation of a zygote.

_____11. Embryonic contribution to the placenta.

MAKING COMPARISONS

Fill in the blanks.

Human Developmental Process	Approximate Time	Event
Fertilization	0 hour	Restores the diploid number of chromosomes
#1	2, 3 weeks	Blastocyst becomes a three-layered embryo
Early cleavage	#2	Formation of the two-cell stage
#3	#4	Activates the egg
Implantation	#5	#6
#7	0 hour	Establishes the sex of the offspring
#8	2.5 weeks	Neural plate begins to form
#9	#10	Morula is formed, reaches uterus

MAKING CHOICES

Place your answer(s) in the space provided. Some questions may have more than one correct answer.

_____ 1. The liver, pancreas, trachea, and pharynx
    a. arise from Hensen's node.
    b. are formed prior to the primitive streak.
    c. are derived from endoderm.
    d. are derived from ectoderm.
    e. are derived from mesoderm.

_____ 2. A hollow ball of several hundred cells
    a. is a morula.
    b. is a blastula.
    c. surrounds a fluid-filled cavity.
    d. is the gastrula.
    e. contains a blastocoel.

_____ 3. Which of the following organ systems is the first to form during organogenesis?
    a. circulatory system
    b. reproductive system
    c. excretory system
    d. nervous system
    e. muscular system

_____ 4. The duration of pregnancy is referred to as the
    a. parturition.
    b. induction cycle.
    c. gestation period.
    d. neonatal period.
    e. prenatal period.

_____ 5. The fertilization envelope (membrane)
    a. promotes binding of sperm to egg.
    b. facilitates egg-sperm recognition.
    c. prevents polyspermy.
    d. expedites entry of sperm into egg.
    e. blocks entrance of sperm.

_____ 6. The cavity within the embryo that is formed during gastrulation is the
    a. archenteron.
    b. gastrulacoel.
    c. blastocoel.
    d. blastopore.
    e. neural tube.

_____ 7. An embryo is called a fetus
    a. when the brain forms.
    b. when the limbs forms.
    c. when its heart begins to beat.
    d. after two months of development.
    e. once it is firmly implanted in the endometrium.

_____ 8. Embryonic cells are arranged in three distinct germ layers (endoderm, mesoderm, ectoderm) during
    a. cleavage.
    b. gastrulation.
    c. blastulation.
    d. organogenesis.
    e. blastocoel formation.

_____ 9. Amniotic fluid
    a. replaces the yolk sac in mammals.
    b. is secreted by the allantois.
    c. is secreted by an embryonic membrane.
    d. fills the space between the amnion and chorion.
    e. fills the space between the amnion and embryo.

_____ 10. Recognition of species compatibility between a sperm and an egg is due to the recognition protein
    a. acrosin.
    b. actin.
    c. vitelline.
    d. bindin.
    e. pellucidin.

_____11. The organ of exchange between a placental mammalian embryo and its mother develops from

    a. chorion and amnion.

    b. chorion and allantois.

    c. chorion and uterine tissue.

    d. amnion and uterine tissue.

    e. amnion and allantois.

_____12. Eggs that have a large amount of yolk concentrated at one pole

    a. are isolecithal.

    b. are telolecithal.

    c. have a metabolically active animal pole.

    d. have a vegetal pole that is metabolically active.

    e. do not have food for the embryo in the egg.

_____13. _____ gives rise to skeletal tissues, muscle, and the circulatory system.

    a. Ectoderm

    b. Mesoderm

    c. Endoderm

    d. One of the germ layers

    e. The blastocoel

_____14. Fertilization

    a. occurs in the upper portion uterus.

    b. is immediately followed by the cortical reaction

    c. is immediately preceded by the acrosome reaction.

    d. occurs about 7 days prior to implantation.

    e. restores polyploidy to the egg.

_____15. Which of the following is mismatched?

    a. capacitation : sperm maturation

    b. blastodisc : vegetable pole

    c. spiral cleavage : diagonal cytokinesis

    d. notochord : mesoderm

    e. primitive groove : blastopore

_____16. Calcium ions

    a. are involved in egg depolarization.

    b. are involved in the cortical reaction.

    c. are involved in capacitation.

    d. stimulate morphogenesis.

    e. stimulate the formation of Hensen's node.

_____17. Mammalian zygotes

    a. have a homogenous cytoplasm.

    b. exhibit regulative development.

    c. exhibit mosaic development.

    d. follow rigid developmental patterns.

    e. rely on surface proteins for tissue formation.

_____18. The archenteron

    a. opens to the outside via the blastopore.

    b. opening becomes the anus in deuterstomes.

    gut.

    c. arises from the oral groove.

    d. is lined by tissue arising from endoderm.

    e. is the newly formed cavity of the developing

_____19. In humans, the umbilical cord

    a. contains two arteries.

    b. contains one vein.

    c. contains two veins.

    d. contains one artery.

    e. blood vessels are formed from blood vessels of the allantois.

VISUAL FOUNDATIONS

Color the parts of the illustration below as indicated. Also label the blastocyst and umbilical cord.

RED	☐ umbilical artery		BROWN	☐ trophoblast cells
GREEN	☐ inner cell mass		TAN	☐ uterine epithelium
YELLOW	☐ yolk sac		PINK	☐ embryo
BLUE	☐ amniotic cavity and amnion		VIOLET	☐ maternal blood and blood vessel
ORANGE	☐ chorionic cavity and chorion		GREY	☐ placenta

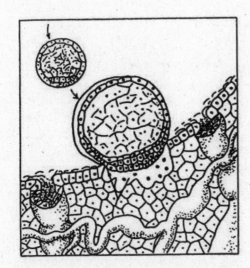
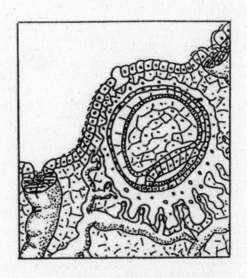
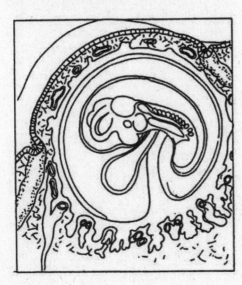
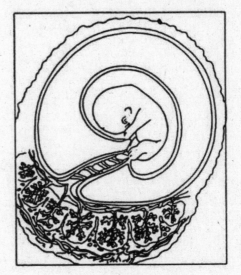

Animal Behavior

Animal behavior consists of those movements or responses an animal makes in response to signals from its environment; it tends to be adaptive and homeostatic. The behavior of an organism is as unique and characteristic as its structure and biochemistry. An organism adapts to its environment by synchronizing its behavior with cyclic change in its environment. Although behaviors are inherited, they can still be modified by experience. Behavioral ecology focuses on the interactions between animals and their environments and on the survival value of their behavior. Social behaviors, that is, interactions between two or more animals of the same species, have advantages and disadvantages. Advantages include confusing predators, repelling predators, and finding food. Disadvantages include competition for food and habitats, and increased ease of disease transmission. Some species that engage in social behaviors form societies. Some societies are loosely organized, whereas others have a complex structure. A system of communication reinforces the organization of the society. Animals communicate in a wide variety of ways; some use sound, some use scent, and some use pheromones. Some invertebrate societies exhibit elaborate and complex patterns of social interactions, such as the bees, ants, wasps, and termites. Vertebrate societies are far less rigid.

REVIEWING CONCEPTS

Fill in the blanks.

INTRODUCTION

1. Behavior refers to the responses an organism makes to _____ from its environment.

2. An animal's behavior is the product of _____ acting on its phenotype and indirectly on the genotype.

3. _____ is the study of behavior in natural environments from an evolutionary perspective.

BEHAVIOR AND ADAPTATION

4. An understanding of behavior requires consideration of

    (a) _____ causes which immediate causes of behavior such the involvement of genetics and (b) _____ causes which are the why questions of behavior.

Behaviors have benefits and costs

5. _____ is an individual's reproductive success as measured by the number of viable offspring it produces

Genes interact with environment

6. (a)_____ behavior is inborn, genetically programmed behavior, whereas (b)_____ behavior is behavior that has been modified in response to environmental experience.

Behavior depends of physiological readiness

7. Behavior is influenced primarily by two systems of the body, namely, the (a) _____ and (b) _____ systems.

Many behavior patterns depend on motor programs

8. Coordinated sequences of muscle actions are called _____.

9. Fixed action pattern (FAP) behaviors are elicited by a _____.

LEARNING: CHANGING BEHAVIOR AS A RESULT OF EXPERIENCE

10. Learning is defined as a persistent change in behavior due to _____.

11. Animals that are poisonous display bright colors. This coloration is called _____.

An animal habituates to irrelevant stimuli

12. _____ is a type of learning in which an animal learns to ignore a repeated, irrelevant stimulus.

Imprinting occurs during an early critical period

13. Imprinting establishes a parent-offspring bond during a critical period early in development, ensuring that the offspring _____ the mother.

In classical conditioning, a reflex becomes associated with a new stimulus

14. In classical conditioning, an animal makes an association between an irrelevant stimulus and a normal body response. It occurs when a (a) _____ becomes a substitute for an (b) _____ stimulus.

In operant conditioning, spontaneous behavior is reinforced

15. In operant conditioning, the behavior of the animal is rewarded or punished after it performs a behavior discovered by chance. Repetitive rewards bring about (a) _____, and consistent punishment induces (b) _____.

16. Operant conditioning plays a role in the development of some behaviors that appear to be _____.

Animal cognition is controversial

17. Some animals have a form of cognition called _____ _____, which is the ability to adapt past experiences that may involve different stimuli to solve a new problem.

Play may be practice behavior

18. Hypotheses for the ultimate causes of play behavior include (a) _____, (b) _____, and (c) _____.

BIOLOGICAL RESPONSES TO ENVIROMENTAL STIMULI

Biological rhythms regulate many behaviors

19. Daily cycles of activity are known as (a) _____.

(b) _____ animals are most active during the day and

(c) _____ animals are most active at dawn and dusk.

20. In mammals, the master clock is located in the

_____ in the hypothalamus.

21. The _____ gland secretes melatonin a hormone that promotes sleep in humans.

Environmental signals trigger physiological responses that lead to migration

22. _____ is the periodic long-distance travel from one location to another.

23. _____ refers to travel by migrating animals in a specific direction.

24. (a) _____ requires compass sense and

(b) _____ sense.

FORAGING BEHAVIOR

25. Foraging involves locating and selecting food, as well as

(a) _____ and (b) _____food.

26. _____ refers to an animal's ability to obtain food in the most efficient manner.

COST AND BENEFITS OF SOCIAL BEHAVIOR

27. The interaction of two or more animals, usually of the same species defines

_____.

28. A _____ is an actively cooperating group of individuals belonging to the same species and often closely related.

Communication is necessary for social behavior

29. Among nonhuman mammals, only _____ are known to match sounds in communicating.

30. _____ are chemical signals that convey information between members of a species.

31. Chemoreceptor cells in the nose of animals comprise the _____ organ.

Dominance hierarchies establish social status

32. Once it is established, dominance hierarchy is one way to prevent _____ within a population.

33. In some animals, the hormone _____ is the apparent cause of their aggressive behaviors.

Many animals defend a territory

34. Most animals have a _____, a geographical area that they seldom leave.

Some insect societies are highly organized

35. Instructions for the honeybee society are (a) _____ and

(b) _____.

SEXUAL SELECTION

Animals of the same sex compete for mates

36. In _____, individuals of the same sex compete for mates.

Animals select quality mates

37. In _____ females select mates on the basis of some physical trait or some resource offered to them.

38. A _____ is a small display area in which the males of some species of insects, birds, and bats compete for females.

39. _____ ensure that the male is a member of the same species, and it provides the female with a means of evaluating the male.

Sexual selection favors polygynous mating systems

40. (a)_____ is a mating system in which males fertilize the eggs of many females during a breeding season, whereas (b)_____ is a mating system in which a female mates with several males.

Some animals care for their young

41. _____ is the contribution each parent makes in producing and rearing offspring.

HELPING BEHAVIOR

Altruistic behavior can be explained by inclusive fitness

42. The _____ is defined as the probability that two individuals inherit the same uncommon allele from a recent common ancestor.

43. Indirect selection, also called _____, is a form of natural selection that increases inclusive fitness through the breeding success of close relatives

Helping behavior may have alternative explanations

44. One example of kin selection includes _____, keeping watch for predators and warning of threats.

Some animals help nonrelatives

45. _____ is a type of behavior in which one animal helps another with no immediate benefit; however, at some later time the animal that was helped repays the debt.

CULTURE IN VERTEBRATE SOCIETIES

Some vertebrates transmit culture

46. Culture is behavior that is (a) _____, learned from other members of the group, and

    (b)_____.

47. Culture is not _____.

Sociobiology explains human social behavior in terms of adaptation

48. Sociobiology focuses on the evolution of social behavior through

    _____.

BUILDING WORDS

Use combinations of prefixes and suffixes to build words for the definitions that follow.

Prefixes	The Meaning		Suffixes	The Meaning
mono-	one		-andr(y)	male
pher-	to carry			
poly-	many, much, multiple, complex		-gyn(y)	female
socio-	social, sociological, society		-(h)ormone	to excite

Prefix	Suffix	Definition
_____	-biology	1. The school of ethology that focuses on the evolution of social behavior through natural selection.
_____	_____	2. A mating system in which a female mates with many males.
_____	_____	3. A mating system in which a male mates with many females.
_____	-gamy	4. Mating with a single partner during a breeding season.
_____	_____	5. Chemical signal secreted into the environment.

MATCHING

Terms:

a. Altruistic behavior
b. Biological clock
c. Circadian rhythm
d. Crepuscular animal
e. Diurnal animal
f. Ethology
g. Habituation
h. Imprinting
i. Innate behaviors
j. Migration
k. Nocturnal animal
l. Optimal foraging
m. Pheromone
n. Sign stimulus
o. Territoriality

For each of these definitions, select the correct matching term from the list above.

____ 1. Means by which activities of plants or animals are adapted to regularly-recurring changes.

____ 2. An animal that is most active during the day.

____ 3. Behaviors that are inherited and typical of the species.

____ 4. A substance secreted by organisms into the external environment that influences the development or behavior of other members of the same species.

____ 5. The theory that animals feed in a manner that maximizes benefits and/or minimizes costs.

____ 6. The process by which organisms become accustomed to a stimulus and cease to respond to it.

____ 7. A form of rapid learning by which a young bird or mammal forms a strong social attachment to an individual or object within a few hours after hatching or birth.

____ 8. Behavior in which one individual appears to act in such a way as to benefit others rather than itself.

____ 9. A daily cycle of activity around which the behavior of many organisms is organized.

____10. The study of animal behavior.

____11. An animal that is most active during the time around dusk and dawn.

____12. A simple signal that triggers a specific response.

MAKING COMPARISONS

Fill in the blanks.

Classification of Behavior	Example of the Behavior
Innate (FAP)	Egg rolling in graylag goose
#1	Wasp responding to cone arrangement to locate nest
#2	Dog salivating at sound of bell
#3	Child sitting quietly for praise
#4	Birds tolerant of human presence
#5	Chicks learning the appearance of the parent
#6	Primate stacking boxes to reach food

MAKING CHOICES

Place your answer(s) in the space provided. Some questions may have more than one correct answer.

_____ 1. A behavior that resembles a simple reflex in some ways and a volitional behavior in others, and is triggered by a sign stimulus is best described as a/an

    a. releaser.
    b. conditioned reflex.
    c. conditioned stimulus.
    d. behavioral pattern (formerly FAP).
    e. unconditioned reflex.

_____ 2. If your dog perks up its ears when you clap your hands, but stops perking its ears after you have repeatedly clapped your hands for a period of time, your dog has displayed a form of

    a. learning.
    b. operant conditioning.
    c. classical conditioning.
    d. sensory adaptation.
    e. habituation.

_____ 3. A change in your behavior derived from experience in your environment is the result of

    a. a FAP.
    b. learning.
    c. inherited behavioral characteristics.
    d. redirected behavior.
    e. habituation and/or sensitization.

_____ 4. Suppose you decide to perform an experiment with a type of lizard that instantly attacks any lizard with green scales on its sides. You place a live lizard with green scales on its sides in a cage next to a styrofoam block with green paint on its sides. Your experimental lizard ignores the spotted lizard, preferentially attacking the styrofoam block. The styrofoam-attacking lizard is responding to a/an

    a. inducer.
    b. sign stimulus.
    c. conditioned reflex.
    d. displacement syndrome.
    e. redirected behavioral syndrome.

_____ 5. Which of the following is/are used by a worker honey bee to indicate the distance of a food source?

    a. round dance
    b. waggle dance
    c. angle relative to a north–south axis
    d. angle relative to gravity
    e. angle relative to sun

_____ 6. The study of the behavior of animals in their natural environments from an evolutionary perspective is

    a. behavioral ecology.
    b. behavioral genetics.
    c. social behaviorist.
    d. behavioral ecology.
    e. neurobiology.

_____ 7. Genetically programmed behavior has been variously termed
    a. innate behavior.   d. instinct.
    b. inborn behavior.   e. patterned behavior.
    c. determinism.

_____ 8. Learning
    a. utilizes past experiences.   d. relies on inherited behavior.
    b. occurs during a critical period.   e. is adaptive.
    c. involves persistent changes in behavior.

_____ 9. You feed a group of just hatched chickens on a daily basis until they are adults. These chicks
    a. are using cognition.   d. will associate you with food.
    b. will imprint on their mother.   e. are following a genetically determined behavior.
    c. are demonstrating operant conditioning.

_____10. Motor programs
    a. may be mainly innate.   d. may have little flexibility in how they are carried out.
    b. may be invoked by a specific color and pattern.   e. may involve learning.
    c. are subject to extinction when a specific signal is not present.

_____ 11. The suprachiasmatic nucleus
    a. is located in the hypothalamus.   d. communicates with the pineal gland.
    b. receives input from the retina.   e. generates an internal 24 hour clock.
    c. is connected to the spinal cord.

_____ 12. Which of the following is not correctly paired?
    a. navigation : integration of distance and time   d. pineal gland : endocrine system
    b. circadian : 48 hours   e. critical period : first month of life
    c. lek : a nest

□

CHAPTER 53

Introduction to Ecology: Population Ecology

This chapter focuses on the study of populations as functioning systems. A population is all the members of a particular species that live together in the same area. Populations have properties that do not exist in their component individuals or in the community of which the population is a part. These properties include birth rates, death rates, growth rates, population density, population dispersion, age structure, and survivorship. Population ecology deals with the number of individuals of a particular species that are found in an area and how and why those numbers change over time. Populations must be understood in order to manage forests, field crops, game, fishes, and other populations of economic importance. Population growth is determined by the rate of addition of new individuals through birth and immigration and the loss of individuals through emigration and death of organisms. Population growth is limited by density-dependent and density-independent factors. Each species has its own survival strategy that enables it to survive. Most organisms fall into one of two survival strategies: One emphasizes a high rate of natural increase and the other emphasizes maintenance of a population near the carrying capacity of the environment. The principles of population ecology apply to humans as well as other organisms. Environmental degradation is related to population growth and resource consumption.

REVIEWING CONCEPTS

Fill in the blanks.

INTRODUCTION

1. Interactions among organisms are called (a) _____, while interactions between organisms and their nonliving, physical environment are called (b)_____.

FEATURES OF POPULATIONS

2. Populations of organisms have properties that individual organisms do not have. Some properties of importance are (a) _____

    (b) _____, (c) _____

    (d) _____, (e) _____,

    and (f) _____.

3. _____ considers the number of individuals of a species that are found in an area along with the dynamics of the population.

4. The study of changes in populations is known as _____.

Density and dispersion are important features of populations

5. _____ is the number of individuals of a species per unit of habitat area or volume at a given time.

6. Individuals within a population may exhibit a characteristic pattern of dispersion. For example, (a)_____ exists when individuals in a population are spaced throughout an area in a manner that is unrelated to the presence of others; (b)_____ occurs when individuals are more evenly spaced than would be expected from a random occupation of a given habitat; and (c)_____ occurs when individuals are concentrated in specific parts of the habitat.

CHANGES IN POPULATION SIZE

7. Ultimately, changes in the size of a population result from the difference between (a) _____ and (b) _____.

8. The rate of change in a population is expressed as:

   $\Delta N / \Delta t = N(b - d)$, where ΔN is the (a) _____, Δt is the (b)_____, b is the birth rate or (c)_____, and d is the death rate or (d)_____.

Dispersal affects the growth rate in some populations

9. (a)_____ occurs when individuals enter a population and thus increase in size; (b)_____ occurs when individuals leave a population and thus decrease its size.

Each population has a characteristic intrinsic rate of increase

10. The maximum rate at which a population of a given species could increase under ideal conditions when resources are abundant and its population density is low is known as its _____.

11. _____ is the accelerating population growth rate that occurs when optimal conditions allow a constant per capita growth rate.

No population can increase exponentially indefinitely

12. The largest population that can be sustained for an indefinite period by a particular environment is that environment's _____.

13. Population growth curves take a variety of forms, their shapes are primarily a function of biotic potential and limiting factors. The _____ growth curve shows the population's initial exponential increase, followed by a leveling out as the carrying capacity of the environment is approached.

FACTORS INFLUENCING POPULATION SIZE

Density-dependent factors regulate population size

14. Examples of density-dependent factors are (a) _____, (b) _____, and (c) _____.

15. Density-dependent factors are an example of a _____ system.

16. Density-dependent limiting factors are most effective at _____ (high or low?) population densities, and therefore they tend to stabilize populations.

17. At high population densities, it is not uncommon for the (a) _____ to go up and the (b) _____ to go down.

18. Competition that occurs within a population is (a) _____ and competition that occurs between two different populations is

    (b) _____.

19. In (a) _____, also called contest competition, certain dominant individuals obtain an adequate supply of the limited resource at the expense of other individuals in the population. In

    (b) _____, also called scramble competition, all the individuals in a population "share" the limited resource more or less equally.

Density-independent factors are generally abiotic

20. Some examples of density-independent factors that affect population size are

    (a) _____, (b) _____, and

    (c) _____.

LIFE HISTORY TRAITS

21. (a)_____ species expend their energy in a single, immense reproductive effort. (b)_____ species reproduce during several breeding seasons throughout their lifetimes.

22. _____ refers to a species potential capacity to produce offspring.

23. Each species has a life history strategy. (a)_____ are usually opportunists found in variable, temporary, or unpredictable environments where the probability of long-term survival is low.

    (b)_____ tend to be found in relatively constant or stable environments, where they have a high competitive ability.

Life tables and survivorship curves indicate mortality and survival

24. _____ is the probability that a given individual in a population or cohort will survive to a particular age.

25. A curve in which mortality is greatest at an early age is called a

    (a) _____ curve. A curve in which the probability of survival decreases more rapidly with increasing age is called a

    (b) _____ curve.

METAPOPULATIONS

26. Good habitats, called (a) _____, are areas where local reproductive success is greater than local mortality. Lower-quality habitats, called (b) _____, are areas where local reproductive success is less than local mortality.

HUMAN POPULATIONS

27. Zero population growth is the point at which the (a) _____ equals the (b) _____.

Not all countries have the same growth rate

28. The amount of time that it takes for a population to double in size is its

    (a) _____. In general, the shorter this time, the

    (b) _____ (more or less?) developed the country.

29. The number of children that a couple must produce to replace themselves is

    known as the (a) _____. This

    value is usually (b) _____ children in highly developed countries

    and (c) _____ children in developing countries.

30. The average number of children born to a woman during her lifetime is called

    the (a) _____. On a worldwide basis, this

    value is currently (b) _____ (higher or lower?) than the replacement

    value.

The age structure of a country helps predict future population growth

31. To predict the future growth of a population, it is important to know its age

    structure, which is the (a) _____ and (b) _____ of

    people at each age in a population.

**Environmental degradation is related to population growth and resource
   consumption**

32. People overpopulation occurs when the environment is worsening from the

    effect of having too many people. Consumption overpopulation occurs when

    each individual in a population consumes too large a share of resources. The

    (a) _____ (former or latter?) is the current problem in many

    developing nations, and the (b) _____ (former or latter?) is the

    current problem in most affluent, highly developed nations.

BUILDING WORDS

Use combinations of prefixes and suffixes to build words for the definitions that follow.

Prefixes	The Meaning		suffixes	The Meaning
bio-	life		-ic	pertaining to
ec-	dwelling		-ology	study of
inter-	between, among			
intra-	within			
intrins-	internally			

Prefix	Suffix	Definition
_____	-specific competition	1. Competition for resources within a population.
_____	-specific competition	2. Competition for resources among populations.
_____	-sphere	3. The entire zone of air, land, and water at the surface of the earth that is occupied by living things.
_____	_____	4. The study of how living organisms and the physical environment interact.
_____	_____	5. The rate of increase in a populations size under ideal conditions.

MATCHING

Terms:

a.	Intrinsic rate of increase	g.	Ecology	m.	Population
b.	Carrying capacity	h.	Emigration	n.	Population crash
c.	Contest competition	i.	Exponential growth	o.	Random dispersion
d.	Density-dependent factor	j.	Immigration	p.	Replacement-level fertility
e.	Density-independent factor	k.	Mortality	q.	Uniform dispersion
f.	Doubling time	l.	Natality		

For each of these definitions, select the correct matching term from the list above.

_____ 1. A pattern of spacing in which individuals in a population are spaced unpredictably.

_____ 2. Growth that occurs at a constant rate of increase over a period of time.

_____ 3. The rate at which organisms produce offspring.

_____ 4. The maximum number of organisms that an environment can support.

_____ 5. Any factor, such as climate, that does not depend on the density of populations.

_____ 6. The number of offspring a couple must produce in order to "replace" themselves.

_____ 7. The movement of individuals out of a population.

_____ 8. The maximum rate of increase of a species that occurs when all environmental conditions are optimal.

_____ 9. The amount of time it takes for a population to double in size, assuming that its current rate of increase does not change.

_____10. An abrupt decline in a population from a high to very low population density.

_____11. The study of the interactions between living things and their environment, both physical and biotic.

_____12. A group of organisms of the same species that live in the same geographical area at the same time.

_____13. Dominant individuals obtain an adequate supply of resources at the expense of others in the population.

_____14. Occurs when cacti in the desert are evenly spaced to maximize the available resources.

MAKING COMPARISONS

Fill in the blanks.

Examples of Environmental Factors That Affect Population Size	Is Factor Density-Dependent or Density-Independent?	Description of Factor
Predation	Density-dependent	As the density of a prey species increases, predators are more likely to encounter an individual of the prey species
Killing frost	#1	#2
Contest competition	#3	#4
Disease	#5	#6
Hurricane	#7	#8

MAKING CHOICES

Place your answer(s) in the space provided. Some questions may have more than one correct answer.

Questions 1-5 pertain to the following equation. Use the list of choices below to answer them.

$$\frac{\Delta N}{\Delta t} = N(b - d)$$

a. growth only   d. ΔN g. value "b"
b. rate of growth   e. Δt h. value "d"
c. change only   f. $b - d$ i. $\Delta N / \Delta t$

_____ 1. Natality.

_____ 2. Mortality.

_____ 3. Population change.

_____ 4. Rate of population change.

_____ 5. The parameter derived from the equation.

_____ 6. The equation $dN/dt = rN$ is used to derive the
a. biotic potential (r_m).   d. per capita growth rate (r).
b. instantaneous growth rate.   e. growth rate at carrying capacity.
c. decline in growth due to limiting factors.

_____ 7. During the exponential growth phase of bacterial growth, the population
    a. increases moderately.
    b. increases dramatically.
    c. decreased moderately.
    d. decreases dramatically.
    e. does not change substantially.

_____ 8. A cultivated field of wheat would most likely display
    a. uniform dispersion.
    b. random dispersion.
    c. clumped dispersion.
    d. no dispersion pattern.
    e. a form of dispersion not found in nature.

_____ 9. In the logistic equation, if environmental limits or resistance are sustained, "K" is the
    a. growth rate.
    b. birth rate.
    c. mortality rate.
    d. carrying capacity.
    e. equivalent of the lag phase.

_____10. A grove of trees that originated from one seed displays
    a. uniform dispersion.
    b. random dispersion.
    c. clumped dispersion.
    d. no dispersion pattern.
    e. a form of dispersion not found in nature.

_____11. Population growth is most frequently controlled by which *one* of the following?
    a. environmental limits
    b. predation
    c. disease
    d. behavioral modifications
    e. resource depletion

_____12. Organisms referred to as r-strategists usually
    a. are mobile animals.
    b. have a high "r" value.
    c. inhabit variable environments.
    d. survive well in the long term.
    e. develop slowly.

_____13. The redwood stands in California are examples of organisms that
    a. use the K-strategy.
    b. use the r-strategy.
    c. often experience local extinction.
    d. live in a relatively stable environment.
    e. pioneer new habitats.

_____14. The number of individuals of a species per unit of habitat area is the _____ for that species.
    a. distribution curve
    b. distribution pattern
    c. population density
    d. ΔN
    e. $b - d$

_____15. Which of the following is incorrectly paired?
    a. population density : spacing of individuals over a given area
    b. S-shaped growth curve : logistic population growth
    c. density dependent factors : negative feedback
    d. cohort : group of individuals of the same age
    e. J-shaped curve : exponential population growth

_____16. The carrying capacity (K)
    a. may change over time. populations.
    b. affects herbivores.
    c. is affected by climate.
    d. has a primary affect on the growth of density independent
    e. affects r-selected populations the most.

_____17. Species that expend their energy in a single, immense reproductive effort
    a. are an iteroparous species.
    b. are likely to be invertebrates.
    c. are a semelparous species.
    d. grow and die in a single season.
    e. are exemplified by the agave plant.

_____18. A population that is expanding would have an age structure diagram
    a. that is widest at the top.
    b. that is widest in the middle.
    c. that is widest at the bottom.
    d. that is even throughout.
    e. that is narrowest at the top.

CHAPTER 54

Community Ecology

A community consists of an association of populations of different species that live and interact in the same place at the same time. A community and its abiotic environment together compose an ecosystem. The interactions among species in a community include competition, predation, and symbiosis. Every organism has its own ecological niche, that is, its own role within the structure and function of the community. Its niche reflects the totality of its adaptations, its use of resources, and its lifestyle. Communities vary greatly in the number of species they contain. The number is often high where the number of potential ecological niches is great, where a community is not isolated or severely stressed, at the edges of adjacent communities, and in communities with long histories. A community develops gradually over time, through a series of stages, until it reaches a state of maturity; species in one stage are replaced by different species in the next stage. Most ecologists believe that the species composition of a community is due primarily to abiotic factors.

REVIEWING CONCEPTS

Fill in the blanks.

INTRODUCTION

1. An association of populations of different species living and interacting in the same place at the same time comprises a _____.

2. A biological community and its abiotic environment together comprise an _____.

COMMUNITY STRUCTURE AND FUNCTIONING

3. Species may interact in a positive way. Such interactions are referred to as _____.

Community interactions are often complex and not readily apparent

4. The three main types of interactions that occur among species in a community are (a) _____, (b) _____, and (c) _____.

The niche is a species' ecological role in the community

5. The totality of adaptations by a species to its environment, its use of resources, and the lifestyle to which it is suited comprises its _____.

6. The potential ecological niche of a species is its (a) _____ niche, while the lifestyle that a species actually pursues and the resources that it actually uses make up its (b) _____ niche.

7. Any environmental resource that, because it is scarce or unfavorable, tends to restrict the ecological niche of a species is called a _____.

Competition is intraspecific or interspecific

8. Competition among individuals within a population is called

(a) _____ competition and when it is between members of different populations it is (b) _____ competition.

9. According to the _____, two species cannot indefinitely occupy the same niche in the same community.

10. The reduction in competition for environmental resources among coexisting species as a result of each species' niche differing from the others in one or more ways is called _____.

11. _____ is the divergence in traits in two similar species living in the same geographic area.

Natural selection shapes the body forms and behaviors of both predator and prey

12. _____ refers to the independent evolution of two interacting species.

13. Conspicuous colors or patterns that advertise a species' unpalatability to potential predators are known as (a) _____ or

(b) _____.

14. _____ coloration help an animal hide by blending into their physical surroundings.

15. _____ is a defense strategy in which a defenseless species is protected from predation by its resemblance to a species that is dangerous in some way.

16. _____ mimicry is used as a defense when both species are harmful.

Symbiosis involves a close association between species

17. The three forms of symbiosis are (a) _____

(b) _____, and (c) _____.

18. When a parasite causes disease or in some cases death it is called a

_____.

19. Parasites may be either (a) _____parasites or (b) _____parasites.

STRENGTH AND DIRECTION OF COMMUNITY INTERACTIONS

Other species of a community dependent on or are greatly affected by keystone species

20. Keystone species determine the nature of an entire community, having an impact that is out of proportion to their abundance because they often affect the amount of (a) _____, (b) _____, or (c) _____.

Dominant species influence a community as a result of their greater size or abundance

21. Dominant species have an influence on a community that is out of proportion to their abundance. They greatly affect a community because they are

_____.

Ecosystem regulation occurs from the bottom up and the top down

22. Bottom-up processes are based on _____.

23. A trophic level is an organism's position _____.

24. Bottom-up processes predominate in aquatic ecosystems where

    (a) _____ and (b) _____ is limiting.

COMMUNITY BIODIVERSITY

25. Species diversity a measure of (a) _____,
    the number of species within a community, and
    (b)_____, a measure of the relative importance
    of each species within a community based on abundance, productivity, or size.

Ecologists seek to explain why some communities have more species than others

26. Species richness is inversely related to the _____
    of a community.

27. Isolated island communities are generally much _____ (more?
    less?) diverse than are communities in similar environments found on
    continents.

28. An _____ is a transitional zone where two or more communities
    meet.

29. The change in species composition produced at ecotones is known as the
    _____.

Species richness may promote community stability

30. Traditionally, ecologists assumed that community stability is a consequence of
    community _____.

COMMUNITY DEVELOPMENT

31. _____ succession is the change in species composition over
    time in a soilless habitat that was not previously inhabited by organisms.

32. _____ succession is the change in species composition that
    takes place after some disturbance removes the existing vegetation; soil is
    already present.

Disturbance influences succession and species richness

33. The relatively stable, often long term stage at the end of ecological succession
    is the _____ which is solely determined by
    climate.

34. Species richness is thought to be greatest at moderate levels of
    _____ of the community.

Ecologists continue to study community structure

35. The cooperative view of community that stresses the interaction of the
    members is known as the _____ model of a community.

36. The view that biological interactions are less important than abiotic factors in producing communities, that, in fact, a community may be a classification category with no reality, is the _____ model of communities.

BUILDING WORDS

Use combinations of prefixes and suffixes to build words for the definitions that follow.

Prefixes	The Meaning	Suffixes	The Meaning
carn-	meat	-ation	the process of
crypt-	hidden	-ic	pertaining to
eco-	home	-gen	production of
herb-	plant	ion	process of
inter-	between, among	-(i)vor(e)	to eat
intra-	within	-sit(e)	food
para-	along side		
patho-	suffering, disease, feeling		
pred-	prey upon		
success-	to follow		

Prefix	Suffix	Definition
_____	_____	1. Any animal that eats flesh.
	-system	2. A community along with its nonliving environment.
_____	_____	3. Any disease-producing organism.
_____	-specific competition	4. Competition for resources within a population.
_____	_____	5. Any animal that eats plants.
_____	-specific competition	6. Competition for resources among populations.
_____	_____	7. The consumption of one species by another.
_____	_____	8. Coloration or markings that help an organism hide by blending in with the environment.
_____	_____	9. In a symbiotic relationship, the organism that receives nutrition from its host.
_____	_____	10. The change in species composition in a community over time.

MATCHING

Terms:

a. Batesian mimicry
b. Commensalism
c. Detritus
d. Ecological niche
e. Ecotone

f. Edge effect
g. Mullerian mimicry
h. Mutualism
i. Parasitism
j. Pathogen

k. Primary consumer
l. Secondary succession
m. Succession
n. Symbiosis
o. Community

For each of these definitions, select the correct matching term from the list above.

_____ 1. The sequence of changes in a plant community over time.

_____ 2. An ecological succession that occurs after some disturbance destroys the existing vegetation.

_____ 3. An organism that is capable of producing disease.

_____ 4. Resemblance of different species, each of which is equally obnoxious to predators.

_____ 5. An intimate relationship between two or more organisms of different species.

_____ 6. A symbiotic relationship between individuals of different species in which one benefits from the relationship and one is harmed.

_____ 7. A transition region between adjacent communities.

_____ 8. A type of symbiosis in which one organism benefits and the other one is neither harmed nor helped.

_____ 9. An assemblage of different species of organisms that live and interact in a defined area or habitat.

_____ 10. The bacteria in your intestine produce vitamin K which is needed by your body for normal blood clotting.

_____ 11. Occurs when a defenseless species is defended by resembling a dangerous species.

MAKING COMPARISONS

Fill in the blanks.

Interaction	Species 1	Effect on Species 1	Species 2	Effect on Species 2
Mutualism of Species 1 and Species 2	Coral	Beneficial	Zooxanthellae	Beneficial
#1	Silverfish	#2	Army ants	#3
#4	Orcas	#5	Salmon	#6
#7	Tracheal mites	#8	Honeybees	#9
#10	*Entamoeba histolytica*	#11	Humans	#12
Mutualism of Species 1 and Species 2	*Rhizobium*	#13	Beans	#14
Competition between Species 1 and Species 2	*Balanus balanoides*	#15	*Chthamalus stellatus*	#16
#17	Orchids	#18	Host tree	#19

MAKING CHOICES

Place your answer(s) in the space provided. Some questions may have more than one correct answer.

_____ 1. The local environment in which a species lives is its
    a. ecosystem.   d. realized niche.
    b. habitat.   e. niche.
    c. fundamental niche.

_____ 2. The actual lifestyle of a species and the resources it uses are known as the
    a. ecosystem.   d. realized niche.
    b. habitat.   e. niche.
    c. fundamental niche.

_____ 3. Resource partitioning results in
    a. extinction of the least adapted species.   d. increased competition among coexisting species.
    b. interspecific breeding.   e. competition for niches among similar species.
    c. reduced competition among coexisting species.

_____ 4. An association of different species of organisms interacting and living together is a/an
    a. ecosystem.   d. pyramid.
    b. community.   e. niche.
    c. food web.

_____ 5. The difference between an organism's potential niche and the niche it comes to occupy may result from
    a. competition.   d. overabundance of resources.
    b. an inverted pyramid.   e. limited space or resources.
    c. factors that exclude it from part of its fundamental niche.

_____ 6. The totality of an organism's adaptations, its use of resources, and the lifestyle to which it is fitted are all embodied in the term
    a. ecosystem.   d. pyramid.
    b. community.   e. niche.
    c. food web.

_____ 7. All food energy in the biosphere is ultimately provided by
    a. producers.   d. consumers.
    b. nitrogen fixers.   e. decomposers.
    c. phosphorus fixers.

_____ 8. The major roles of organisms in a community are
    a. producers.   d. consumers.
    b. nitrogen fixers.   e. decomposers.
    c. phosphorus fixers.

_____ 9. Critically important minerals and other essential substances would be permanently lost to organisms if it were not for the
    a. producers.   d. consumers.
    b. nitrogen fixers.   e. decomposers.
    c. phosphorus fixers.

_____10. The nonliving environment together with different interacting species is a/an
    a. ecosystem.   d. pyramid.
    b. community.   e. niche.
    c. food web.

_____11. Divergence of traits of two similar species living in the same geographical area is known as

    a. coevolution.   d. Batesian mimicry.

    b. aposematic displacement.   e. Mullerian mimicry.

    c. character displacement.

_____12. Both dominant and keystone species

    a. are very common in the community.   d. disproportionally effect bottom-up processes.

    b. have effects on energy in a community.   e. have community-wide effects.

    c. have large effects on a commmunity even though they represent a relatively small biomass.

_____13. An ecological niche

    a. includes an organism's habitat.   d. includes what a species eats.

    b. is difficult to precisely define.   e. is restricted by limiting resources.

    c. represents the totality of adaptations of a species to its environment.

_____14. Predator strategies include

    a. ambush.   d. chemical protection.

    b. pursuit.   e. attracting prey.

    c. generally have larger brains.

_____15. Which of the following relationships are incorrectly paired?

    a. aposematic coloration : warning coloration   d. resource portioning : timing of feeding

    b. secondary succession : no soil   e. cryptic coloration : camouflage

CHAPTER 55

Ecosystems and the Biosphere

This chapter examines the dynamic exchanges between communities and their physical environments. Energy cannot be recycled and reused. Energy moves through ecosystems in a linear, one-way direction from the sun to producer to consumer to decomposer. Much of this energy is converted into less useful heat as the energy moves from one organism to another. Matter, the material of which living things are composed, cycles from the living world to the abiotic physical environment and back again. All materials vital to life are continually recycled through ecosystems and so become available to new generations of organisms. Key cycles include the carbon cycle, the nitrogen cycle, the phosphorus cycle, and the water, or hydrologic, cycle. Life is possible on earth because of its unique environment. The atmosphere protects organisms from most of the harmful radiation from the sun and space, while allowing life supporting visible light and infrared radiation to penetrate. An area's climate is determined largely by temperature and precipitation. Many ecosystems contain fire-adapted organisms.

REVIEWING CONCEPTS

Fill in the blanks.

INTRODUCTION

1. An individual community *and* its abiotic components together make up an

   _____.

2. All of earth's communities combined and their interactions with the abiotic environment comprise the Earth's _____.

ENERGY FLOW THROUGH ECOSYSTEMS

3. Energy enters an ecosystem in the form of (a) _____ energy, which plants can convert to useful (b) _____ energy that is stored in the bonds of molecules. When cellular respiration breaks these molecules apart, energy becomes available to do work. As the work is accomplished, energy escapes the organisms and dissipates into the environment as (c) _____ energy where organisms cannot use it.

4. A complex of interconnected food chains in an ecosystem is a

   _____.

5. Each "link" in a food chain is a trophic level, the first and most basic of which is comprised of the photosynthesizers, or (a) _____. Once "fixed," materials and energy pass linearly through the chain from photosynthesizers to the (b) _____, and then to the (c) _____.

6. Dead organic matter is called (a) _____and it is consumed by (b) _____.

Ecological pyramids illustrate how ecosystems work

7. The ecological pyramid of (a) _____ illustrates how many organisms there are in each trophic level; while the pyramid of (b) _____ is a quantitative estimate of the amount of living material at each level.

8. The pyramid of _____ is a somewhat unique representation since this pyramid, unlike the others, and because of the second law of thermodynamics, can never be inverted.

Ecosystems vary in productivity

9. The rate at which energy is captured during photosynthesis is known as the _____ of an ecosystem.

10. The energy that remains in plant tissues after cellular respiration has occurred is called _____; it represents the rate at which organic matter is actually incorporated into plant tissues to produce growth.

11. It is estimated that humans use between _____ and _____% of annual land-based net primary productivity

Food chains and poisons in the environment

12. The buildup of a toxin in the body of an organism is known as _____.

13. The increase in concentration as a toxin passes through successive levels of the food web is known as _____.

CYCLES OF MATTER IN ECOSYSTEMS

14. The cycling of matter from one organism to another and from living organisms to the abiotic environment and back again is referred to as _____.

15. Four cycles that are of particular importance to organisms are the cycles for

(a) _____, (b) _____,

(c) _____, and (d) _____.

Carbon dioxide is the pivotal molecule in the carbon cycle

16. Carbon is incorporated into the biota, or "fixed," primarily through the process of (a) _____, and in the short term much of it is returned to the abiotic environment by (b)_____. Some carbon is tied up for long periods of time in wood and fossil fuels, which may eventually be released through combustion.

Bacteria are essential to the nitrogen cycle

17. The nitrogen cycle has five steps: (a) _____, which is the conversion of nitrogen gas to ammonia; the conversion of ammonia or ammonium to nitrate, called (b)_____; absorption of nitrates, ammonia, or ammonium by plant roots, called (c)_____; the release of organic nitrogen compounds back into the abiotic environment in the form of ammonia and ammonium ions in waste products and by decomposition, a process known as (d)_____; and finally, to complete the cycle, (e)_____ reduces nitrates to gaseous nitrogen.

18. Nitrogen oxides are necessary for the production of _____ which is a mixture of several air pollutants that injure plant tissues and irritate eyes.

The phosphorus cycle lacks a gaseous component

19. Erosion of (a) _____ releases inorganic phosphorus to the (b) _____ where it is taken up by plant roots.

Water moves among the ocean, land, and atmosphere in the hydrologic cycle

20. The hydrological cycle results from the evaporation of water from both land and sea and its subsequent precipitation when air is cooled. The movement of surface water from land to ocean is called _____.

21. The loss of water vapor from plants into the atmosphere is called

    _____.

22. Water seeps down through the soil to become _____, which supplies much of the water to streams, rivers, ocean, soil, and plants.

ABIOTIC FACTORS IN ECOSYSTEMS

23. The two abiotic factors that probably most affect organisms in ecosystems are
    (a) _____, and (b) _____.

The sun warms Earth

24. Ultimately, all of the sun's energy that is absorbed by the earth's surface and atmosphere is returned to space in the form of

    _____.

25. The solar energy that reaches polar region is (a) _____
    and produces (b) _____.

The atmosphere contains several gases essential to organisms

26. Winds generally blow from (a) _____ (high or low?) pressure areas to
    (b)_____ (high or low?) pressure areas, but they are deflected from these paths by Earth's rotation, a phenomenon known as the
    (c)_____.

The global ocean covers most of Earth's surface

27. The ocean is separated into continents into four sections or "individual" oceans. They are the (a) _____, (b) _____,
    (c) _____, and (d) _____ oceans.

28. The _____ Ocean covers one third of the Earth's surface and contains more than half of Earth's water.

29. Earth's rotation from west to east causes surface ocean currents to swerve to the (a) _____ (left or right?) in the Northern Hemisphere, producing a (b) _____ (clockwise or counterclockwise?) gyre of water currents, and to the (c) _____ (left or right?) in the Southern Hemisphere, producing a (d) _____ (clockwise or counterclockwise?) gyre.

Climate profoundly affects organisms

30. _____ is the average weather conditions, plus extremes, that occur in a given location over an extended period of time.

31. _____ occurs when air heavily saturated with water vapor cools and loses its moisture-holding ability.

32. The dry lands on the sides of the mountains away from the prevailing wind are called _____.

33. Local variations in climate as known as _____.

Fires are a common disturbance in some ecosystems

34. Fires have several effects on organisms. First, they

    (a) _____; second, they

    (b) _____

    _____; and third, they

    (c) _____.

STUDYING ECOSYSTEM PROCESSES

35. _____ is the clearance of large expanses of forest for agriculture or other uses.

BUILDING WORDS

Use combinations of prefixes and suffixes to build words for the definitions that follow.

Prefixes	The Meaning	Suffixes	The Meaning
assimil-	to bring into conformity	-ation	the process of
atmo-	air	-ic	pertaining to
bio-	life	-troph	nutrition
detrit-	worn off	-ule	small
hydro-	water	-us	thing
micro-	small		
nod-	knob		
sapr(o)-	rotten		
troph-	nutrition		

Prefix	Suffix	Definition
_____	-mass	1. The total weight of all living things in a particular habitat.
_____	-sphere	2. The gaseous envelope surrounding the earth; the air.
_____	-logic	3. The water cycle, which includes evaporation, precipitation, and flow to the seas.
_____	-climate	4. A localized variation in climate; the climate of a small area.
_____ _____		5. A feeding level in a food web.
_____ _____		6. An organism that obtains energy by breaking down organic molecules in the remains of all member of the food chain.

_____ _____ 7. Dead organic matter.

_____ _____ 8. Oxygen-excluding swellings on the roots of legumes.

_____ _____ 9. The incorporation of ammonia and nitrate nitrogen into proteins by roots that absorb them.

MATCHING

Terms:

a. Assimilation
b. Bioaccumulation
c. Ecological pyramid
d. Biosphere
e. Ecosystem
f. Food chain

g. Food web
h. Fossil fuel
i. Heterocyst
j. Nitrification
k. Nitrogen fixation
l. Rain shadow

m. Tradewinds
n. Trophic level
o. Westerlies

For each of these definitions, select the correct matching term from the list above.

_____ 1. The distance of an organism in a food chain from the primary producers of a community.

_____ 2. Cells of certain cyanobacteria that are the site of nitrogen fixation.

_____ 3. The interacting system that encompasses a community and its nonliving, physical environment.

_____ 4. The system of interconnected food chains in a community.

_____ 5. The ability of certain microorganisms to bring atmospheric nitrogen into chemical combination.

_____ 6. An area on the downwind side of a mountain range with very little precipitation.

_____ 7. All of the Earth's communities.

_____ 8. A sequence of organisms through which energy is transferred from its ultimate source in a plant; each organism eats the preceding member and is eaten by the following member of the sequence.

_____ 9. A graphical representation of the relative energy value at each trophic level.

_____10. The oxidation of ammonium salts or ammonia into nitrites and nitrates by soil bacteria.

_____11. The buildup of mercury in tuna is one example.

_____12. These winds circulate from 30° N and 30° S of the equator.

MAKING COMPARISONS

Fill in the blanks.

Biogeochemical Cycle	Material Cycled	Description
Phosphorus	Phosphorus	No gas formed; dissolves in water, is taken into plants
#1	#2	As a gas, is taken into plants, converted into glucose; returned to atmosphere by respiration
#3	#4	Converted from a gas into form usable by plants, then fixed in plant tissue, then consumed by animals
#5	Water	#6

MAKING CHOICES

Place your answer(s) in the space provided. Some questions may have more than one correct answer.

_____ 1. The energy that remains as biomass in plants after they have carried out cellular respiration is

    a. significant only in a changing ecosystem.   d. equal to the primary producer's metabolic rate.

    b. called gross primary productivity.   e. evident only in a climax community.

    c. called net primary productivity.

_____ 2. Compared to the overall regional climate, a microclimate

    a. may be significantly different.   d. covers a larger area.

    b. is insignificant to organisms.   e. exists for a brief period.

    c. is more important to resident organisms.

_____ 3. When it is warmest in the United States, it is coolest in

    a. the Southern Hemisphere.   d. Europe.

    b. the Northern Hemisphere.   e. Mexico.

    c. Rio de Janiero.

_____ 4. Activities performed by microorganisms involved in the nitrogen cycle include

    a. liberation of nitrogen from nitrate.   d. fixation of molecular nitrogen.

    b. oxidization of ammonia to nitrates.   e. continuous cycling of nitrogen.

    c. production of ammonia from proteins, urea, or uric acid.

_____ 5. An animal burrow in a hostile desert environment is an example of a/an

    a. subterranean biome.   d. microclimate.

    b. O-horizon.   e. ecosystem.

    c. macroclimate.

_____ 6. Which of the following would be positioned at the top of a typical pyramid?

    a. producers   d. trophic level with the least numbers

    b. consumers   e. trophic level with the least energy

    c. the trophic level with the greatest biomass

_____ 7. The main collector(s) of solar energy used to power life processes is/are the

    a. producers.   d. green leaves.

    b. atmosphere.   e. photosynthetic organisms.

    c. hydrosphere.

_____ 8. Most carbon is fixed _____ and liberated _____

    a. in proteins/as organic compounds.   d. in humus/by cyanobacteria.

    b. in CO_2/as complex compounds.   e. by plants/by plants.

    c. in complex compounds/as CO_2.

_____ 9. Phosphorus enters aquatic communities by means of

    a. decomposers.   d. secondary consumers.

    b. producers.   e. erosion.

    c. primary consumers.

_____ 10. The totality of the Earth's living inhabitants is the

    a. ecosphere.   d. biome.

    b. lithosphere.   e. hydrosphere.

    c. biosphere.

_____ 11. Which of the following is incorrectly paired?

    a. saprotrophs : decomposers   d. food web : food chains

    b. detritivores : dead organic matter   e. ecosystem energy flow : two-way

    c. biological magnification : increase in toxin concentration as it moves through trophic levels

_____12. Ecological pyramids may represent

    a. the number of organisms at each level.

    b. the rate of energy flow through an ecosystem.

    c. the biomass at each level.

    d. the producers, consumers, and decomposers in an ecosystem.

    e. the kilocalories per square meter per year in an ecosystem.

_____13. The oxygen-excluding cells of filamentous cyanobacteria that fix nitrogen are called

    a. nodules.   d. anaerobe cells.

    b. nitrification cells.   e. heterocysts.

    c. assimilocytes.

_____14. Approximatley _____% of the water a plant absorbs is transported to the leaves and is lost there by transpiration.

    a. 85   d. 95

    b. 87   e. 97

    c. 90

_____15. Air currents and the direction of wind flow are affected by

    a. the Coriolis effect.   d. differences in atmospheric pressure.

    b. the Earth's rotation.   e. upwelling.

    c. the position of the continents.

VISUAL FOUNDATIONS

Color the parts of the illustration below as indicated. Label N_2 in the atmosphere, NO_3^- in the soil, and NH_3 in the soil.

RED ☐ nitrogen fixation	BLUE ☐ atmosphere	TAN ☐ animal
GREEN ☐ plants	ORANGE ☐ assimilation	PINK ☐ nitrification
YELLOW ☐ ammonification	BROWN ☐ soil	VIOLET ☐ denitrification

Color the parts of the illustration below as indicated.

RED ☐ equator

GREEN ☐ trade winds

YELLOW ☐ warm air

BLUE ☐ cool air

ORANGE ☐ westerlies

VIOLET ☐ polar easterlies

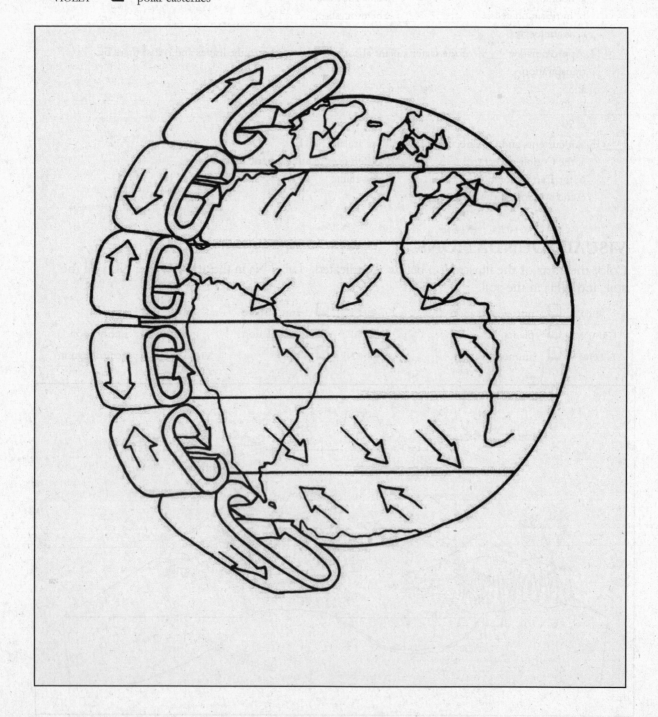

CHAPTER 56

Ecology and the Geography of Life

This chapter examines the characteristics of Earth's major biomes, aquatic ecosystems, and biogeographical realms. Major terrestrial life zones (biomes) are large, relatively distinct ecosystems characterized by similar climate, soil, plants, and animals, regardless of where they occur on earth. Their boundaries are determined primarily by climate. Temperature and precipitation are the major determinants of plant and animal inhabitants. Examples of biomes include tundra; the taiga, or boreal forests; temperate forests; temperate grasslands; chaparral; deserts; tropical grasslands, or savanna; and tropical forests. In aquatic life zones, salinity, dissolved oxygen, nutrient minerals, and light, are the major determinants of plant and animal inhabitants. Freshwater ecosystems include flowing water (rivers and streams), standing water (lakes and ponds), and freshwater wetlands. Marine ecosystems include estuaries, the intertidal zone, the ocean floor, the neritic province, and the oceanic province. All terrestrial and aquatic life zones interact with one another. The study of biogeography helps biologists relate ancient communities to modern communities that exist in a given area. Biologists believe that each species originated only once, and each has a characteristic geographic distribution. The ranges of different species do not include everywhere that they could survive. The world's land areas are divided into six biogeographical realms.

REVIEWING CONCEPTS

Fill in the blanks.

INTRODUCTION

BIOMES

1. Biomes are (a) _____regions that have similar climate, soil, plants, and animals . (b) _____ is the most influential factor in determining the characteristics and boundaries of a biome.

Tundra is the cold, boggy plains of the far north

2. Due to its location, tundra is the (a) _____biome. It is characterized by low-growing vegetation, a short growing season, and a permanently frozen layer of subsoil called (b) _____.

3. Tundra may be found at specific latitudes, but tundra is also found at high _____.

Boreal forest is the evergreen forest of the north

4. Another name for the boreal forest is _____.

5. Most of the boreal forest is not suitable for agriculture because of its
   (a) _____, and
   (b) _____.

6. Boreal forest trees are currently the primary source of the world's _____.

Temperate rain forest has cool weather, dense fog, and high precipitation

7. Coniferous temperate rainforest can be found in

    (a) _____, (b) _____,

    and (c) _____.

8. Temperate rain forest is one of the richest wood producers in the world,
   supplying us with (a) _____ and (b) _____.

Temperate deciduous forest has a canopy of broad-leaf trees

9. Temperate deciduous forest receives approximately _____ to _____
   inches (cm) of rain and is dominated by broad-leaf hardwood trees.

10. Worldwide, temperate forests have been cleared and converted to

    _____.

Temperate grasslands occur in areas of moderate precipitation

11. Temperate grasslands receive approximately _____ to _____ inches
    (cm) of rain.

12. Temperate grasslands are also known as _____.

Chaparral is a thicket of evergreen shrubs and small trees

13. Chaparral is an assemblage of small-leaved shrubs and trees in areas with mild
    Winters with abundant rainfall, combined with extremely dry summers.
    These climatic conditions are referred to as

    _____ climate.

Deserts are arid ecosystems

14. North America has four distinctive deserts. They are the

    (a) _____, (b) _____,

    (c) _____ and (d) _____.

15. During the driest months of the year, many desert animals tunnel
    underground where they remain inactive; this period of dormancy is known as

    _____.

Savanna is a tropical grassland with scattered trees

16. Seasons in tropical grassland are regulated by (a) _____
    and not by (b) _____.

17. The best known savanna, with its large herds of grazing, hoofed animals and
    predators, is the _____ savanna.

18. Desertification of savannas is occurring as a result of

    _____.

There are two basic types of tropical forests

19. Tropical rain forests have about _____ to _____ inches (cm) of rainfall
    annually. They have many diverse species and are noted for tall foliage, many
    epiphytes, and leached soil.

20. Most trees of tropical rain forests are _____
    plants.

AQUATIC ECOSYSTEMS

21. Aquatic ecosystems are classified primarily on

    _____.

22. Aquatic life is ecologically divided into the free-floating (a) _____, the strong swimming (b)_____, and the bottom dwelling (c)_____ organisms.

Freshwater ecosystems are closely linked to land and marine ecosystems

23. Freshwater ecosystems include (a) _____and (b) _____, (c) _____ and (d) _____ as well as (e) _____ and (f) _____.

24. Rivers and streams are flowing-water ecosystems. The kinds of organisms they contain depends primarily on the _____.

25. Lakes and ponds are standing-water ecosystems. Large lakes have three zones: the (a)_____ zone is the shallow water area along the shore; the (b)_____ zone is the open water area away from shore that is illuminated by sunlight; and the (c)_____ zone, which is the deeper, open water area beneath the illuminated portion. Of these, the (d)_____ zone is the most productive.

26. Vertical thermal layering of standing water is called (a)_____. It is usually marked in the summer by an abrupt temperature transition at some depth called the (b)_____. When atmospheric temperatures drop in fall, vertical layers mix, a phenomenon known as (c)_____. In spring, surface waters sink and bottom waters return to the surface. This is called (d) _____.

27. Freshwater wetlands may be (a) _____, where grasslike plants dominate, or (b) _____, where woody trees or shrubs dominate.

28. Freshwater wetlands may be small, shallow ponds called (a) _____, or (b) _____ that are dominated by mosses.

Estuaries occur where fresh water and salt water meet

29. Temperate estuaries usually contain _____, shallow wetlands in which salt-tolerant grasses dominate.

30. The tropical equivalent of salt marshes are _____.

Marine ecosystems dominate earth's surface

31. The main marine habitats are similar to those of fresh water but are modified by tides and currents. The marine habitat near the shoreline between low and high tide is the _____.

32. The benthic environment is divided into zones based upon (a) _____, (b) _____and (c) _____.

33. The abyssal zone is at a depth of ___ to ___ thousand meters or miles.

34. The _____ is open ocean that overlies the continental shelves.

35. The portion of the neritic province that contains sufficient light to support photosynthesis is called the _____ region.

36. The deeper region of the open ocean, beyond the reach of light, is known as the_____.

ECOTONES

37. Within landscapes, ecosystems intergrade with one another at their

_____.

BIOGEOGRAPHY

38. One of the basic tenets of biogeography is that each species originated only once. The particular place where this occurred is known as the species'

_____.

39. Some species have a nearly worldwide distribution and occur on more than one continent or throughout much of the ocean. Such species are said to be

_____.

Land areas are divided into six biogeographic realms

40. The six biogeographical realms are the (a) _____
    (b) _____, (c) _____,
    (d) _____, (e) _____,
    and (f) _____.

BUILDING WORDS

Use combinations of prefixes and suffixes to build words for the definitions that follow.

Prefixes	The Meaning	Suffixes	The Meaning
abys(s)-	bottomless	-al	pertaining to
benth-	depth	-ic	pertaining to
inter-	between, among		
littor-	seashore		
phyto-	plant		
thermo-	heat, hot		
zoo-	animal		

Prefix	Suffix	Definition
_____	-plankton	1. Planktonic plants; primary producers.
_____	-plankton	2. Planktonic protists and animals.
_____	-tidal	3. Pertains to the zone of a shoreline between high tide and low tide.
_____	-cline	4. A layer of water marked by an abrupt temperature transition.
_____	_____	5. A zone of shallw water along the shore of a lake or pond.
_____	_____	6. Refers to the ocean floor or the bottom of a lake or pond.
_____	_____	7. Region or zone of the benthic environment of the ocean ranging from 4000 to 6000 m.

MATCHING

Terms:

a. Benthos
b. Biome
c. Chaparral
d. Desert
e. Desertification

f. Ecotone
g. Endemic species
h. Estuary
i. Euphotic zone
j. Limnetic zone

k. Neritic province
l. Permafrost
m. Savanna
n. Taiga
o. Tundra

For each of these definitions, select the correct matching term from the list above.

_____ 1. Northern coniferous forest biome that stretches across North America and Eurasia.

_____ 2. Upper reaches of the neritic province; region of aquatic habitats lying near enough the surface that sufficient light is present for photosynthesis to take place.

_____ 3. The open water area away from the shore of a lake or pond, extending down as far as sun light penetrates.

_____ 4. Organisms that live on the bottom of oceans or lakes.

_____ 5. Distinctive vegetation type characteristic of certain areas with cool moist winters and long dry summers.

_____ 6. Permanently frozen ground.

_____ 7. Dry areas found in temperate and subtropical or tropical regions.

_____ 8. The conversion of marginal savanna into desert due to severe overgrazing by domestic animals..

_____ 9. A coastal body of water, partly surrounded by land, with access to the open ocean and a large supply of fresh water from rivers.

_____ 10. Localized, native species that are not found anywhere else in the world.

_____ 11. Transition zone where two communities or biomes meet.

_____ 12. Open ocean that overlies the continental shelves.

MAKING COMPARISONS

Fill in the blanks.

Biome	Climate	Soil	Characteristic Plants and Animals
Temperate rain forest	Cool weather, dense fog, high rainfall	Relatively nutrient-poor	Large conifers, epiphytes, squirrels, deer
Tundra	#1	#2	#3
#4	Not as harsh as tundra	Acidic, mineral poor, decomposed needles	Coniferous and deciduous trees, medium to small animals and a few larger species
Temperate deciduous forest	#5	#6	#7
#8	Moderate, uncertain rainfall	#9	Grasses, grazing animals and predators such as wolves
Chaparral	#10	Thin, infertile	#11
#12	Temperate and tropical, low rainfall	#13	Organisms adapted to water conservation, small animals
#14	Tropical areas with low or seasonal rainfall	#15	Grassy areas with scattered trees; hoofed mammals, large predators
Tropical rain forest	#16	#17	#18

MAKING CHOICES

Place your answer(s) in the space provided. Some questions may have more than one correct answer.

_____ 1. One expects to find organisms with adaptations for retaining moisture and clinging securely to the substrate in the

     a. salt marsh.   d. littoral zone.

     b. intertidal zone.   e. estuary.

     c. profundal zone.

_____ 2. The communities containing the most permafrost are called

     a. polar.   d. tundra.

     b. steppes.   e. muskegs.

     c. boreal.

_____ 3. A community with two distinct soil layers abounding in trees, reptiles, and amphibians is

     a. grassland.   d. temperate.

     b. deciduous forest.   e. chaparral.

     c. temperate rain forest.

_____ 4. Insect larvae, water striders, and crayfish are found mainly in

    a. the littoral zone.

    b. the limnetic zone.

    c. the profundal zone.

    d. shallow lakes.

    e. the area just below the compensation point.

_____ 5. Small or microscopic organisms carried about by currents are

    a. littoral.

    b. all producers.

    c. nektonic.

    d. benthic.

    e. planktonic.

_____ 6. Most of midwestern North America is

    a. grassland.

    b. deciduous forest.

    c. temperate rain forest.

    d. temperate.

    e. chaparral.

_____ 7. The area(s) essentially corresponding to the waters above the continental shelf is/are the

    a. littoral zone.

    b. limnetic zone.

    c. profundal zone.

    d. neritic province.

    e. benthic zone.

_____ 8. The two most diverse of communities are the _____ and _____

    a. deciduous forest.

    b. tropical rain forest .

    c. coniferous forest.

    d. coral reef.

    e. grassland.

_____ 9. An aquatic zone at the juncture of a river and the sea is a/an

    a. salt marsh.

    b. intertidal zone.

    c. profundal zone.

    d. littoral zone.

    e. estuary.

_____10. Communities dominated by evergreens and containing acidic, mineral-poor soil is/are

    a. in South America.

    b. called taiga.

    c. temperate coniferous forests.

    d. typically impregnated with a layer of permafrost.

    e. boreal forest.

_____11. The factor that determines the nature of the biogeographic realms more than any other single factor is

    a. soil.

    b. climate.

    c. latitude.

    d. types of animals.

    e. altitude.

_____12. Which biome is found only in the northern hemisphere?

    a. chaparral

    b. desert

    c. tundra

    d. boreal forest (taiga)

    e. temperate grasslands

_____13. Trees that might be found in a boreal forest include

    a. spruces.

    b. aspen.

    c. birch.

    d. firs.

    e. larch.

_____14. Which biome has an annual precipitation range of 30 to 50 inches (75 to 126 cm).

    a. temperate grasslands

    b. savanna

    c. temperate deciduous forest

    d. taiga

    e. temperate rain forest

_____15. Which desert is located in California, Arizona, and Mexico?

    a. Mojave Desert   d. Great Basin Desert

    b. Sonoran Desert   e. Chihuahuan Desert

    c. Death Valley Desert

_____16. Desert soil is rich in

    a. $MgSO_4$

    b. $CaSO_4$ d. minerals.

    c. $CaCO_3$ e. NaCl

_____17. About ___% of the organisms living in a tropical rainforest live in the middle and upper canapoies.

    a. 80   d. 95

    b. 85   e. 97

    c. 90

_____18. Streams and rivers depend upon _____ for much of their energy.

    a. aquatic producers   d. cyanobacteria

    b. detritus   e. land

    c. plankton

_____19. Coral reefs

    a. is comparable to a tropical rain forest.   d. can be found in water poor in nutrients.

    b. have stinging tentacles.   e. provide humans with pharmaceuticals.

    c. are dependent upon a symbiotic relationship with zooxanthellae.

VISUAL FOUNDATIONS

Color the parts of the illustration below as indicated. Label intertidal zone, benthic environment, abyssal zone, hadal zone, and pelagic environment.

GREEN ☐ neritic province

BLUE ☐ oceanic province

TAN ☐ land

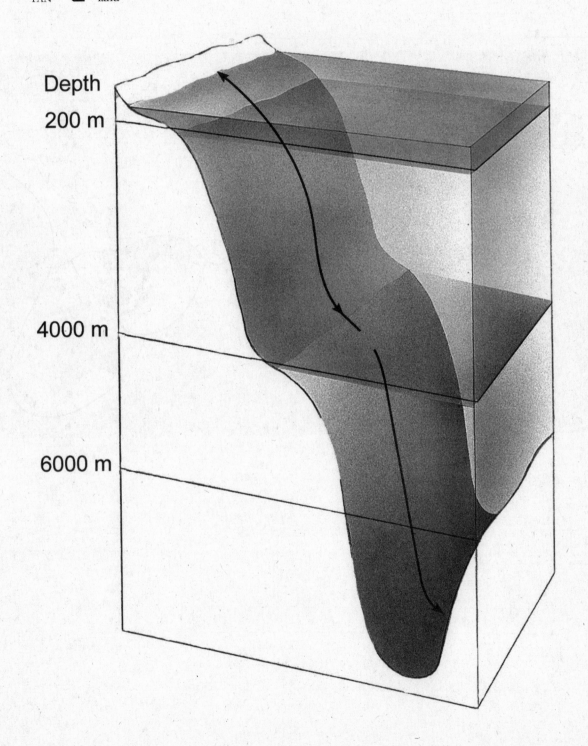

Color and label the biogeography realms of the illustration below as indicated.

GRAY	☐	Nearartic	GREEN	☐	Neotropical
BLUE	☐	Ethiopian	ORANGE	☐	Paleartic
TAN	☐	Austrailian	BROWN	☐	Oriental

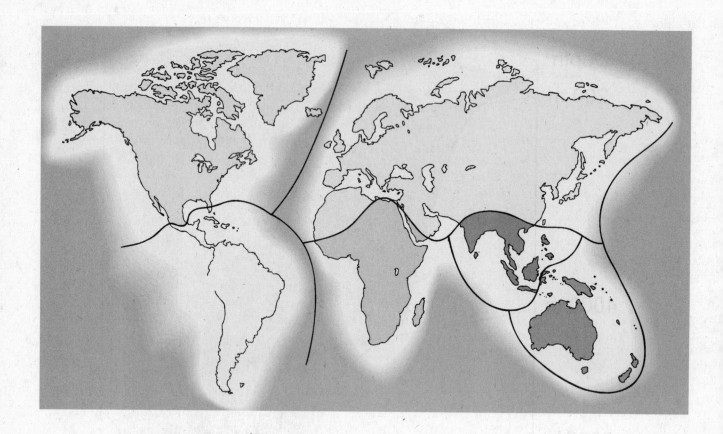

CHAPTER 57

Biological Diversity and Conservation Biology

This chapter examines some of the consequences of the growth in the human population on the biosphere. Although humans have been on earth a comparatively brief span of time, our biological success has been unparalleled. We have expanded our biological range into almost every habitat on earth, and, in the process, have placed great strain on the Earth's resources and resilience, and profoundly affected other life forms. As a direct result of our actions, many environmental concerns exist today. Species are disappearing from earth at an alarming rate. Although forests provide us with many ecological benefits, deforestation is occurring at an unprecedented rate. The production of atmospheric pollutants that trap solar heat in the atmosphere threaten to alter the earth's climate. Global warming may cause a rise in sea level, changes in precipitation patterns, death of forests, extinction of animals and plants, and problems for agriculture. It could result in the displacement of thousands or even millions of people. Ozone is disappearing from the stratosphere at an alarming rate. The ozone layer helps to shield Earth from damaging ultraviolet radiation. If the ozone were to disappear, Earth would become uninhabitable for most forms of life.

REVIEWING CONCEPTS

Fill in the blanks.

INTRODUCTION

THE BIODIVERSITY CRISIS

1. Many people are concerned with _____, the ability to meet humanity's current needs without compromising the ability of future generations to meet their needs.

2. Biological diversity is the variety of living organisms considered at three levels: _____, _____, and _____.

3. When the number of individuals in a species is reduced to the point where extinction seems imminent, the species is said to be (a) _____, and when numbers are seriously reduced so that extinction is feared but thought to be less imminent, the species is said to be (b) _____.

Endangered species have certain characteristics in common

4. Many threatened and endangered species share certain characteristics that make them vulnerable to extinction. Some of these characteristics are having

   (a) _____, requiring a large
   (b) _____, living (c) _____,
   having a (d) _____, needing specialized
   (e) _____ and possessing specialized

(f) _____.

Human activities contribute to declining biological diversity

5. Species become endangered and extinct for variety of reasons, but the two most common reasons are the (a) _____ and (b) _____ of habitat.

6. The breakup of large areas of habitat into small, isolated patches is known as _____.

7. _____, the introduction of a foreign species into an area where it is not native, often upsets the balance among the organisms living in that area.

CONSERVATION BIOLOGY

8. *In situ* conservation attempts to preserve species in _____ by establishing reserves (e.g., parks).

9. *Ex situ* conservation tries to preserve species in _____ such as zoos and seed banks.

In situ conservation is the best way to preserve biological diversity

10. The single best way to protect biological diversity is to protect _____.

11. _____ is a subdivision of ecology studying the connections in a heterogeneous landscape consisting of multiple interacting ecosystems.

12. Strips of habitat connecting isolated habitat patches are called _____.

13. (a)_____, is a process in which the principles of ecology are used to return a degraded environment to one that is more functional and sustainable, is an important part of (b)_____conservation.

Ex situ conservation attempts to save species on the brink of extinction

14. In (a) _____, sperm is collected from a male and used to impregnate a female, while in (b)_____ a female is treated with fertility drugs so she will produce more eggs.

The Endangered Species Act provides some legal protection for species and habitats

15. The Endangered Species Act authorizes the U.S. Fish and Wildlife Service to_____ _____.

International agreements provide some protection of species and habitats

16. The Convention on International Trade in Endangered Species of Wild Flora and Fauna protects_____ _____.

DEFORESTATION

17. Forests on mountains and hillsides help to protect valleys from flooding by
_____.

18. Deforestation may increase global temperature by releasing
_____.

Why are tropical rain forests disappearing?

19. The three most immediate causes of deforestation in tropical rains forests are

(a) _____, (b) _____,

and (c) _____.

Why are boreal forests disappearing?

20. Boreal forests are currently the primary source of the world's
_____.

CLIMATE CHANGE

Greenhouse gases cause climate change

21. Gases that are accumulating in the atmosphere as a result of human activities

include (a) _____, (b) _____. ,

(c) _____, (d) _____ and

(e) _____.

22. Because CO_2 and other gases trap the sun's radiation somewhat like glass does in a greenhouse, the natural trapping of heat is referred to as the
_____.

What are the probable effects of climate change?

23. The probable effects of global warming include changes in

(a) _____, and (b) _____,

as well as effects on (c) _____ and (d) _____.

DECLINING STRATOSPHERIC OZONE

24. According to the National Center for Atmospheric Research, ozone levels over Europe and North America have dropped almost _____% since the 1970s.

Certain chemicals destroy stratospheric ozone

25. Both (a) _____ and (b) _____ containing substances catalyze ozone destruction.

26. The primary chemicals responsible for ozone loss in the stratosphere are a group of chlorine compounds called _____.

27. The thinning of the ozone layer that was discovered over Antarctica occurs annually between _____ and _____.

Ozone depletion harms organisms

28. Excessive exposure to UV radiation is linked to human health problems, such as (a) _____, (b) _____,

and (c) _____.

International cooperation is helping to repair the ozone layer

29. In 1987, the Montreal Protocol was signed by many countries. At the time it stipulated a (a)_____ % reduction in (b)_____ by 1998.

BUILDING WORDS

Use combinations of prefixes and suffixes to build words for the definitions that follow.

Prefixes	The Meaning		Suffixes	The Meaning
bi(o)-	life			
de-	to remove		demic-	pertaining to people or country
en-	in		tic-	pertaining to
strat(o)-	layer			

Prefix	Suffix	Definition
_____	-forestation	1. The removal or destruction of all tree cover in an area.
_____	-sphere	2. The layer of the atmosphere between the troposphere and the mesosphere, containing a layer of ozone that protects life by filtering out much of the sun's ultraviolet radiation.
_____	_____	3. Species that are found on an island and nowhere else in the world.
_____	_____	4. A type of pollution that occurs when a foreign species is introduced into an area where it was not found previously.

MATCHING

Terms:

a. Artificial insemination
b. Biological diversity
c. Endangered species
d. Ex situ conservation
e. Extinction
f. Genetic diversity
g. Greenhouse effect
h. Greenhouse gases
i. Host mothering
j. In situ conservations
k. Keystone species
l. Ozone
m. Slash-and-burn agriculture
n. Subsistence agriculture
o. Threatened species

For each of these definitions, select the correct matching term from the list above.

_____ 1. The warming of the Earth resulting from the retention of atmospheric heat caused by the build-up of certain gases, especially carbon dioxide.

_____ 2. A species whose population is low enough for it to be at risk of becoming extinct.

_____ 3. The number and variety of living organisms.

_____ 4. The disappearance of a species from a given habitat.

_____ 5. Carbon dioxide, methane, nitrous oxide, CFCs, and ozone.

_____ 6. A method of controlled breeding in zoos.

_____ 7. Efforts to preserve biological diversity in the wild.

_____ 8. The production of enough food to feed oneself and one's family.

_____ 9. A species whose numbers are so severely reduced that it is in imminent danger of becoming extinct.

_____10. The layer in the upper atmosphere that helps shield the Earth from damaging ultraviolet radiation.

_____11. Species conservation in which species are conserved in human controlled settings.

_____12. Species conservation in which eggs are collected from a female of a rare species, fertilized artificially and implanted in a female from a closely related species.

MAKING COMPARISONS

Fill in the blanks.

Issue	Why It Is An Issue	Cause of Problem	Solution
Declining biological diversity	Diversity contributes to a sustainable environment; lost opportunities, lost solutions to future problems	Commercial hunting, habitat destruction, efforts to eradicate or control species, commercial harvesting	*In situ* conservation, *ex situ* conservation, habitat protection
#1	Forests are needed for habitat, watershed protection, prevention of erosion	#2	#3
#4	Threatens food production, forests, biological diversity; could cause submergence of coastal areas, change precipitation patterns	#5	#6
Ozone depletion	Ozone layer protects living organisms from exposure to harmful amounts of UV radiation	#7	#8

REFLECTIONS

The subject matter covered in this chapter is of critical importance to the continuance of life on earth as we know it. This chapter points out some of the dangers humanity faces if we continue some of our habits. Biologists and other responsible citizens can make a difference. One might even say that we have a responsibility to make a difference. A few genuinely concerned individuals will assume leadership roles in bringing about the social, cultural, and political changes required to protect the environment. All of us can participate in assuring that our children and grandchildren will inherit a clean, healthy, and productive environment.

We encourage you to reflect on what your personal role could be to bring about needed reforms. What might you contribute to the welfare of the planet? What changes are needed to ensure that future generations can enjoy a reasonably good quality of life? Think about your lifestyle and aspirations. Are they compatible with a long-term investment in the future of the planet?

VISUAL FOUNDATIONS

Color the parts of the illustration below as indicated. Also label greenhouse gases in atmosphere.

RED   ☐   radiated heat redirected back to Earth

GREEN   ☐   Earth

YELLOW   ☐   solar energy

BLUE   ☐   atmosphere

ORANGE   ☐   heat escaping to space

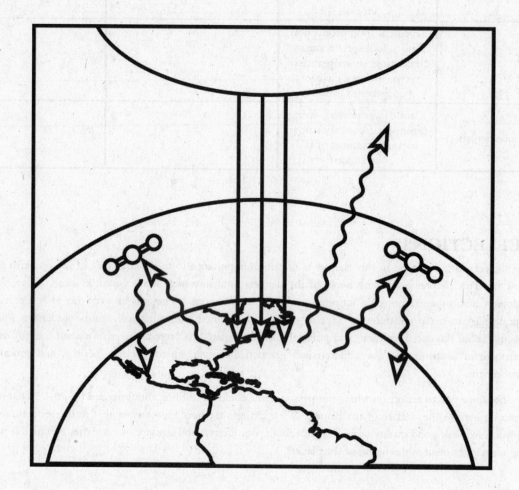

Color the parts of the illustration below as indicated. Also label normal ozone levels, and reduced ozone levels.

RED ☐ ozone molecules

GREEN ☐ oxygen molecules

BLUE ☐ troposphere

YELLOW ☐ stratosphere

ORANGE ☐ UV solar radiation

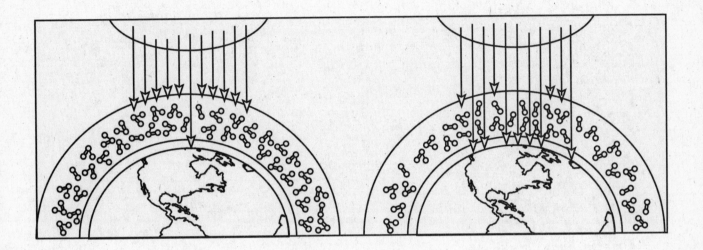

ANSWERS

□

REVIEWING CONCEPTS, BUILDING WORDS, MATCHING, MAKING COMPARISONS, MAKING CHOICES, VISUAL FOUNDATIONS

Chapter 1
Reviewing Concepts: **1.** (a)decreasing biological diversity, (b)diminishing natural resources, (c) expanding human population, (d)prevention and cure of diseases, (e) global climate change **2.** (a) evolution, (b) information transfer, (c) energy transfer **3.** composed of cells, grow and develop, regulate metabolic processes, respond to stimuli, reproduce, evolve, and adapt **4.** unicellular/multicellular **5.** Prokaryote/eukaryote **6.** size/number **7.** development **8.** chemical activities **9.** homeostasis **10.** stimuli **11.** Louis Pasteur **12.** sexually or asexually **13.** adaptations **14.** Reductionism **15.** Emergent properties **16.** cell **17.** biosphere **18.** population **19.** genes **20.** Watson and Crick **21.** proteins **22.** metabolism **23.** cellular respiration **24.** producers, consumers, and decomposers **25.** producers or autotrophs **26.** Evolution **27.** Taxonomy **28.** Species **29.** genus and species **30.** orders **31.** domain **32.** (a)Archaea, (b)Bacteria, and (c)Eukarya **33.** protists **34.** Evolutionary processes **35.** *On the Origin of Species By Means of Natural Selection* **36.** (a)variation, (b)survive, (c)competition, (d) survive **37.** (a) mutations (b) DNA **38.** gene pool **39.** environmental changes (or pressures) **40.** scientific method **41.** deductive **42.** inductive **43.** Alexander Fleming **44.** phenomenon or problem **45.** (a)consistent with well-established facts, (b)capable of being tested, (c)falsifiable **46.** Proven **47.** models **48.** variable **49.** double-blind study **50.** not all cases of what is being studied can be observed **51.** Conclusions **52.** theory **53.** the Big Bang and evolution of major groups of organisms **54.** paradigm **55.** systems biologists **56.** genetic research, stem cell research, cloning, human and animal experimentation
Building Words: **1.** biology **2.** ecology **3.** asexual **4.** homeostasis **5.** ecosystem **6.** biosphere **7.** autotroph **8.** heterotroph **9.** prokaryote **10.** Photosynthesis **11.** hypothesis
Matching: **1.** o **2.** f **3.** m **4.** c **5.** l **6.** n **7.** a **8.** e **9.** b **10.** k
Making Comparisons: **1.** population **2.** Lions, elephants, wildebeests, hyena, etc living in the Kalahari desert **3.** All animal populations and physical features in the Kalahari desert **4.** biosphere
Making Choices: **1.** e **2.** a, d **3.** b **4.** d **5.** c,e **6.** a **7.** a,b,e **8.** b **9.** a,c,d **10.** d **11.** c **12.** c,d **13.** c **14.** a,b,c,e **15.** a-e
Visual Foundations: **1.** d **2.** b,c,d **3.** e

Chapter 2
Reviewing Concepts: **1.** water, many simple acids and bases, and simple salts **2.** chemical reactions (ordinary chemical means) **3.** carbon, hydrogen, oxygen, nitrogen **4.** (a)protons, neutrons, electrons, (b)protons and neutrons, (c)electrons **5.** (a)protons, (b)neutrons, (c)electrons **6.** atomic number **7.** eight **8.** dalton **9.** atomic mass

551

10. (a)atomic mass (or numbers of neutrons), (b)atomic number (or numbers of protons) **11.** radioisotopes **12.** electron shell (or energy level) **13.** valence **14.** chemical properties **15.** two or more different elements **16.** molecule **17.** chemical **18.** structural **19.** atomic masses of its constituent atoms **20.** mole **21.** (a)mol, (b)dissolve, (c)liter **22.** (a)reactants, (b)products **23.** (a)left, (b)right **24.** (a)chemical bonds (b)valence electrons **25.** single, double, triple **26.** (a)nonpolar bonds, (b)polar bonds **27.** (a)cations, (b)anions **28.** (a)solvent (b)solute **29.** hydration **30.** (a)hydrogen, (b)oxygen or nitrogen, (c)negative **31.** (a)weak, (b)strong **32.** electric charges **33.** redox **34.** (a)oxidized, (b)reduced **35.** 70 **36.** (a)hydrogen bonding (b)surface tension **37.** adhesion **38.** capillary action **39.** (a)hydrophilic, (b)hydrophobic **40.** hydrophobic interactions **41.** heat of vaporization **42.** calorie (cal) **43.** specific heat **44.** (a)H^+ and an anion, (b)OH^- and a cation, **45.** (a)0-14, (b)logarithmic, (c)7 **46.** (a)base, (b)acid **47.** buffer **48.** electrolytes
Building Words: **1.** neutron **2.** equilibrium **3.** tetrahedron **4.** hydration **5.** hydrophobic **6.** nonelectrolyte **7.** hydrophilic
Matching: **1.** l **2.** c **3.** g **4.** b **5.** f **6.** o **7.** a **8.** j **9.** e **10.** d **11.** i
Making Comparisons: **1.** CO_2 **2.** Ca **3.** iron (III) oxide **4.** CH_4 **5.** Cl⁻ **6.** potassium ion **7.** H^+ **8.** magnesium **9.** NH_4^+ **10.** H_2O **11.** $C_6H_{12}O_6$ **12.** HCO_3^- **13.** carbonic acid **14.** Sodium chloride **15.** hydroxide ion **16.** NaOH
Making Choices: **1.** e **2.** a **3.** b **4.** a **5.** e **6.** a,b **7.** e **8.** c,d **9.** a **10.** c,d,e **11.** a,e **12.** d **13.** b **14.** e **15.** a **16.** a,b,d **17.** a **18.** a,b,d,e **19.** a,b
Visual Foundations: 1a. Carbon-12 **1b.** Carbon-14 **2.** e

Chapter 3

Reviewing Concepts: **1.** smaller molecules **2.** carbohydrates, lipids, proteins, nucleic acids **3.** four **4.** single, double, triple **5.** (a)hydrocarbons, (b)hydrogen, (c)unbranched, (d)ring **6.** structural **7.** geometric **8.** enantiomers **9.** (a) properties (b)functional **10.** (a)hydroxyl (b)polar **11.** (a)carboxyl (b)acidic **12.** (a)amino (b)basic **13.** monomers **14.** hydrolysis **15.** (a)condensation reactions (b)polymers **16.** carbon, hydrogen, oxygen **17.** (a)glucose (b)ring **18.** glycosidic **19.** (a)sucrose (b)lactose **20.** simple sugars (monosaccharides) **21.** starch **22.** amylose **23.** amyloplasts **24.** glycogen **25.** (a)cellulose (b)cell walls **26.** chitin **27.** (a)glycoprotein, (b)glycolipids **28.** (a)structure (b)insoluble **29.** carbon, hydrogen, oxygen **30.** (a)glycerol, (b)fatty acids **31.** (a)fats (b)storage **32.** van der Waals interactions **33.** (a)liquid (b) double bonds **34.** (a)hydrogenation (b)saturated fats **35.** amphipathic **36.** (a)glycerol, (b)fatty acids, (c)phosphate, (d)organic **37.** (a)isoprene units (b) Vitamin A **38.** (a)carbon, (b)six, (c)five **39.** cholesterol **40.** fatty acids **41.** (a)amino and carboxyl, (b)R group (radical) **42.** essential **43.** peptide **44.** amino acids **45.** (a)carboxyl (b)amino **46.** (a)conformation (b)function **47.** primary, secondary, tertiary, quaternary **48.** polypeptide chains **49.** (a)hydrogen (b) α-helix and β-pleated sheet **50.** (a)side chains (b)shape **51.** polypeptides **52.** molecular chaperones **53.** function **54.** denaturation **55.** Alzheimer's, Huntington's, mad cow **56.** nucleotide **57.** (a)purine or pyrimidine, (b)sugar, (c)phosphate **58.** phospodiester linkages **59.** ATP
Building Words: **1.** isomer **2.** macromolecule **3.** monomer **4.** hydrolysis **5.** monosaccharide **6.** hexose **7.** disaccharide **8.** polysaccharide **9.** amyloplast **10.** monoacylglycerol **11.** diacylglycerol **12.** triacylglycerol **13.** amphipathic **14.** dipeptide **15.** polypeptide **16.** pentose
Matching: **1.** i **2.** c **3.** l **4.** n **5.** a **6.** o **7.** g **8.** k **9.** m **10.** e
Making Comparisons: **1.** phosphate **2.** organic phosphates **3.** $R—CH_3$ **4.** hydrocarbon; nonpolar **5.** amino **6.** amines **7.** thiols **8.** hydoxyl **9.** alcohols **10.** polar because electronegative oxygen attracts covalent electrons

Making Choices: **1.** c,e **2.** b,d **3.** c,d,e **4.** b,d **5.** c,e **6.** b,c,d **7.** a,c,d **8.** b,c **9.** a,c,d,e **10.** b,c,e **11.** a,b,d **12.** b,c,d,e **13.** a,b,c,d **14.** a,b,d **15.** a,c
Visual Foundations: **1.** b,d **2.** b **3.** b **4.** c

Chapter 4

Reviewing Concepts: **1.** (a)cells (b)single **2.** cells are the basic living units of organization and function in all organisms, and all cells come from other cells **3.** homeostasis **4.** plasma membrane **5.** organelles **6.** 1/1000 **7.** surface area-to-volume ratio **8.** sizes and shapes **9.** (a)magnification, (b)resolution **10.** (a)wavelengths (b)visible light **11.** staining **12.** ultrastructure **13.** (a)electrons, (b)pass, (c)TEM **14.** cell fractionation **15.** (a)prokaryotic (b)nucleoid **16.** (a)cell wall (b)flagella **17.** (a)ribosome (b)smaller **18.** eukaryotic **19.** cell nucleus **20.** (a)localized (b)simultaneously **21.** endomembrane system **22.** (a)nuclear envelope (b)nuclear pores **23.** nuclear lamina **24.** chromatin **25.** (a)nucleolus (b)RNA **26.** RNA and proteins **27.** endoplasmic reticulum **28.** (a)phospholipids, cholesterol, steroid hormones, (b) calcium ions, (c) detoxification **29.** ribosomes **30.** (a)transport vesicles (b)proteasome **31.** cisternae **32.** glycoproteins **33.** digestive enzymes **34.** Golgi complex **35.** tonoplast **36.** Central, food, contractile **37.** hydrogen peroxide **38.** (a)fatty acid breakdown (b)detoxification **39.** prokaryotes **40.** (a) aerobic respiration (b)cristae **41.** apoptosis **42.** free radicals **43.** chloroplasts **44.** thylakoid **45.** (a)stroma, (b)thylakoids **46.** strength, shape, and movement **47.** microfilaments and microtubules **48.** alpha and beta tubulin **49.** (a)microtubule associated proteins (b)motor **50.** (a)centrosome (b)MTOC **51.** (a)centrioles, (b)centrosome, (c)cell division **52.** (a)flagella, (b)cilia, (c)movement **53.** basal body **54.** cell cortex **55.** myosin **56.** intermediate filaments **57.** glycocalyx **58.** collagen, fibronectin, integrins **59.** cellulose
Building Words: **1.** chlorophyll **2.** chromoplasts **3.** cytoplasm **4.** cytoskeleton **5.** glyoxysome **6.** leukoplast **7.** lysosome **8.** microfilaments **9.** microtubules **10.** microvilli **11.** myosin **12.** peroxisome **13.** proplastid **14.** prokaryotes **15.** cytosol **16.** eukaryote
Matching: **1.** b **2.** h **3.** f **4.** m **5.** n **6.** c **7.** e **8.** l **9.** d **10.** j **11.** o
Making Comparisons: **1.** ribosomes **2.** prokaryotic, most eukaryotic **3.** cytoplasm **4.** modifies, sorts, and packages proteins **5.** cytoplasm **6.** ATP synthesis (aerobic respiration) **7.** eukaryotic **8.** nucleus (and cytoplasm during some stages of mitosis and meiosis) **9.** eukaryotic **10.** centrioles **11.** cell division, microtubule formation **12.** eukaryotic (animals primarily) **13.** cell wall **14.** encloses cell, including the plasma membrane **15.** plasma membrane **16.** boundary of cell cytoplasm **17.** prokaryotic, eukaryotic **18.** cytoplasm **19.** captures light energy (site of photosynthesis)
Making Choices: **1.** c **2.** d,e **3.** a-e **4.** c,e **5.** d,e **6.** a,b,d,e **7.** a,b,c,d **8.** a-e **9.** a,c,d,e **10.** b,c,e **11.** a,b,d,e **12.** a,d **13.** a,b,c,e **14.** e **15.** a,d **16.** a, c, d
Visual Foundations: **1.** prokaryotic **2.** plasma membrane, cell wall **3.** eukaryotic plant cell **4.** prominent vacuole, chloroplast, mitochondrion, internal membrane system (excluding the nucleus) **5.** prominent vacuole, chloroplast **6.** eukaryotic animal cell **7.** centriole, lysosome, mitochondrion, internal membrane system (excluding the nucleus) **8.** centriole

Chapter 5

Reviewing Concepts: **1.** plasma membrane **2.** cadherins **3.** phospholipids **4.** (a)hydrophilic, (b)hydrophobic **5.** amphipathic **6.** (a)fluid mosaic, (b)lipid bilayer, (c)proteins **7.** (a)head groups, (b)fatty acid chains **8.** hydrocarbon chains **9.** unsaturated fats **10.** cholesterol **11.** plasma membrane **12.** (a)integral membrane proteins (b)transmembrane proteins **13.** peripheral membrane proteins **14.** (a)free

ribosomes, (b)cytoplasm   15. (a)RER ribosomes (b)glycoproteins (c)Golgi complex
16. protein   17. (a)integrins (b)transport (c)receptor (d)signal transduction 18.
(a)recognition (b)junctions 19. selectively permeable 20. nonpolar 21. (a)ions (b)large
22. (a)carrier proteins (b)shape change 23. (a)channel proteins (b)gated 24. aquaporins
25. energy   26. size, shape, electrical charge   27. (a)higher (b)lower   28. osmosis
29. osmotic pressure   30. isotonic 31. (a)hypertonic (b)out (c)shrink 32. (a)hypotonic
(b)swell (c)into 33. (a)turgor pressure (b)hypotonic (c)plasmolysis (d)hypertonic
34. (a)higher, (b)lower 35. (a)transport (b)channel, carrier 36. (a)concentration gradient
(b)ATP 37. fewer   38. negatively 39. (a)electrical gradient (b)electrochemical gradient
40. (a)uniporters, (b)symporters (c)antiporters   41. indirect active transport 42. energy
43. vesicle   44. (a)phagocytosis, (b)pinocytosis 45. Receptor-mediated endocytosis
46. (a)desmosomes, tight junctions, gap junctions (b)plasmodesmata   47. (a)desmosomes
(b)adhering junctions 48. blood-brain barrier   49. (a)proteins or connexin, (b)cylinders
50. gap junctions 51. desmotubule
Building Words: 1. endocytosis   2. exocytosis   3. hypertonic   4. hypotonic
5. isotonic 6. phagocytosis 7. pinocytosis 8. desmosome 9. plasmodesma
Matching: 1. g 2. i 3. f 4. b 5. a 6. k 7. e 8. m 9. l 10. h
Making Comparisons: 1. cell ingests large particles such as bacteria and other food;
literally means "cell eating"   2. Facilitated diffusion   3. endocytosis   4. carrier-
mediated transport   5. cell junctions anchored by intermediate filaments   6. the active
transport of materials out of the cell by fusion of cytoplasmic vesicles with the plasma
membrane   7. diffusion of water across a selectively permeable membrane   8. active
transport   9. cell takes in materials dissolved in droplets of water; literally means "cell
drinking" 10. desmosomes
Making Choices: 1. a,e 2. a,c,e 3. c 4. c,e 5. b,c,d 6. c 7. b 8. d 9. b,d,e
10. b,e 11. b,c,d,e 12. b,d 13. a 14. c,d
Visual Foundations: 1. fluid mosaic model   2. transmembrane proteins 3.
cholesterol

Chapter 6
Reviewing Concepts: 1. communicate   2. secreting chemical signals   3. biofilm
4. cell signaling   5. target cells   6. signal transduction 7. neurotransmitters
8. paracrine regulation 9. prostaglandins   10. receptors 11. (a)ligand (b)hydrophilic
(c)hydrophobic 12. (a)domains (b) docking (c)tail 13. cryptochromes   14. (a)high,
(b)decreasing 15. (a)low, (b)increasing 16. (a)ion channel-linked receptors or ligand-
gated channels, (b)chemical (c)electrical 17. (a)G protein linked (b)signal transduction
(c)pharmaceutical   18. (a)enzyme-linked receptors   (b)kinase (c)phosphorylation
19. (a)transcription factors (b)small,hydrophobic 20. plasma membrane 21. (a)signaling
pathway/cascade (b)amplify 22. (a)phosphorylation (b)dephosphorylation   23. ligand-
gated chloride ion channels   24.signal transduction   25. second messengers
26. ions/small molecules   27. cyclic AMP (cAMP)   28. cytoplasmic   29. (a)cAMP,
(b)protein kinase A, (c)proteins 30. (a)phosphilipase C (b)IP$_3$/DAG (c)PKC (d)calcium
31. low   32. (a)kinases, (b)phosphorylate   33. intracellular receptors
34. (a)intracellular signaling molecules, (b)signaling complexes   35. (a)integrins
(b)inside-out 36. ion channels open/close, altered enzyme activity, gene turned on/off
37. (a)ras proteins (b)GTP (c)cancer 38. (a)MAP kinase /ERK (b)division and
development 39. signal amplification   40. (a)signal transduction, (b)inactive 41.
(a)conserved (b)G proteins, protein kinases, phosphatases 42. prokaryotes
Building Words: 1. paracrine 2. neurotransmitter 3. phytochrome 4. intracellular
5. triphosphate 6. diphosphate 7. diacylglycerol 8. intercellular
Matching: 1. h 2. n 3. j 4. i 5. f 6. d 7. c 8. g 9. p 10. k 11. o 12. a
13. b 14. l 15. m

Making Comparisons: 1. phosphate group   2. cAMP   3. protein   4. ATP
5. phosphorylated protein 6. G protein   7. signaling molecule 8. plasma membrane
Making Choices: 1. c   2. e   3. b   4. a   5. c   6. a,b,d   7. e   8. a,b,c,e   9. a
10. b,e   11. c,d   12. a,c,d,e   13. a,b,d   14. d   15. d,e   16. a,b,c,d   17. a,c,e   18. a-e

Chapter 7

Reviewing Concepts: 1. photosynthetic plants   2. chemical   3. work
4. (a)potential energy, (b)kinetic energy, (c) chemical   5. thermodynamics   6. (a)created
or destroyed, (b)transferred   7. heat   8. randomness or disorder   9. bond energy
10. total bond energies   11. entropy   12. (a)enthalpy, (b)entropy   13. exergonic
reactions   14. ΔG 15. positive   16. . (a)high, (b)lower   17. dynamic equilibrium
18. (a)release, (b)require input of   19. more   20. adenine, ribose, three phosphates
21. phosphorylation   22. (a)catabolic, (b)anabolic   23. large   24. acceptor molecule
25. break existing bonds   26. accelerate/speed up   27. enzyme substrate complex
28. induced fit   29. -ase   30. active site   31. coenzyme   32. apoenzyme
33. temperature, pH, ion concentration   34. temperature, pH, ion concentration
35. metabolic pathway 36. inhibits   37. reversible   38. competitive 39. non
competitive   40. PABA
Building Words: 1. kilocalorie   2. anabolism   3. catabolism   4. exergonic
5. endergonic   6. allosteric
Matching: 1. b   2. k   3. i   4. l   5. f   6. d   7. j   8. c   9. m   10. e
Making Comparisons: 1. catabolic   2. exergonic   3. anabolic   4. endergonic
5. anabolic   6. endergonic   7. catabolic   8. exergonic   9. anabolic   10. endergonic
11. catabolic   12. exergonic
Making Choices: 1. b,c,e   2. a–e   3. a,c   4. a,b,e   5. a,b,d   6. b,c,e   7. b,d,e
8. b,c,d   9. c   10. none   11. c   12. d

Chapter 8

Reviewing Concepts: 1. catabolism   2. anabolism   3. adenosine triphosphate (ATP)
4. (a) cellular respiration, (b) nutrients, (c) ATP 5. (a) aerobic, (b) anaerobic, (c)
fermentation   6. aerobic respiration   7. (a)carbon dioxide and water, (b)energy
8. glycolysis, formation of acetyl coenzyme A, citric acid cycle, and electron transport
chain and chemiosmosis 9. (a) dehydrogenations (b) decarboxylations 10. cytosol (or
cytoplasm)   11. glyceraldehyde-3-phosphate (PGAL)   12. pyruvate   13. (a)two, (b)
substrate-level phosphorylation, (c)two   14. (a)two, (b)two   15. (a)three, (b)two, (c)one
16. two   17. (a)NAD$^+$ and FAD, (b)the electron transport chain (c)oxidative
phosphorylation 18. (a)electron transport chain, (b)molecular oxygen 19. (a)proton, (b)
intermembrane space, (c)matrix 20. (a)protons (b)ATP synthase 21. chemiosmosis 22.
(a)oxidative phosphorylation   (b) substrate-level phosphorylation   23. NADH
24. oxygen   25. fatty acids   26. deamination   27. glycerol and fatty acid   28. beta
oxidation   29. anaerobic respiration   30. (a)organic molecule (b) NAD+   31. ethyl
alcohol 32. lactate   33. (a)two, (b)36-38
Building Words: 1. aerobe   2. anaerobe   3. dehydrogenation   4. decarboxylation
5. deamination   6. glycolysis
Matching: 1. c   2. d   3. h   4. i   5. a   6. g   7. k   8. l   9. f   10. m
Making Comparisons: 1. fructose-6-phosphate   2. phosphofructokinase   3. –1
ATP   4. dihydroxyacetone phosphate   5. aldolase   6. zero   7. 3-phosphoglycerate
8. +2 ATP   9. 2-phosphoglycerate   10. zero   11. phosphoenolpyruvate   12. pyruvate
kinase
Making Choices: 1. a,b,c   2. b,c,d,e   3. a,b,c,d   4. c   5. b,e   6. c,e   7. a,d   8. c,d
9. a,b,e   10. a,b   11. a-e   12. a,b,c

Visual Foundations: **1.** citrate synthase   **2.** aconitase   **3.** isocitrate dehydrogenase
4. α-ketoglutarate dehydrogenase   **5.** succinyl CoA synthetase   **6.** succinate
dehydrogenase **7.** fumarase **8.** malate dehydrogenase

Chapter 9

Reviewing Concepts: **1.** (a)carbon dioxide and water, (b)light (solar), (c) chemical   **2.**
(a)wavelength (b)photons (c)shorter wavelengths **3.** (a)absorbs (b)fluorescence,
(c)electron acceptor   **4.** (a)mesophyll (b)stroma, **5.** thylakoids   **6.** blue and red
7. (a)chlorophyll *a*, (b)chlorophyll *b*, (c)carotenoids   **8.** (a)absorption spectrum,
(b)spectrophotometer **9.** (a)action spectrum, (b)accessory pigments **10.** (a)chlorophyll
(b)light energy (c)carbon dioxide (d)glucose (e)oxygen **11.** (a)synthesis (b)thylakoids
12. (a)NADPH and ATP, (b)carbohydrate (c)stroma **13.** (a)light energy (b)ATP and
NADPH **14.** chlorophyll *a* **15.** antenna complexes **16.** reaction center **17.** electron
transfer reactions   **18.** (a) P700 (b)P680   **19.** reaction center   **20.** (a)ferredoxin
(b)NADP+ **21.** antenna complex   **22.** photolysis **23.** (a)energy (b)protons (c)ATP
24. (a)thylakoid membrane (b)PSI   **25.** cyclic electron transport   **26.**
photophosphorylation **27.** (a)thylakoid lumen (b)ATP synthase **28.** electron transport
29. carbon fixation **30.** stroma **31.** (a) ribulose biphosphate (RuBP), (b)rubisco
(c)phosphoglycerate (PGA) **32.** (a)CO_2, (b)NADPH, (c)ATP **33.** RuBP **34.** (a)RuBP
carboxylate, (b)oxygen **35.** CO_2 **36.** oxaloacetate   **37.** (a)stomata (b)desert
38. (a)heterotrophs, (b)autotrophs **39.** (a)organic molecules (b)carbon dioxide
40. photolysis of water
Building Words: **1.** mesophyll   **2.** photosynthesis   **3.** photolysis
4. photophosphorylation   **5.** chloroplast   **6.** chlorophyll   **7.** autotroph
8. heterotrophy **9.** photorespiration
Matching: **1.** m **2.** n **3.** h **4.** e **5.** a **6.** c **7.** k **8.** f **9.** j **10.** d **11.** o **12.** l
Making Comparisons: **1.** anabolic   **2.** catabolic **3.** site of electron transport
4. thylakoid membrane **5.** H_2O **6.** glucose or other carbohydrate **7.** terminal
hydrogen acceptor **8.** O_2
Making Choices: **1.** a **2.** b **3.** b **4.** a,b,d **5.** a,b,d **6.** a,c,d,e **7.** a,c,d **8.** a
9. c **10.** c,e **11.** a,c,e **12.** c **13.** b **14.** a,c,d **15.** c **16.** b

Chapter 10

Reviewing Concepts: **1.** growth, repair, reproduction   **2.** chromatin   **3.** 20,000 **4.**
(a)circular (b)chromosomes (c)packaged **5.** nucleosomes **6.** (a)30nm fiber (b)scaffolding
proteins (c) condensin **7.** (a)chromosome (b)genes **8.** 46   **9.** cell cycle **10.** interphase
11. (a)first gap, (b)synthesis, (c)second gap **12.** (a)nuclear (b)cytoplasmic **13.** (a)identical
(b)centromere   **14.** kinetochore   **15.** (a)mitotic spindle (b)MTOC (c)centrioles
16. spindle microtubules **17.** metaphase plate **18.** (a)kinetochore (b)polar **19.** (a)sister
(b)chromosome **20.** (a)depolymerize (b)push **21.** arrival of the chromosomes at the
poles   **22.** (a)contractile ring (b)cleavage furrow (c)cell plate   **23.** chromosomes
24. binary fission **25.** cell-cycle checkpoints **26.** (a)phosphorylating (b)Cdks (c)cyclins
27. (a)cytokinins (b)growth factors **28.** (a)asexual (b)clones **29.** (a)sexual, (b)diploid,
(c)zygote **30.** size, shape, and the position of their centromeres **31.** four **32.** meiosis I
and meiosis II   **33.** synapsis   **34.** (a)prophase, (b)genetic variability   **35.** anaphase
36. (a)four, (b)two   **37.** (a)mitosis, (b)meiosis   **38.** (a)gametes, (b)gametogenesis
39. spermatogenesis **40.** oogenesis **41.** (a)alternation (b)sporophyte (c)gametophyte
Building Words: **1.** chromosome   **2.** interphase   **3.** haploid   **4.** diploid
5. centromere **6.** gametogenesis **7.** spermatogenesis **8.** oogenesis **9.** gametophyte
10. sporophyte **11.** metaphase **12.** anaphase **13.** prophase
Matching: **1.** n **2.** l **3.** f **4.** c **5.** o **6.** b **7.** g **8.** j **9.** i **10.** d **11.** h

Making Comparisons: **1.** M phase **2.** Telophase I and II **3.** does not occur **4.** anaphase I **5.** tetrads form **6.** does not occur **7.** telophase **8.** telophase I, telophase II **9.** anaphase **10.** anaphase II **11.** chromosomes line up at the midplane of the cell **12.** metaphase I, metaphase II **13.** S phase of interphase **14.** S phase of interphase

Making Choices: **1.** a,b **2.** c,d **3.** b **4.** b,c,e **5.** a,d **6.** a-d **7.** b,f **8.** b **9.** h **10.** d **11.** b **12.** c **13.** d **14.** b **15.** d **16.** g **17.** h **18.** b **19.** e **20.** a,i **21.** a,g,h **22.** c **23.** j **24.** b **25.** g **26.** d

Chapter 11

Reviewing Concepts: **1.** Gregor Mendel **2.** (a)phenotype (b)genotype **3.** (a)first filial, (b)F_2 or second filial **4.** (a)dominant, (b)recessive **5.** (a)genes (b)DNA (c)RNA and protein **6.** segregate **7.** locus **8.** (a)heterozygous (b)homozygous **9.** Punnett square **10.** (a)genotype (b)homozygous dominant or heterozygous **11.** homozygous recessive **12.** two loci **13.** meiosis **14.** linearly arranged **15.** (a)zero, (b)one **16.** product rule **17.** product **18.** sum rule **19.** (a)past events, (b)future events **20.** linked **21.** crossing over **22.** (a)recombination (b)map unit **23.** autosomes **24.** (a)SRY (b)testosterone **25.** (a)XY male, (b)XX female **26.** (a)males (b)X **27.** (a)hyperactive, (b)inactivation **28.** Barr body **29.** (a)incomplete dominance (b)intermediate **30.** codominance **31.** (a)three or more, (b)locus (c)two **32.** pleiotropy **33.** epistasis **34.** polygenic inheritance **35.** norm of reaction

Building Words: **1.** polygene **2.** homozygous **3.** heterozygous **4.** dihybrid **5.** monohybrid **6.** hemizygous **7.** genotype **8.** phenotype

Matching: **1.** d **2.** a **3.** j **4.** b **5.** k **6.** h **7.** f **8.** c

Making Comparisons: **1.** tall plant with yellow seeds (TT Yy)) **2.** tall plant with green seeds (TT yy) **3.** tall plant with yellow seeds (Tt Yy) **4.** tall plant with green seeds (Tt yy) **5.** tY **6.** tall plant with yellow seeds (Tt Yy) **7.** short plant with yellow seeds (tt YY) **8.** short plant with yellow seeds (tt Yy) **9.** ty **10.** tall plant with yellow seeds (Tt Yy) **11.** tall plant with green seeds (Tt yy) **12.** short plant with green seeds (tt yy)

Making Choices: **1.** b,c,d,e,f **2.** j **3.** both j **4.** RrTT **5.** l,m,n,r,s,t **6.** g **7.** c/n and d/r **8.** g **9.** b/m and e/s **10.** b,e **11.** b,c,e **12.** d **13.** a-d **14.** b **15.** b or d **16.** a,c,d **17.** d **18.** c,d **19.** a,e **20.** c

Chapter 12

Reviewing Concepts: **1.** inheritance **2.** (a)protein (b)4 **3.** transformation **4.** bacteriophages **5.** pentose sugar (deoxyribose), phosphate, nitrogenous base **6.** (a)adenine and guanine, (b)cytosine and thymine **7.** (a)covalent (b)phosphodiester linkage **8.** (a)phosphate (b)hydroxyl **9.** X-ray diffraction **10.** (a)antiparallel nucleotide strands (b)sugar-phosphate backbone (c)bases **11.** (a)opposite (b)antiparallel **12.** (a)two, (b)three **13.** complementary base pairing **14.** template **15.** density **16.** mutation **17.** (a)origins of replication (b)helicase (c)single-strand binding proteins **18.** (a)supercoiling (b)topoisomerase **19.** DNA polymerases **20.** (a)RNA primer (b)DNA primase **21.** (a)5, (b)3, (c)Okazaki (d)DNA ligase **22.** (a)bidirectional (b)2 **23.** DNA polymerase **24.** mismatch repair **25.** nuclease, DNA polymerase, and DNA ligase **26.** (a)short, noncoding (b)telomerase **27.** apoptosis **28.** cell aging and apoptosis

Building Words: **1.** avirulent **2.** antiparallel **3.** semiconservative **4.** telomere

Matching: **1.** l **2.** d **3.** k **4.** i **5.** g **6.** h **7.** b **8.** e **9.** c **10.** d

Making Comparisons: **1.** Found substance in heat-killed bacteria that "transformed" living bacteria **2.** Hershey-Chase **3.** found that the ratios of guanine to cytosine, purines to pyrimidines, and adenine to thymine were very close to 1 **4.** Inferred from x-ray crystallographic films of DNA patterns that nucleotide bases are stacked like rungs in a ladder **5.** Avery, MacLeod, and McCarty

Making Choices: **1.** a,c,e **2.** a,c **3.** c,d **4.** b **5.** d **6.** a,c **7.** b,e **8.** d **9.** b,c,e **10.** a,c,e **11** a **12.** b **13.** e **14.** d **15.** a-e **16.** d **17.** d **18.** a-c

Chapter 13

Reviewing Concepts: **1.** DNA bases **2.** recessive mutant allele **3.** single-stranded, contains ribose, uses uracil **4.** mRNA, tRNA, and rRNA **5.** amino acid **6.** codon **7.** (a)anticodon (b) tRNA (c)mRNA **8.** reading frame **9.** (a)64, (b)61, (c)3 **10.** evolved early in ancient life **11.** methionine and tryptophan **12.** wobble hypothesis **13.** RNA polymerases **14.** (a)5' to 3' (b)3' to 5' **15.** (a)upstream, (b)downstream **16.** promoter **17.** (a)elongation (b)3' end **18.** termination **19.** (a)leader sequence (b)start codon **20.** trailing sequences **21.** precursor mRNA **22.** (a)5' end (b) poly-A tail (c)3' end **23.** (a)spliceosome (b)introns (c)exons **24.** peptide bonds **25.** aminoacyl-tRNA synthetases **26.** (a)one (b)three (c)P (d)A (e)E **27.** (a)initiation (b)small **28.** AUG **29.** (a)methionine (b)large (c)initiation **30.** (a)elongation (b)GTP **31.** peptidyl transferase **32.** (a)translocation (b)5' to 3' **33.** release factor **34.** molecular chaperones **35.** polyribosome **36.** (a)protein domains (b)exons **37.** (a)snRNA (b)SRP RNA **38.** sno-RNA **39.** (a)gene expression (b)siRNA and miRNA **40.** a specific RNA or polypeptide **41.** reverse transcriptase **42.** nucleotide sequence **43.** (a)point or base-substitution, (b) silent (c)missense, (d)nonsense **44.** inserted or deleted **45.** movable **46.** retrotransposons **47.** hot spots **48.** (a)mutagens (b)carcinogens

Building Words: **1.** anticodon **2.** ribosome **3.** polysome **4.** mutagen **5.** carcinogen

Matching: **1.** g **2.** j **3.** h **4.** m **5.** b **6.** c **7.** k **8.** d **9.** f **10.** e **11.** n **12.** l

Making Comparisons: **1.** 5'—ACC—3' **2.** 3'—TGG —5' **3.** 3'—CGC—5' **4.** 3'—CGC—5' **5.** 3'—ACA—5' **6.** 5'—UAA—3', 5'—UAG—3', 5'—UGA—3' **7.** no tRNA molecule binds to a stop codon **8.** 3'—TAC—5' **9.** 5'—UTC—3'

Making Choices: **1.** d **2.** c **3.** b,d **4.** c,e **5.** a,c **6.** a,c,d **7.** b,c **8.** a,c,e **9.** d **10.** a **11.** b **12.** b,c **13.** b,d **14.** d,e

Chapter 14

Reviewing Concepts: **1.** control of the amount or mRNA transcribed, the rate of translation of mRNA, and the activity of the protein product **2.** transcriptional-level control **3.** differential expression **4.** constitutive genes **5.** operon **6.** promoter region **7.** operator **8.** inducer **9.** corepressor **10.** (a)repressors (b)positive **11.** promoter **12.** posttranscriptional controls **13.** feedback inhibition **14.** operons **15.** (a)heterochromatin, (b)euchromatin (c)histones **16.** (a)methylation (b)inactivate (c)genomic imprinting (d)epigenetics **17.** (a)tandemly repeated gene sequences (b)gene amplification **18.** TATA box **19.** (a)enhancers (b)silencers **20.** (a)transcription factors (b)domain **21.** RNA processing **22.** alternative splicing **23.** mRNA **24.** chemical modification **25.** kinases **26.** phosphatases **27.** (a)ubiquitin (b)proteasome

Building Words: **1.** euchromatin **2.** heterochromatin **3.** corepressor **4.** heterodimer **5.** homodimer

Matching: **1.** f **2.** a **3.** j **4.** k **5.** d **6.** b **7.** g **8.** c **9.** e **10.** l

Making Comparisons: **1.** genes are inducible only during specific phases of life cycle **2.** molecular chaperones **3.** phenotypic expression is determined by whether an allele is inherited from the male or female parent **4.** posttranslational controls **5.** adjusts the rate of synthesis in a metabolic pathway **6.** enhancers **7.** regulate the rate of mRNA translation

Making Choices: **1.** b,e   **2.** a   **3.** a,d   **4.** c   **5.** e   **6.** e   **7.** b   **8.** b,d   **9.** a,b,c,e   **10.** a,d,e   **11.** a,d,e   **12.** b,c,d   **13.** b,e

Chapter 15
Reviewing Concepts: **1.** genetic engineering **2.** biotechnology **3.** restriction enzymes **4.** plasmids   **5.** palindromic **6.** (a)sticky ends (b)DNA ligase   **7.** origin of replication, restriction sites, and selection genes **8.** cosmid cloning vectors   **9.** genomic DNA library **10.** probes   **11.** reverse transcriptase   **12.** (a)DNA polymerase, (b)heat   **13.** Taq **14.** (a)positive pole   (b)size   **15.** (a)Southern blot (b)Northern   (c)Western **16.** polymorphism   **17.** (a)chain termination method (b)genome   **18.** (a)structural (b)functional (c)comparative (d)metagenomics   **19.** expressed sequence tags **20.** RNAi **21.** gene targeting **22.** mutagenesis screening   **23.** DNA microarrays   **24.** (a)function (b)evolution   **25.** (a)bioinformatics, (b)pharmacogenetics, (c) proteomics   **26.** gene therapy **27.** (a)drugs (b)vaccines   **28.** polymorphic   **29.** (a)fertilized egg or stem cell (b)viruses **30.** pharming   **31.** genetically modified crops   **32.** risk assessment
Building Words: **1.** transgenic   **2.** retrovirus   **3.** bacteriophage   **4.** kilobase   **5.** biotechnology
Matching: **1.** i   **2.** e   **3.** a   **4.** h   **5.** f   **6.** j   **7.** g   **8.** c   **9.** l   **10.** b   **11.** d   **12.** k
Making Comparisons: **1.** restriction enzymes   **2.** a segment of DNA that is homologous to a part of a DNA sequence that is being studied **3.** cloning   **4.** amplifying DNA by alternate heat denaturization and use of heat-resistant DNA polymerase to replicate DNA strands *in vitro* **5.** plasmids   **6.** DNA fragments comprising the total DNA of an organism   **7.** proteomics
Making Choices: **1.** a   **2.** d   **3.** c   **4.** d   **5.** c,e   **6.** b   **7.** a   **8.** c,e   **9.** b,e   **10.** b   **11.** d,e   **12.** c,d   **13.** b,c   **14.** c,d,e   **15.** a,c   **16.** b

Chapter 16
Reviewing Concepts: **1.** nucleotides in DNA   **2.** karyotype   **3.** (a)FISH (b)fluorescent dyes   **4.** pedigree analysis   **5.** 2.9 billion   **6.** noncoding regulatory elements, repetitive sequences, and gene segments of unknown function **7.** genome wide associations scans **8.** 500   **9.** gene targeting   **10.** polyploidy   **11.** (a)aneuploidy, (b)trisomic, (c)monosomic (d)nondisjunction   **12.** maternal age   **13.** Klinefelter syndrome   **14.** Turner syndrome **15.** XYY karyotype **16.** Translocation   **17.** deletion **18.** fragile sites **19.** epigenetic inheritance **20.** inborn errors of metabolism   **21.** enzyme **22.** hemoglobin   **23.** ion transport   **24.** mucous   **25.** (a) lipid, (b) brain **26.** autosomal dominant allele   **27.** factor VIII **28.** (a)cognitive abilities (b)males **29.** (a)mutant, (b)normal **30.** viral vectors **31.** amniocentesis   **32.** chorionic villus sampling (CVS) **33.** pre-implantation genetic diagnosis   **34.** preventive medicine **35.** carriers **36.** probability **37.** (a)all (b)all (c) recessive **38.** consanguineous mating **39.** normal **40.** (a)discriminating (b)genetic tests **41.** privacy and confidentiality
Building Words: **1.** cytogenetics   **2.** polyploidy   **3.** trisomy   **4.** translocation   **5.** aneuploidy
Matching: **1.** a   **2.** j   **3.** f   **4.** h   **5.** l   **6.** e   **7.** i   **8.** c   **9.** b   **10.** g
Making Comparisons: **1.** valine substitutes for glutamic acid in hemoglobin **2.** abnormal RBCs block small blood vessels.   **3.** Klinefelter syndrome   **4.** sex chromosome nondisjunction during meiosis   **5.** absence of an enzyme that converts phenylalanine to tyrosine; instead phenylalanine is converted into toxic phenylketones that accumulate   **6.** Hemophilia A   **7.** X-linked recessive trait   **8.** lack enzyme that breaks down a brain membrane lipid **9.** blindness, severe retardation **10.** Huntington's disease   **11.** autosomal dominant allele **12.** 47 chromosomes because they have extra copies of chromosome 21 **13.** retardation, abnormalities

Making Choices: **1.** b **2.** c **3.** a,d,e **4.** b,d **5.** c,d, **6.** b,c,d **7.** a,b,c **8.** b,e **9.** b,c,e **10.** e **11.** c,e **12.** a,d
Visual Foundations: **1.** c **2.** a **3.** d

Chapter 17

Reviewing Concepts: **1.** developmental genetics **2.** immunoflourescence **3.** (a)cell determination, (b)cell differentiation **4.** (a)morphogenesis, (b)pattern formation **5.** nuclear equivalence **6.** transcriptional **7.** totipotent **8.** clones **9.** cell cycles **10.** transgenic organisms **11.** pluripotent **12.** transcription factors **13.** (a)human reproductive (b) human therapeutic **14.** characteristics that allow for the efficient analysis of biological processes **15.** mutants **16.** (a)egg, larval, pupal, (b)adult (c)metamorphosis **17.** maternal effect **18.** (a)segmentation genes, (b)homeotic genes **19.** (a)homeobox (b)homeodomain **20.** hermaphrodites **21.** founder cells **22.** mosaic **23.** induction **24.** (a)apoptosis (b)caspases **25.** chimera **26.** regulative **27.** IGF **28.** SEPALLATA **29.** (a)tumor (b)metastasis **30.** oncogenes **31.** tumor suppressor gene or antioncogene
Building Words: **1.** morphogenesis **2.** oncogene **3.** totipotent **4.** pluripotent **5.** neoplasm
Matching: **1.** j **2.** k **3.** b **4.** i **5.** e **6.** d **7.** c **8.** a **9.** m **10.** g
Making Comparisons: **1.** ectoderm **2.** epidermis **3.** ectoderm **4.** neuron **5.** germ line cells **6.** lungs **7.** endoderm **8.** mesoderm
Making Choices: **1.** c,e **2.** c,e **3.** a,c,d **4.** c **5.** a,d **6.** c,e **7.** b **8.** c **9.** d **10.** c,e **11.** a **12.** b **13.** a,c,d **14.** b,c,e

Chapter 18

Reviewing Concepts: **1.** gradual divergence **2.** links all fields of the life sciences into a unified body of knowledge **3.** populations **4.** species **5.** Bioremediation **6.** scale of nature **7.** Leonardo da Vinci **8.** Jean Baptiste de Lamarck (Lamarck) **9.** Lamarck **10.** (a)H.M.S. Beagle, (b)Galapagos Islands **11.** artificial selection **12.** Thomas Malthus **13.** adaptation **14.** Alfred Russell Wallace **15.** *Origin of Species by Means of Natural Selection* **16.** (a) overproduction, (b) variation, (c) limits on population growth (a struggle for existence), (d) differential reproductive success ("survival of the fittest") **17.** (a) Darwin, (b) Mendel **18.** mutations **19.** (a) natural selection, (b) chance **20.** has evolved through time **21.** (a) rapid covering, (b) decay **22.** rain forests **23.** 195,000 **24.** index fossils **25.** biogeography **26.** continental drift **27.** homologous **28.** convergent evolution **29.** homoplastic features **30.** vestigial structures (organs) **31.** amino acid **32.** DNA sequencing **33.** years
Building Words: **1.** homologous **2.** biogeography **3.** homoplastic
Matching: **1.** b **2.** d **3.** f **4.** m **5.** n **6.** o **7.** j **8.** e **9.** i **10.** k **11.** a
Making Comparisons: **1.** homologous features **2.** homoplastic features **3.** the presence of useless structures is to be expected as a species adapts to a changing mode of life **4.** areas of the world that have been separated from the rest of the world for a long time have organisms unique to that area **5.** evidence that all life is related **6.** differential reproductive success **7.** fossil record **8.** convergent evolution **9.** the order of nucleotide bases in DNA is useful in determining evolutionary relationships
Making Choices: **1.** a,c,d **2.** a,b,c,d,e **3.** c **4.** b **5.** a **6.** b,e **7.** a **8.** e **9.** c,d **10.** c **11.** d **12.** c **13.** a,b,c,e **14.** a,b,c,e **15.** d **16.** e

Chapter 19

Reviewing Concepts: **1.** genetic variability **2.** all the alleles for all of the loci present **3.** phenotype frequency **4.** allele frequencies must change over successive generations

5. (a)0, (b)1 **6.** (a)$p^2 + 2pq + q^2 = 1$ (b) at genetic equilibrium   **7.** large population, genetic isolation, no net mutations, no natural selection, random mating   **8.** codominant   **9.** visible phenotype **10.** microevolution **11.** genetically similar individuals within a population **12.** genetic fitness   **13.** assortative mating **14.** (a) at the loci involved in mate choice, (b) in the entire genome **15.** mutation   **16.** genetic drift **17.** (a) decrease, (b) increase **18.** (a) bottleneck, (b) genetic diversity **19.** a small number of individuals from a larger population establish a colony in a new area **20.** gene flow   **21.** stabilizing selection **22.** directional selection   **23.** disruptive selection   **24.** phenotypes **25.** balanced polymorphism   **26.** heterozygote advantage   **27.** frequency-dependent selection **28.** neutral variation   **29.** geographic variation

Building Words: **1.** microevolution   **2.** phenotype

Matching: **1.** n   **2.** k   **3.** h   **4.** f   **5.** e   **6.** b   **7.** e   **8.** a   **9.** j   **10.** m   **11.** l

Making Comparisons: **1.** yes   **2.** yes   **3.** no   **4.** yes   **5.** yes   **6.** yes   **7.** no   **8.** yes **9.** no

Making Choices: **1.** b,c   **2.** a,e   **3.** a,c   **4.** c,d   **5.** d   **6.** c   **7.** a,c,d   **8.** a,c   **9.** a **10.** c,e   **11.** e   **12.** n   **13.** i   **14.** y   **15.** r   **16.** m

Chapter 20

Reviewing Concepts: **1.** (a) 4-100 million, (b) 99   **2.** (a) Linnaeus, (b) morphological species concept   **3.** (a) reproductively, (b) gene pool   **4.** sexually reproducing organisms **5.** subspecies **6.** genetic integrity **7.** interspecific   **8.** (a) temporal isolation, (b) behavioral or sexual isolation, (c) mechanical isolation, (d) gametic isolation   **9.** hybrid inviability **10.** hybrid sterility   **11.** hybrid breakdown **12.** lysin   **13.** (a) populations, (b) reproductively isolated **14.** geographically isolated   **15.** allele frequency **16.** (a) a change in ploidy, (b) a change in ecology   **17.** plants **18.** (a) polyploidy, (b) allopolyploidy **19.** hybrid zone   **20.** reinforcement **21.** stability   **22.** punctuated equilibrium **23.** gradualism   **24.** preadaptations **25.** (a) allometric growth, (b) paedomorphosis **26.** (a) adaptive zones, (b) adaptive radiation   **27.** background extinction   **28.** mass **29.** (a) microevolutionary, (b) macroevolutionary

Building Words: **1.** paedomorphic   **2.** polyploidy   **3.** allopatric   **4.** allopolyploidy **5.** macroevolution   **6.** allometric

Matching: **1.** j   **2.** p   **3.** t   **4.** a   **5.** n   **6.** q   **7.** m   **8.** i   **9.** o   **10.** e   **11.** k

Making Comparisons: **1.** hybrid breakdown   **2.** gametes of similar species are incompatible   **3.** hybrid sterility   **4.** similar species have structural differences in their reproductive organs   **5.** hybrid inviability   **6.** temporal isolation **7.** two different species do not attempt to mate because they live in different parts of the environment **8.** prezygotic barrier

Making Choices: **1.** a   **2.** e   **3.** a,b,e   **4.** c   **5.** d   **6.** e   **7.** b,d   **8.** a,c,e   **9.** c,e   **10.** e **11.** a,c,e   **12.** c,d   **13.** b,d   **14.** a,c,d,e   **15.** a,b

Chapter 21

Reviewing Concepts: **1.** chemical evolution **2.** 4.6   **3.** little or no free oxygen, energy, chemical building blocks, time   **4.** (a) prebiotic soup, (b) sulfur world   **5.** (a) Oparin, (b) Haldane   **6.** (a) Miller, (b) Urey, (c) hydrogen (H_2), methane (CH_4), ammonia (NH_3), water   **6.** protobionts **7.** microspheres   **8.** (a) RNA, (b)ribosomes   **9.** (a) microfossils, (b) 3.5   **10.** (a) anaerobic, (b) prokaryotic   **11.** autotrophs **12.** cyanobacteria **13.** obligate anerobes **14.** oxygen **15.** 2.2   **16.** serial endosymbiosis **17.** index fossils **18.** eons   **19.** epochs   **20.** Proterozoic   **21.** Ediacaran   **22.** 542 million   **23.** (a) Ordovician, (b) ostracoderm   **25.** Devonian   **26.** 251 million   **27.** (a) Triassic, (b) Jurassic, (c) Cretaceous **28.** (a) Paleogene period, (b) Neogene period, (c) Quaternary period

Building Words: **1.** autotroph   **2.** protobiont   **3.** heterotroph   **4.** precambrian
5. aerobe   **6.** anaerobe
Matching: **1.** n   **2.** g   **3.** b   **4.** q   **5.** l   **6.** f   **7.** e   **8.** k   **9.** 9   **10.** h   **11.** i   **12.** m
Making Comparisons: **1.** Ordovician   **2.** Paleozoic   **3.** Triassic   **4.** Mesozoic
5. Cretaceous   **6.** Mesozoic   **7.** Cambrian   **8.** Paleozoic   **9.** Quaternary
10. Cenozoic   **11.** Devonian   **12.** Paleozoic   **13.** Paleogene   **14.** Cenozoic
15. Carboniferous   **16.** Paleozoic
Making Choices: **1.** d   **2.** b,c   **3.** d   **4.** c   **5.** b,c   **6.** a,d   **7.** a-e   **8.** c,d,e   **9.** c,e
10. a,b,c,d   **11.** d   **12.** b,d   **13.** a-e   **14.** b,e   **15.** a,b,d,e   **16.** e   **17.** b,d   **18.** c

Chapter 22

Reviewing Concepts: **1.** placental mammals   **2.** arboreal past   **3.** a) flexible, b) have
five digits   **4.** 56 million   **5.** integrate visual information from both eyes simultaneously **6.**
(a) prosimii, (b) tarsiiformes, (c) anthropoidea **7.** a) Africa, b) Asia   **8.** of their brain   **9.**
prehensile tails   **10.** hominoids   **11.** gibbons (*Hylobates*), orangutans (*Pongo*), gorillas
(*Gorilla*), chimpanzees (*Pan*), humans (*Homo*) **12.** brachiation   **13.** knuckle walking   **14.**
foramen magnum   **15.** supraorbital ridges   **16.** 6-7 million   **17.** Africa   **18.** *Ardipithecus*
19. sexual dimorphism   **20.** *Australopithecus afarensis*   **21.** Tanzania   **22.** 2.5 million **23.**
(a) *Homo ergaster* (b) *H. erectus*   **24.** Indonesia **25.** 1.7 million   **26.** *H. ergaster*   **27.** Spain
28. Neander Valley in Germany   **29.** dead end   **30.** out of Africa   **31.** (a)98, (b) 99   **32.**
development of hunter-gatherer societies, development of agriculture and the Industrial
Revolution   **33.** 10,000 **34.** 2.6 billion **35.** environment
Building Words: **1.** hominoid   **2.** quadrupedal   **3.** bipedal   **4.** arboreal
5. supraorbital
Matching: **1.** e   **2.** g   **3.** a   **4.** j **5.** l   **6.** k   **7.** g   **8.** c   **9.** d   **10.** f
Making Comparisons: **1.** *Australopithecus afarensis*   **2.** Neandertal   **3.** *Ardipithecus
ramidus*   **4.** *Australopithecus africanus*   **5.** 2.5 mya   **6.** more modern hominid features
than *Australopithecus*, fashioned primitive tools;   **7.** 6-7 mya   **8.** 1.7 mya   **9.** larger brain
than *H. habilis*, made more sophisticated tools **10.** 1.2 mya **11.** *Australopithecus anamensis*
Making Choices: **1.** c,d,e   **2.** a,b,e   **3.** a,e   **4.** a,d,e   **5.** c   **6.** e   **7.** b   **8.** b   **9.** a
10. d,e   **11.** d   **12.** b,c   **13.** c   **14.** d   **15.** a,b,c,d,e   **16.** e

Chapter 23

Reviewing Concepts: 1. biological diversity or biodiversity   **2.** systematics
3. taxonomy   **4.** binomial system   **5.** (a) genus, (b) specific epithet   **6.** (a) family, (b)
class, (c) phylum   **7.** domain   **8.** taxon   **9.** (a) Plantae, (b) Animalia   **10.** Protista
11. Fungi   **12.** (a) Prokaryotes, (b) Bacteria, (c) Archea   **13.** clade   **14.** cladogram
15. vertical gene transfer   **16.** horizontal gene transfer or lateral gene transfer   **17.**
evolutionary relationships   **18.** population   **19.** convergent evolution   **20.**
homoplasy   **21.** shared ancestral characters   **22.** shared derived characters   **23.**
combination   **24.** bar code   **25.** molecular systematics   **26.** homologous   **27.** (a)
amino acid sequences, (b) nucleotide sequences   **28.** monophyletic   **29.** paraphyletic
30. polyphyletic   **31.** shared characters   **32.** (a) cladistics, (b) evolutionary
systematics   **33.** (a) outgroup, (b) ingroup   **34.** derived   **35.** ancestral   **36.** most
recent common ancestor   **37.** ancestor-descendent   **38.** parsimony   **39.**
maximum likelihood
Building Words: **1.** subphylum   **2.** paraphyletic   **3.** Polyphyletic
Matching:
1. Classification   **2.** Taxonomy   **3.** Genus   **4.** Phylum   **5.** Taxon   **6.** Clade   **7.** Phenetics
8. Horizontal gene transfer   **9.** Derived character   **10.** Parsimony

Making Comparisons: **1.** Archaea   **2.** Archaea   **3.** Eukarya   **4.** Protista
5. Eukarya **6.** Fungi **7.** Eukarya **8.** Animalia **9.** Eukarya **10.** Plantae
Making Choices: **1.** c **2.** d **3.** d **4.** c,e **5.** a,d **6.** e **7.** a **8.** b,d **9.** b **10.** e
Visual Foundations:
1. 2,3,4,5,6 **2.** 4,5,6 **3.** This taxon shares the most common ancestor , E.   **4.** This taxon
includes some but not all of the descendents of a common ancestor, A.   **5.** This taxon
does not share the same recent common ancestor.

Chapter 24
Reviewing Concepts: **1.** disease **2.** 20, 300   **3.** nucleic acid (DNA or RNA)
4. capsid   **5.** viron **6.** capsid **7.** Capsomers (protein subunits in the capsid) **8.** host cell's
plasma membrane **9.** host range   **10.** (a) transcriptional, (b) translational **11.** polyhedral
head **12.** (a) lytic, (b) lysogenic **13.** virulent   **14.** attachment (absorption), penetration,
replication and synthesis, assembly, release   **15.** restriction enzymes   **16.** temperate
17. (a) prophage, (b) lysogenic cells   **18.** lysogenic conversion **19.** pathogens **20.** (a) a
capsid, (b) an envelope **21.** single-stranded RNA   **22.** the plant wall has been damaged
23. bioterrorism   **24.** retroviruses   **25.** reverse transcriptase **26.** cellular origin **27.**
coevolution   **28.** regressive **29.** subviral **30.** RNA **31.** (a) nucleus, (b) gene regulation
32. transmissible spongiform encephalopathies or TSEs
Building Words: **1.** bacteriophage   **2.** viroid **3.** capsid **4.** obligate **5.** retrovirus
6. pathogen
Matching: **1.** h **2.** m **3.** k **4.** f **5.** d **6.** c **7.** i **8.** g **9.** j **10.** l
Making Comparisons: **1.** yes **2.** no **3.** no **4.** no **5.** yes **6.** yes **7.** yes **8.** yes **9.**
yes **10.** no **11.** yes
Making Choices: **1.** a-e **2.** a,c,d **3.** a,d,e **4.** a **5.** a,b,d **6.** b,d,e **7.** a **8.** a,b,c,e
9. a,b **10.** c **11.** d,e

Chapter 25
Reviewing Concepts: **1.** disease   **2.**(a) 50, (b) 35, (c) 15 **3.**(a) cocci, (b) diplococci,
(c) streptococci, (d) staphylococci   **4.** bacilli   **5.** (a)spirillum, (b)spirochete   **6.** vibrio
7. Nucleoid or nucleus area **8.** peptidoglycan **9.** gram-positive **10.** gram-positive **11.**
other microorganisms   **12.** the host's white blood cells **13.** (a) proteins, (b) flagella
14. reproduction   **15.** (a)rotating flagella, (b)chemotaxis **16.** plasmids **17.** (a)binary,
fission,(b) budding,(c) fragmentation   **18.** transformation   **19.** transduction
20. conjugation **21.** mutations **22.** autotrophs **23.** (a) chemotrophs, (b) phototrophs
24. decomposers **25.** (a)aerobic, (b)facultative anaerobes, (c)obligate anaerobes **26.** (a)
amino acids, (b) nucleic acids   **27.** other organisms   **28.** (a) size, (b) shape **29.** (a)
Bacteria, (b) Archaea **30.** peptidoglycan **31.** eukaryotes **32.** the sequence of entire
genomes **33.** extreme thermophiles   **34.** methanogens   **35.** (a) mutualism, (b)
commensalism, (c) parasitism **36.** (a) biofilms, (b) dental plaque **37.** bacteria **38.** (a)
nitrogen, (b) rhizobial bacteria **39.** antibiotics **40.** bioremidation **41.** microbiota **42.**
Koch's postulates **43.** (a) adhere to a specific cell type, (b) multiply, (c) produce toxic
substances **44.** R factors **45.** bioflims
Building Words: **1.** chemotaxis   **2.** exotoxin   **3.** endotoxin   **4.** methanogen
5. endospore **6.** pathogen **7.** anaerobe **8.** prokaryote **9.** Eukaryote **10.** obligate
11. halophile
Matching: **1.** i **2.** a **3.** h **4.** e **5.** b **6.** c **7.** j **8.** d **9.** n **10.** k **11.** l **12.** f
Making Comparisons: **1.** absent   **2.** present   **3.** present   **4.** absent (except in
mitochondria and chloroplasts)   **5.** absent   **6.** absent   **7.** present   **8.** present
9. absent **10.** absent **11.** absent **12.** absent **13.** present

Making Choices: **1.** a,b,c,e **2.** b,d,e **3.** d,e **4.** a,b,d **5.** e **6.** a,d **7.** a-e **8.** b **9.** b,c,d,e **10.** a **11.** c **12.** a,b

Chapter 26

Reviewing Concepts: **1.** Eukarya, (b) nucleus **2.** unicellular, (b) free-living **3.** plankton **4.** symbiotic **5.** serial endosymbiosis **6.** apicomplexa **7.** (a) monophyletic, (b) polyphyletic **8.** mitochondria **9.** oral groove **10.** (a) functional mitochondria, (b) Golgi complex **11.** (a) termites, (b) wood-eating cockroaches **12.** photosynthetic **13.** pellicle **14.** photosynthetic **15.** mitochondrion **16.** *Trypanosoma brucei* **17.** flattened vesicles **18.** flagella, (b) cellulose **19.** (a) zooxanthellae, (b) carbohydrates, (c) coral reefs **20.** locomotion **21.** apical complex **22.** sporozoites **23.** *Plasmodium* **24.** (a)micronuclei, (b)macronucleus **25.** conjugation **26.** mycelium **27.** (a)zoospores, (b)oospores **28.** silica **29.** (a) radial, (b) bilateral **30.** diatomaceous earth **31.** seaweeds **32.** (a) gametes, (b) zoospores **33.** alternation of generation **34.** unicellular **35.** (a) silica, (b) calcium carbonate **36.** test **37.** marine **38.** chalk **39.** index fossils **40.** microtubules **41.** holdfast **42.** (a) agar, (b) carrageenan **43.** (a) mitosis, (b) cell division **44.** fragmentation **45.** motile spores with flagella **46.** posterior flagellum **47.** two flagella **48.** asymmetrical **49.** pseudopodia **50.** *Entamoeba histolytica* **51.** sporangia **52.** (a)swarm cell, (b)myxamoeba **53.** amoebas, (b) plasmodial slime molds **54.** microvilli

Building Words: **1.** protozoa **2.** pellicle **3.** pseudopodium **4.** micronucleus **5.** macronucleus

Matching: **1.** b **2.** k **3.** g **4.** i **5.** d **6.** c **7.** f **8.** e **9.** j **10.** l **11.** m **12.** o

Making Choices: **1.** a-d **2.** c,e **3.** c,d,e **4.** a,b,e **5.** b **6.** c **7.** b **8.** a,c,e **9.** b **10.** a **11.** a,d **12.** a,c,d **13.** e **14.** c **15.** a,c,d,e

Chapter 27

Reviewing Concepts: **1.** green algae **2.** (a)chlorophylls *a* and *b* and accessory pigments, yellow and orange carotenoids, (b)starch, (c) cellulose **3.** maternal tissues **4.** (a)waxy cuticle, (b)stomata **5.** carbon dioxide **6.** gametangia **7.** alternation of generations **8.** (a) gametophyte generation, (b) sporophyte generation **9.** (a)antheridia, (b)archegonia **10.** (a)zygote (fused gametes), (b)spores **11.** stoneworts (charophytes) **12.** (a)mosses and other bryophytes, (b)bryophytes and seedless vascular plants, (c)gymnosperms, (d)angiosperms (flowering plants) **13.** lignin **14.** mosses, liverworts, hornworts **15.** rhizoids **16.** archegonium **17.** sporophyte **18.** protonema **19.** (a)*Sphagnum*, (b)peat mosses **20.** thallus **21.** archegonia and antheridia **22.** gemmae **23.** Anthocerophyta **24.** indeterminate growth **25.** photoperiodism **26.** monophyletic **27.** bryophytes **28.** (a) microphyll, (b) megaphyll, (c) megaphyll **29.** (a) ferns, (b) club mosses **30.** coal deposits **31.** (a)rhizome, (b)fronds **32.** (a)sporangia, (b)sori, (c)prothallus **33.** (a)upright stems, (b) dichotomous **34.** fungus **35.** Australia **36.** silica **37.** (a) homospory, (b) heterospory, (c) microspores, (d) megaspores **38.** apical meristem **39.** megafossils

Building Words: **1.** xanthophyll **2.** gametangium **3.** archegonium **4.** gametophyte **5.** sporophyte **6.** microphyll **7.** megaphyll **8.** sporangium **9.** homospory **10.** heterospory **11.** megaspore **12.** microspore **13.** bryophyte **14.** bryology **15.** thallus **16.** strobilus

Matching: **1.** d **2.** c **3.** j **4.** l **5.** b **6.** h **7.** a **8.** c **9.** p **10.** k **11.** e **12.** f

Making Comparisons: **1.** vascular **2.** sporophyte **3.** vascular **4.** sporophyte **5.** vascular **6.** sporophyte **7.** vascular **8.** naked seeds **9.** nonvascular **10.** seedless, reproduce by spores **11.** gametophyte **12.** seedless, reproduce by spores **13.** seedless, reproduce by spores **14.** sporophyte

Making Choices: **1.** a,c,e **2.** a-e **3.** a,c,d **4.** b,d **5.** b,c,e **6.** c,d **7.** b,d **8.** a,e **9.** a,c **10.** a,c,d **11.** a,e **12.** b,d **13.** a-e **14.** d **15.** b **16.** a,c,d,e

Chapter 28

Reviewing Concepts: **1.** seeds **2.** pinon pine **3.** (a) gymnosperms, (b) angiosperms **4.** (a) xylem, (b) phloem **5.** Coniferophyta, Ginkgophyta, Cycadophyta, Gnetophyta **6.** resin **7.** monoecious **8.** (a) *Pinus*, (b) sporophytes **9.** (a) sporophylls, (b) microsporocytes, (c) male gametophytes (pollen grains) **10.** (a) megasporangia, (b) megaspores, (c) female gametophyte **11.** pollen tube **12.** (a) water, (b) air **13.** on land **14.** Cycadophyta **15.** dioecious **16.** Ginkgophyta **17.** Gnetophyta **18.** vessel elements **19.** (a) vascular, (b) sexually **20.** vessel elements **21.** palms, grasses, orchids, irises, onions, lilies **22.** oaks, roses, mustards, cacti, blueberries, sunflowers **23.** (a)three, (b)one, (c)endosperm **24.** (a)four or five, (b)two, (c)cotyledons **25.** (a)sepals, petals, stamens, carpels; (b)stamens; (c)carpels; (d)perfect **26.** (a) sepals, (b) leaflike **27.** (a)anther, (b)ovary **28.** (a)megaspores, (b)embryo sac **29.** (a)microspores, (b)male gametophyte (pollen grain) **30.** (a) zygote, (b) endosperm tissue **31.** fruit **32.** (a)cross-fertilization, (b)genetic variation **33.** (a) food, (b) water storage **34.** leaves or leaflike structures **35.** (a)progymnosperms, (b)seed ferns **36.** gymnosperms **37.** conifers

Building Words: **1.** gymnosperm **2.** angiosperm **3.** monoecious **4.** dioecious **5.** fertilization **6.** endosperm

Matching: **1.** a **2.** e **3.** f **4.** i **5.** k **6.** m **7.** h **8.** l **9.** b **10.** c **11.** n

Making Comparisons: **1.** herbaceous or woody **2.** endosperm **3.** cotyledons **4.** mostly broader than in monocots **5.** 1 cotyledon **6.** floral parts in multiples of 4 or 5 **7.** parallel venation **8.** netted venation

Making Choices: **1.** b,c **2.** a **3.** b **4.** b **5.** a,c,d,e **6.** c,d,e **7.** a,e **8.** b,e **9.** a-e **10.** b,c **11.** d **12.** a,d **13.** a-e **14.** d **15.** c

Chapter 29

Reviewing Concepts: **1.** mycologists **2.** decomposers **3.** heterotrophs **4.** less **5.** extracellularly **6.** lipid droplets or glycogen **7.** a cell wall **8.** (a) chitin (b) Arthropoda **9.** yeasts **10.** hyphae **11.** (a) mycelium (b) molds **12.** spores **13.** (a) fruiting body, (b) spores) **14.** (a) buds, (b) spores **15.** haploid **16.** pheromones **17.** (a)dikaryotic, (b)monokaryotic **18.** cellulose **19.** flagellated protist **20.** animals **21.** Chytridiomycota, Zygomycota, Ascomycota, Basidiomycota, Glomeromycota **22.** monophyletic **23.** sexual stage **24.** (a)thallus, (b)rhizoids **25.** flagellated **26.** *Rhizopus stolonifer* **27.** heterothallic **28.** (a) opportunistic (b) compromised immune systems **29.** mitochondria **30.** mycorrhizae **31.** arbuscules **32.** (a)asci, (b)conidia, (c)conidiophores **33.** Ascocarps **34.** budding **35.** (a)basidium, (b)basidiospores **36.** (a)button, (b)basidiocarp **37.** (a) carbon dioxide, (b)minerals **38.** (a) cellulose (b) lignin **39.** (a) endomycorrhizal (b) ectomycorrhizae **40.** (a)green alga, (b) cyanobacterium, (c) both (d) ascomycete **41.** (a) crustose, (b) foliose, (c) fruticose **42.** Soredia **43.** lignin **44.** (a) yeasts, (b) fruit sugars, (c) sugar from starch in grains, (d) carbon dioxide

45. (a) *Penicillium*, (b) *Aspergillus* **46.** *Amanita* **47.** psilocybin **48.** (a)*Penicillium notatum*, (b)penicillin **49.** ergot **50.** (a) coal tars, (b) petroleum **51.** (a) malaria, (b) *Anopehles* **52.** aflatoxins **53.** cutinase **54.** haustoria

Building Words: **1.** coenocytic **2.** monokaryotic **3.** heterothallic **4.** conidiophore **5.** homothallic **6.** mycelium **7.** plasmogamy **8.** pheromone **9.** rhizoid **10.** basidiocarp

Matching: **1.** j **2.** k **3.** b **4.** c **5.** o **6.** l **7.** e **8.** d **9.** h **10.** f **11.** m

Making Comparisons: **1.** Zygomycota **2.** Ascomycota **3.** Basidiomycota **4.** Glomeromycota

Making Choices: **1.** a,c,d,e **2.** c,e **3.** a-e **4.** e **5.** a-e **6.** a,c,e **7.** c,d,e **8.** a,d **9.** b **10.** b,d **11.** a-e **12.** c **13.** a,c,d **14.** a-e **15.** a,e **16.** a,b,c **17.** b,d,e **18.** a,b,d **19.** a,b,c,e **20.** a-e **21.** e **22.** a-e **23.** d

Chapter 30

Reviewing Concepts: **1.** 99 **2.** monophyletic **3.** collagen **4.** heterotrophs **5.** (a) fluid, (b) salt **6.** less constant **7.** hypotonic **8.** dehydrate **9.** internal fertilization **10.** choanoflagellate **11.** parsimony **12.** Cambrian radiation or Cambrian explosion **13.** Hox **14.** (a) similarities, (b) morphology **15.** (a) bilateral symmetry, (b) cephalization **16.** (a)anterior, (b)posterior, (c)dorsal, (d)ventral **17.** (a)medial, (b)lateral, (c)cephalic, (d)caudal **18.** germ layers **19.** acoelomates **20.** (a)pseudocoelomates, (b)coelomates **21.** (a)mouth, (b)anus **22.** (a)radial cleavage, (b)spiral cleavage **23.** metozoa **24.** eumetazoa **25.** (a) Lophotrochozoa, (b) Ecydsozoa **26.** segmentation

Building Words: **1.** pseudocoelom **2.** ectoderm **3.** mesoderm **4.** protostome **5.** deuterostome **6.** schizocoely **7.** enterocoely **8.** blastula **9.** gastrodermis **10.** cleavage

Matching: **1.** m **2.** j **3.** a **4.** e **5.** h **6.** c **7.** p **8.** n **9.** o **10.** i

Making Comparisons: **1.** radial **2.** spiral **3.** indeterminate **4.** determinate **5.** anus **6.** mouth **7.** enterocoelous (enterocoely) **8.** schizocoelous (shizocoely)

Making Choices: **1.** c **2.** b,c,d **3.** c **4.** d,e **5.** a **6.** b **7.** a,e **8.** e **9.** d **10.** a,c,d,e **11.** a,b,d,e **12.** a-e **13.** b,c,d,e **14.** b,c

Chapter 31

Reviewing Concepts: **1.** 99 **2.** (a) adaptations that facilitated food capture, (b) escape from predators, (c) reproduction **3.** do not form tissues **4.** choanoflagellates **5.** (a) asconoid, (b) syconoid, (c) leuconoid **6.** (a)spongocoel, (b)osculum **7.** (a) mesohyl, (b) spicules **8.** (a)Hydrozoa, (b)Scyphozoa, (c)Anthozoa, (d) Cubozoa **9.** cnidocytes **10.** nematocysts **11.** (a) ectoderm (epidermis), (b) endoderm (gastrodermis), (c)mesoglea **12.** (a) freshwater, (b) sexually **13.** medusa **14.** sea turtles **15.** Planula **16.** biradial **17.** (a) cnidocytes, (b) mouth, (c) anal pores **18.** (a) bilateral symmetry, (b) three **19.** (a) Turbellaria, (b) Trematoda, (c) Monogenea, (d) Cestoda **20.** auricles **21.** pharynx **22.** scolex (head) **23.** proglottid **24.** everted from the anterior end of the body **25.** Mantle **26.** (a) radula, (b) hemocoel **27.** (a)trochophore, (b)veliger **28.** eight **29.** (a) eyes, (b) tentacles **30.** insects **31.** torsion **32.** (a)shell, (b)calcium carbonate **33.** (a)tentacles, (b)ten, (c)eight **34.** "head-foot" **35.** (a)septa, (b)setae

36. (a) marine, (b) parapodia   **37.** (a) crop, (b) gizzard   **38.** (a) hemoglobin, (b) metanephridia, (c) moist skin   **39.** (a) Brachiopoda, (b) Phoronida, (c) Bryozoa   **40.** (a) brain, (b) sense organs, (c) flame cells

41. increase in cell size **42.** (a) decomposers, (b) predators of smaller organisms **43.** 80 **44.** (a)exoskeleton, (b)chitin **45.** tracheae **46.** trilobites **47.** (a) Chilopoda, (b) one **48.** (a) Diplopoda, (b) two

49. Chelicerata **50.** (a)cephalothorax, (b)six, (c) four **51.** book lungs **52.** spinnerets **53.** (a) Crustacea, (b) mandibles, (c) maxillae, (d)biramous, (e) two **54.** barnacles **55.** Decapoda **56.** (a)maxillae, (b)maxillipeds, (c)chelipeds, (d)walking legs **57.** (a)reproductive, (b)swimmeretes **58.** (a) Insecta, (b) Hexapoda **59.** (a)three, (b)one or two, (c)one, (d)Malpighian tubules

Building Words: 1. exoskeleton   **2.** bivalve   **3.** biramous   **4.** uniramous   **5.** trilobite **6.** cephalothorax   **7.** hexapod   **8.** arthropod   **9.** spongocoel   **10.** choanocytes   **11.** hydrostatic   **12.** auricle(s)

Matching: 1. s   **2.** j   **3.** q   **4.** f   **5.** d   **6.** c   **7.** l   **8.** e   **9.** r   **10.** i   **11.** u   **12.** n   **13.** t

Making Comparisons: 1. hydras, jellyfish, coral   **2.** biradial symmetry; diploblastic; gastrovascular cavity with mouth and anal pores   **3.** Platyhelminthes   **4.** biradial symmetry; triploblastic; simple organ systems; gastrovascular cavity with one opening **5.** Mollusca   **6.** soft body with dorsal shell; muscular foot; mantle covers visceral mass; most with radula   **7.** some marine worms, earthworms, leeches   **8.** segmented body; most with setae   **9.** Rotifera   **10.** crown of cilia; constant cell number   **11.** cylindrical, threadlike body; pseudocoelom   **12.** centipedes, crabs, lobsters, spider, insects **13.** segmented body; jointed appendages; exoskeleton; some with compound eyes; insects with tracheal tubes   **14.** Echinodermata   **15.** bilateral larva, pentaradial adult; triploblastic; organ systems; complete digestive tube

Making Choices: 1. a,b,d   **2.** c   **3.** c,e   **4.** e   **5.** a   **6.** b   **7.** a,e   **8.** c,e   **9.** d **10.** a,c,e   **11.** c   **12.** b,c,e   **13.** a   **14.** d   **15.** b,d,e   **16.** a,d   **17.** a,b,d   **18.** b,c,e **19.** a,c,e   **20.** b,d,e   **21.** d   **22.** b   **23.** a,b   **24.** b,c

Chapter 32

Reviewing Concepts: 1. filter   **2.** (a) echinoderms, (b) chordates   **3.** Proterozoic **4.** (a) radial, (b) indeterminate   **5.** (a)bilateral, (b) pentaradial, (c)coelom   **6.** (a) endoskeleton (internal skeleton), (b) pedicellariae   **7.** Early Cambrian   **8.** (a) Crinoidea, (b) oral   **9.** Asteroidea   **10.** central disk **11.** tube feet   **12.** carnivores (active predators)   **13.** Ophiuroidea   **14.** (a) locomotion, (b) to collect and handle food   **15.** (a) Echinoidea, (b) absent, (c) test, (d) spines   **16.** (a) Holotghuroidea, (b) tube feet **17.** Chordata   **18.** (a) notochord, (b) dorsal, tubular nerve cord, (c) postanal tail, (d) endostyle **19.** tadpoles   **20.** tunic   **21.** *Brachiostoma*   **22.** filtering particles from water   **23.** Tunicate larva **24.** somites   **25.** (a)vertebral column, (b)cranium, (c)cephalization   **26.** (a) 6, (b) 4, (c) Amphibia, (d) Reptilia, (e) Aves, (f) Mammalia   **27.** ostracoderms   **28.** Myxini   **29.** lampreys **30.** fins **31.** gill arch skeleton   **32.** *Chondrichthyes*   **33.** placoid scales   **34.** (a) lateral line organs, (b)electroreceptors **35.** (a)oviparous, (b)ovoviviparous, (c)viviparous   **36.** (a)ray-finned, (b)sarcopterygians **37.** swim bladders

38. lung fish **39.** lobed fins of fishes **40.** (a) *Tiktaalik,* (b) movable neck, (c) rigs that supported lungs
41. (a)Urodela, (b)Anura, (c)Apoda **42.** paedomorphosis **43.** (a) skin, (b)three **44.** amniotic egg
45. (a)two, (b)one **46.** (a)three, (b)ectothermic **47.** (a) Testudines, (b) Squamata, (c) Sphenodonta, (d) Crocodilia, (e) Archosaurs **48.** horny **49.** overlap **50.** pit organ **51.** iguanas **52.** (a) long, slender snout, (b) large fourth tooth **53.** asymmetrical **54.** endothermy **55.** the two clavicles fuse
56. (a)four, (b) endothermic, (c)uric acid **57.** (a)crop, (b)proventriculus, (c)gizzard **58.** (a) hair, (b) mammary glands, a pair of temporal openings in the skull, (c) differentiation of teeth, (d) three middle ear bones **59.** (a) viviparous, (b) placenta **60.** (a)therapsids, (b)Triassic period **61.** (a) arboreal, (b) nocturnal **62.** monotremes **63.** (a) marsupials, (b)marsupium **64.** eutherians
Building Words: **1.** agnathan **2.** Anura **3.** Apoda **4.** Chondrichthyes **5.** tetrapod **6.** Urodela **7.** echinoderm **8.** endoskeleton **9.** notochord **10.** somite **11.** cephalization **12.** endotherm
Matching: **1.** k **2.** i **3.** l **4.** a **5.** o **6.** q **7.** n **8.** r **9.** p **10.** f **11.** c
Making Comparisons: **1.** sharks, rays, skates, chimeras **2.** cartilage **3.** jawed, gills, marine and fresh water, placoid scales, well-developed sense organs **4.** Actinopterygii **5.** bone **6.** salamanders, frogs and toads, caecilians **7.** three chambered heart **8.** Aves **9.** four chambered heart **10.** light hollow bone with air spaces **11.** mammalia **12.** bone **13.** mostly tetrapods, possess hair and mammary glands, endothermic, highly developed nervous system
Making Choices: **1.** d **2.** d **3.** b,c **4.** c **5.** a,b **6.** c,e **7.** d **8.** a-e **9.** a,b **10.** b,e **11.** b,c,d **12.** c,e **13.** a,d **14.** d **15.** c,e **16.** d,e **17.** a,d **18.** a-e **19.** e **20.** a-e **21.** d,e **22.** e **23.** b,d,e

Chapter 33
Reviewing Concepts: **1.** (a) annuals, (b) biennials, (c) perennials **2.** (a) deciduous, (b) evergreens
3. (a) root, (b) shoot **4.** (a) tissue, (b) simple tissues, (c) complex tissues **5.** organs **6.** (a) parenchyma, (b) collenchyma, (c) sclerenchyma **7.** (a) photosynthesis, (b) storage, (c) secretion **8.** differentiate **9.** support **10.** (a) sclereids, (b) fibers **11.** hemicelluloses and pectins **12.** lignin **13.** (a)water and dissolved minerals, (b)parenchyma cells or xylem parenchyma, (c)fibers **14.** (a) apoptosis, (b) hollow
15. (a) food, (b) sieve tube elements **16.** plasmodesmata **17.** (a)epidermis and periderm, (b)cuticle, (c)stomata **18.** trichomes **19.** (a) protection, (b) cork **20.** meristems **21.** grow **22.** (a) primary,
(b) secondary **23.** (a)area of cell division, (b)area of cell elongation, (c)area of cell maturation **24.** (a) leaf primordia, (b) bud primordia **25.** (a)vascular cambium, (b)cork cambium **26.** bark **27.** preprophase band
28. (a) increase in number, (b) increase in size **29.** differential gene expression **30.** different concentrations of signaling molecules
Building Words: **1.** trichome **2.** biennial **3.** epidermis **4.** stoma **5.** deciduous **6.** cuticle
Matching: **1.** o **2.** m **3.** a **4.** n **5.** c **6.** k **7.** b **8.** h **9.** f **10.** g
Making Comparisons: **1.** ground tissue **2.** stems and leaves **3.** sclerenchyma **4.** structural support **5.** xylem **6.** vascular tissue **7.** conducts sugar in solution **8.** extends throughout plant body **9.** dermal tissue **10.** protection of plant body, controls gas exchange and water loss on stems, leaves **11.** covers body of herbaceous plants **12.** periderm **13.** protection of plant body

Making Choices: **1.** d **2.** a **3.** e **4.** a,d **5.** d **6.** c,d,e **7.** c **8.** a,b,e **9.** c **10.** a,b,c,e **11.** b **12.** d **13.** b **14.** a-e **15.** a,b **16.** b,d,e **17.** a,c,d,e

Chapter 34

Reviewing Concepts: **1.** (a) blade, (b) petiole, (c) stipules **2.** (a) simple, (b) compound **3.** (a) alternate leaf arrangement, (b) opposite leaf arrangement, (c) whorled leaf arrangement **4.** (a) parallel, (b) monocots, (c) netted, (d) eudicots **5.** epidermis **6.** (a) cuticle, (b) cutin **7.** subsidiary cells **8.** (a) mesophyll, (b) palisade mesophyll, (c) spongy mesophyll **9.** (a) xylem, (b) phloem, (c) bundle sheath **10.** (a) transparent, (b) photosynthesis **11.** (a)carbon dioxide, (b)oxygen **12.** (a) rigidity, (b) flaccid (limp) **13.** pigment **14.** yellow **15.** (a) malic acid, (b) starch **16.** (a) proton pumps, (b) H$^+$ **17.** (a) sucrose, (b) starch **18.** light and darkness **19.** (a) CO$_2$ concentration in the leaf, (b) water stress, (c) a circadian rhythm **20.** soil **21.** (a) prevent plant from overheating, (b) distribute essential minerals throughout plant **22.** sweating **23.** guttation **24.** shedding leaves **25.** ethylene **26.** (a)carotenoids, (b)anthocyanins **27.** (a)abscission zone, (b)fibers **28.** middle lamella **29.** spines **30.** tendrils **31.** (a)passive, (b)active

Building Words: **1.** abscission **2.** circadian **3.** mesophyll **4.** Transpiration **5.** guttation

Matching: **1.** b **2.** j **3.** g **4.** c **5.** h **6.** f **7.** d **8.** m **9.** i **10.** a

Making Comparisons: **1.** allows for efficient capture of sunlight **2.** helps reduce or control water loss, enabling plants to survive the dry terrestrial environment **3.** stomata **4.** allows light to penetrate to photosynthetic tissue **5.** air space in mesophyll tissue **6.** bundle sheaths and bundle sheath extensions **7.** transports water and minerals from roots **8.** phloem in veins

Making Choices: **1.** a **2.** d **3.** b,c,e **4.** c **5.** b,c,e **6.** d **7.** a,d,e **8.** a,b,e **9.** a-e **10.** a,b,c **11.** c **12.** a,e **13.** e **14.** a-e **15.** b,c,e **16.** c,e **17.** a,e **18.** b

Chapter 35

Reviewing Concepts: **1.** (a) roots, (b) leaves, (c) stems **2.** (a) support, (b) conduction (internal transport), (c) produce new stem tissue **3.** primary **4.** herbaceous eudicot **5.** vascular bundles **6.** (a) xylem and phloem, (b) vascular cambium **7.** (a) xylem, (b) phloem **8.** lateral meristems **9.** (a)vascular cambium, (b)cork cambium, (c)periderm **10.** radially (at right angles to the normal division) **11.** (a) xylem (wood), (b) phloem (inner bark) **12.** crushed **13.** (a) rays, (b) parenchyma cells **14.** (a) periderm, (b) epidermis **15.** bark **16.** cork cells **17.** cork parenchyma **18.** (a) bud scales, (b) leaves **19.** bud scale scars **20.** leaf scars **21.** (a)heartwood, (b)sapwood **22.** (a) hard, (b) soft **23.** (a) fibers, (b) vessel elements **24.** (a)springwood, (b)late summerwood **25.** natural physical process (rather than by a heart) **26.** (a) pushed, (b) pulled, (c) pulled **27.** (a)water potential, (b)less, (c)higher (less negative), (d)lower (more negative) **28.** negative **29.** transpiration **30.** cohesion, (b) adhesion **31.** (a) mineral ions, (b) decrease **32.** (a) source, (b) sink **33.** (a) decrease, (b) osmosis, (c) increase

Building Words: **1.** translocation **2.** periderm **3.** internode **4.** adhesion **5.** cohesion **6.**

Matching: **1.** a **2.** g **3.** h **4.** b **5.** c **6.** l **7.** m **8.** j **9.** i **10.** d **11.** k **12.**

Making Comparisons: **1.** secondary phloem **2.** conducts dissolved sugar **3.** produced by cork cambium **4.** storage **5.** cork cells **6.** periderm **7.** vascular

cambium   **8.** produces secondary xylem and secondary phloem   **9.** a lateral meristem, usually arises from parenchyma   **10.** produces periderm (secondary growth)
Making Choices: **1.** c   **2.** d   **3.** a,b,c,d   **4.** a,d   **5.** a,b,d   **6.** b,d,e   **7.** a-e   **8.** a,b,d   **9.** a,c,e   **10.** b   **11.** e   **12.** a,c,d,e   **13.** e   **14.** b,d,e   **15.** a   **16.** e

Chapter 36

Reviewing Concepts: **1.** (a) anchoring, (b) absorption of water, (c) minerals, conduction and storage
2. (a)taproot, (b)fibrous   **3.** (a)taproot, (b)fibrous   **4.** adventitious   **5.** (a) root cap, (b) apical meristem
6. root hairs   **7.** (a) parenchyma, (b) collenchyma   **8.** plasmodesmata   **9.** (a) endodermis, (b) Casparian strip   **10.** aquaporins   **11.** (a) pericyle, (b) xylem, (c) phloem
12. (a) cortex, (b) endodermis, (c) pericycle
13. vascular cambium   **14.** roots   **15.** (a) periderm, cork parenchyma   **16.** nodes   **17.** (a) prop roots, (b) buttress roots   **18.** (a) epiphytes, (b) aerial   **19.** mycorrhizae   **20. 15.** cell signaling   **21.** rock   **22.** clay
23. (a) protons (H^+), (b) cation exchange   **24.** (a) sand and silt, (b) clay, (c) air and water
25. humus
26. air   **27.** (a) leaching, (b) illuviation   **28.** castings   **29.** acid precipitation   **30.** (a) 90, (b) 60
31. (a) macronutrients, (b) micronutrients   **32.** (a) as free K^+, (b) turgidity, (c) stomata
33. Hydroponics
34. nitrogen, phosphorus, potassium   **35.** soil erosion   **36.** salinization
Building Words: **1.** hydroponics   **2.** macronutrient   **3.** micronutrient
4. endodermis   **5.** Mycorrhiza
6. nodule
Matching: **1.** i   **2.** g   **3.** d   **4.** a   **5.** c   **6.** k   **7.** j   **8.** e   **9.** m   **10.** l   **11.** n   **12.** o
13. h
Making Comparisons: **1.** area of cell division that causes an increase in length of the root   **2.** root hairs   **3.** protects root   **4.** storage   **5.** endodermis   **6.** pericycle
7. xylem   **8.** conducts dissolved sugars
Making Choices: **1.** a,c,e   **2.** b,e   **3.** c   **4.** b,d   **5.** a,b,d   **6.** b   **7.** a,b,d,e   **8.** c
9. c   **10.** c   **11.** C
12. e   **13.** c   **14.** a,e   **15.** a,b,d   **16.** a,d   **17.** e   **18.** a,c,d

Chapter 37

Reviewing Concepts: **1.** (a) fertilization, (b) different   **2.** (a) asexually, (b) similar   **3.** (a) alternation of generation, (b) gametophyte generation, (c) sporophyte generation   **4.** flowering   **5.** (a) sepals, (b) petals, (c) stamens, (d) carpels   **6.** (a)calyx, (b)corolla   **7.** ovules   **8.** (a)microsporocytes, (b)microspores, (c)pollen grain, (d)sperm cells   **9.** (a)anther, (b)stigma   **10.** (a) inbreeding, (b) outcrossing   **11.** self-incompatability   **12.** (a) showy petals, (b) scent   **13.** (a)blue or yellow, (b)red   **14.** *bee's purple*   **15.** (a) red, (b) orange, (c) yellow, (d) visible light   **16.** (a) night, (b) dull white, (c) fermented fruit   **17.** female bee   **18.** insect   **19.** flowers   **20.** (a)style, (b)ovule   **21.** endosperm   **22.** (a) embryo, (b) nutrients   **23.** suspensor
24. (a) radicle, (b) cotyledons   **25.** plummule   **26.** (a) simple, (b) aggregate, (c) multiple, (d) accessory   **27.** (a)simple, (b)berries, (c)drupes   **28.** (a)follicle, (b)legumes, (c)capsules
29. (a)aggregate, (b)multiple   **30.** (a) accessory, (b) receptacle, (c) floral tube   **31.** wind, animals, water, explosive dehiscence
32. imbibition   **33.** scarification   **34.** coleoptile   **35.** (a) rhizomes, (b) tubers, (c) bulbs, (d) corms, (e) stolons   **36.** suckers   **37.** dispersed   **38.** sexual reproduction

Building Words: **1.** endosperm **2.** hypocotyl **3.** coevolution **4.** plumule **5.** dormancy **6.** germinate
7. coleoptile
Matching: **1.** d **2.** g **3.** l **4.** b **5.** a **6.** k **7.** j **8.** m **9.** c **10.** f **11.** i **12.**
Making Comparisons: **1.** dry: does not open; nut **2.** stony wall does not split open at maturity **3.** single seeded; fully fused to fruit wall **4.** simple **5.** fleshy: berry **6.** accessory **7.** multiple **8.** formed from the ovaries of many flowers that fuse together and enlarge after fertilization **9.** aggregate **10.** many separate ovaries from a single flower **11.** simple **12.** fleshy, drupe **13.** hard, stony pit; single seeded
Making Choices: **1.** c **2.** e **3.** a **4.** b **5.** b,d,e **6.** d **7.** b,d,e **8.** b,c,d **9.** e **10.** d **11.** d **12.** e **13.** b,d,e **14.** a,c **15.** c **16.** a,b,d,e **17.** c,d,e **18.** a,c,e **19.** a,d **20.** a-e

Chapter 38
Reviewing Concepts: **1.** hormone **2.** directional response **3.** (a) phototropic, (b) blue **4.** photoreceptor
5. (a) gravitropism, (b) thigmotropism **6.** (a) auxins, (b) gibberellins, (c) cytokinins, (d) ethylene, (e) abscisic acid, (f) brassinosteroids **7.** (a) enzyme-linked receptors, (b) enzymatic **8.** coleoptile
9. indoleacetic acid (IAA) **10.** (a) polar, (b) shoot apical meristem, (c) root, (d) **11.** apical dominance
12. fruit **13.** Fungus **14.** (a) flowering, (b) germination **15.** (a) cell division, (b) differentiation **16.** senescence **17.** thigmomorphogenesis **18.** dormancy **19.** environmental stress **20.** Gibberellins
21. dwarf **22.** florigen **23.** photoperiodism **24.** long-night plants **25.** (a)short-day (long-night), (b)long-day (short-night), (c)intermediate-day, (d)day-neutral **26.** photoperiodism **27.** shade avoidance
28. (a)red, (b)Pr, (c)Pfr **29.** potassium **30.** the time of day **31.** (a) prostaglandins, (b) lipids
Building Words: **1.** phototropism **2.** gravitropism **3.** phytochrome **4.** Photoperiodism **5.** senescence **6.** promoter
Matching: **1.** b **2.** a **3.** m **4.** d **5.** c **6.** f **7.** o **8.** e **9.** p **10.** l **11.** h **12.** k
Making Comparisons: **1.** auxin **2.** cytokinin **3.** gibberellin **4.** ethylene **5.** gibberellin and cytokinin **6.** abscisic acid
Making Choices: **1.** a,c **2.** b **3.** a-e **4.** c **5.** a **6.** a,d **7.** a **8.** b **9.** e **10.** a-e **11.** d **12.** a,c,d,e
13. e **14.** a,c **15.** d **16.** b,c **17.** a-e

Chapter 39
Reviewing Concepts: **1.** ratio of its surface area to its volume. **2.** (a) number, (b) size **3.** complex organs systems **4.** (a) tissue, (b) organs, (c) organ systems **5.** (a) layer, (b) tightly, (c) basement membrane **6.** (a) protection, (b) absorption, (c) secretion, (d) sensation **7.** (a) squamous, (b) cuboidal, (c) columnar **8.** (a) simple, (b) stratified **9.** pseudostratified **10.** (a) exocrine glands, (b) endocrine glands **11.** (a) intercellular substance, (b) microscopic fibers, (c) matrix **12.** intercellular substance
13. (a) collagen, (b) elastin, (c) collagen and glycoprotein **14.** (a) fibers, (b) protein, (c) carbohydrate

15. macrophages   16. (a) loose connective tissue, (b) move   17. collagen
18. (a)tendons, (b)ligaments   19. skeleton   20. (a) cartilage, (b) bone   21. (a)
chondrocytes, (b) collagen fibers, (c) lacunae
22. (a) osteocytes, (b) lacunae, (c) vascularized   23. canaliculi   24. (a) osteons, (b)
lamellae, (c) Haversian canal   25. plasma   26. transport oxygen   27. defend the body
against disease-causing microorganisms   28. red bone marrow   29. muscle fiber   30.
(a)actin, (b) myosin, (c)myofibrils   31. (a)skeletal, (b)smooth, (c)cardiac   32. (a) neurons,
(b) glial   33. synapses   34. nerve   35. (a) cell body, (b) dendrite(s), (c) axon(s)
36. (a) integumentary, (b) skeletal, (c) muscular, (d) digestive, (e) cardiovascular, (f)
immune (lymphatic), (g) respiratory, (h) urininary, (i) nervous, (j) endocrine, (k)
reproductive   37. homeostasis   38. stressors
39. (a) sensor, (b) intergrator (control center)   40. opposite (negative)   41. intensifies   42.
lower   43. six
44. (a)torpor, (b)hibernation, (c)estivation
Building Words: 1. multicellular   2. pseudostratified   3. fibroblast   4. intercellular
5. macrophage   6. chondrocyte   7. osteocyte   8. myofibril   9. homeostasis   10.
dendrite   11. synapsis   12.
Matching: 1. q   2. b   3. o   4. d   5. m   6. g   7. c   8. j   9. i   10. e   11. p   12. s
Making Comparisons: 1. simple squamous epithelium   2. air sacs of lungs, lining of
blood vessels   3. epithelial tissue   4. some respiratory passages, ducts of many glands
5. secretion, movement of mucus, protection   6. connective tissue   7. food storage,
protection of some organs, insulation   8. bone   9. support and protection of internal
organs, calcium reserve, site of skeletal muscle attachments   10. connective tissue
11. within heart and blood vessels of the circulatory system   12. transport of oxygen,
nutrients, waste product and other materials   13. walls of the heart   14. skeletal muscle
15. nervous tissue
Making Choices: 1 a   2. a,b,c,e   3. d   4. a,b,e   5. c   6. a,d   7. a,c,e   8. b,e   9. d
10. b   11. c   12. c,e   13. a,c,d   14. a,c   15. c   16. b   17. d   18. a,c,d   19. c   20.
c,d,e   21. a,c,d,e   22. d,e   23. a-e
24. a,c,d,e

Chapter 40

Reviewing Concepts: 1. MRSA   2. protection   3. skeleton   4. (a) environment, (b)
lifestyle   5. cuticle   6. (a) lubricants, (b) adhesives   7. (a) skin, (b) structures that develop
from the skin.   8. (a) feathers, (b) hair
9. (a) nails, (b) hair, (c) sweat glands, (d) oil glands, (e) sensory receptors   10. (a)
epidermis, (b) strata   11. stratum basale   12. (a)keratin, (b)strength, (c) flexibility,
(d) a diffusion barrier   13. (a)connective tissue, (b) collagen, (c) subcutaneous   14.
mechanical forces   15. fluid-filled body compartments
16. muscle contraction 17. (a) hydrostatic skeleton, (b) longitudinally, (c) circularly,
(d) outer, (e) inner   18. septa   19. exoskeleton   20. (a) chitin, (b) molting   21. (a)
endoskeletons, (b) chordates   22. calcium salts   23. transmits muscle forces   24. (a)
axial, (b) appendicular   25. (a) skull, (b) vertebral column, (c) rib cage   26. (a)
pectoral girdle, (b) pelvic girdle, (c) limbs   27. (a) periostium, (b) tendons,
(c) ligaments   28. (a) epiphyses, (b) diaphysis, (c) metaphysis, (d) epiphyseal line, (e)
compact bone   29. (a)endochondral, (b)intramembranous   30. (a) joints, (b)
immovable joints,
(c) slightly movable joints, (d) freely movable joints   31. (a)actin, (b)myosin   32. (a)
striated (skeletal), (b) smooth   33. asynchronous   34. tendons   35.
(a)antagonistically   36. (a) agonist (b) antagonist   37. long, cylindrical cell   38. (a)
myofibrils, (b) myofilaments (filaments)   39. (a) actin myofilaments, (b) myosin

myofilaments **40.** sarcomere **41.** (a)acetylcholine (neurotransmitter) , (b)action potential **42.** (a) sacroplasmic reticulum, (b) troponin, (c) actin
43. center **44.** ATP **45.** creatine phosphate **46.** glycogen **47.** (a) slow-oxidative, (b) aerobic respiration, (c) myoglobin **48.** (a) fast-glycolytic, (b) glycolysis **49.** fast-oxidative **50.** motor unit **51.** simple twitch **52.** muscle tone (state of tonus) **53.** (a) smooth, (b) cardiac
Building Words: **1.** epidermis **2.** periosteum **3.** endochondral **4.** osteoblast **5.** osteoclast **6.** myofilament **7.** keratin **8.** chitin **9.** axial **10.** sarcomere **11.** sarcolemma
Matching: **1.** i **2.** f **3.** l **4.** o **5.** g **6.** b **7.** e **8.** p **9.** j **10.** a **11.** m **12.** k **13.** h
Making Comparisons: **1.** chitin **2.** exoskeleton **3.** external covering jointed for movement, does not grow so animal must periodically molt **4.** echinoderms **5.** calcium salts **6.** cartilage or bone **7.** endoskeleton **8.** internal, composed of living tissue and grows with the animal **9.** fast **10.** fast
11. slow **12.** fast **13.** aerobic respiration **14.** aerobic respiration **15.** low **16.** high
Making Choices: **1.** c **2.** a,b,c **3.** d **4.** c **5.** a **6.** e **7.** c,d **8.** a **9.** d **10.** c,e
11. a **12.** a,b,d **13.** b,c **14.** a,c **15.** b,e **16.** b,e **17.** d **18.** a **19.** a,c,e **20.** b,c
21. a,b,c,e **22.** b,c,d
23. a,c **24.** b,d,e **25.** c

Chapter 41
Reviewing Concepts: **1.** stimuli **2.** (a) endocrine, (b) nervous **3.** (a) reception, (b) afferent, (c) interneurons **4.** (a) efferent neurons, (b) effectors **5.** motor **6.** (a) dendrites, (b) cell body, (c) axon
7. (a) terminal branches, (b) synaptic terminals, (c) neurotransmitter **8.** (a) Schwann cells, (b) myelin sheath, (c) nodes of Ranvier **9.** neurogenesis **10.** (a) nerves, (b) tracts (pathways) **11.** (a) ganglia, (b) nuclei **12.** neuroglia **13.** (a) astrocytes, (b) oligodendrocytes, (c) ependymal cells, (d) microglia
14. membrane potential **15.** (a) resting potential, (b) -70 mV **16.** (a) differences in the concentrations of specific ions inside and outside of the cell, (b) selective permeability of the plasma membrane to these ions
17. (a) positive, (b) negative) **18.** (a) inside, (b) outside **19.** (a) sodium-potassium pumps, (b) 3, (c) 2
20. (a) excitatory, (b) inhibitory **21.** (a) threshold, (b) action potential **22.** (a) voltage-activated sodium channels, (b) sodium, (c) depolarization **23.** (a) sodium channel inactivation gates, (b) voltage-activated potassium channels, (c) repolarization **24.** repolarization **25.** absolute refractory **26.** all-or-none
27. (a) salutatory, (b) node of Ranvier **28.** synapse **29.** (a)presynaptic, (b)postsynaptic
30. (a) electrical synapses, (b) chemical synapses, (c) neurotransmitter molecules, (d) synaptic cleft **31.** (a)acetylcholine, (b) brain, (c) autonomic nervous system **32.** adrenergic neurons **33.** (a) norepinephrine, (b) epinephrine, (c) dopamine **34.** (a) serotonin, (b) histamine **35.** (a) endorphins, (b) enkephalins **36.** endorphins
37. synaptic vesicles **38.** (a) excitatory postsynaptic potential (EPSP), (b) inhibitory postsynaptic potential (IPSP) **39.** graded potentials **40.** (a) summation, (b) temporal summation, (c) spatial summation
41. (a) converging, (b) diverging
Building Words: **1.** interneuron **2.** neuroglia **3.** neurotransmitter **4.** multipolar **5.** postsynaptic **6.** presynaptic
Matching: **1.** d **2.** j **3.** f **4.** b **5.** i **6.** k **7.** a **8.** h **9.** n **10.** g **11.** m **12.** o
Making Comparisons: **1.** -70mV **2.** stable **3.** sodium-potassium pump active, sodium and potassium channels open **4.** threshold potential **5.** voltage-activated

sodium ion channels open   **6.** varies, but is about +35mV   **7.** wave of depolarization (rise: depolarization; fall: repolarization)   **8.** depolarization   **9.** neurotransmitter-receptor combination opens sodium ion channels   **10.** hyperpolarization

Making Choices: **1.** b,c,e   **2.** b   **3.** c,d,e   **4.** c,d   **5.** b   **6.** a,d   **7.** c   **8.** b,c   **9.** d   **10.** c,e   **11.** a

12. e   **13.** e   **14.** a,d   **15.** a,b,d,e   **16.** a,b,c   **17.** c,e   **18.** b,d   **19.** b,d   **20.** b   **21.** a   **22.** a   **23.** a-e

Chapter 42

Reviewing Concepts: **1.** (a) nerve net, (b) cnidarians   **2.** nerve ring   **3.** (a) ladder-type, (b) cerebral ganglia   **4.** ganglia   **5.** (a) 100 billion, (b) learning ability   **6.** (a) central nervous (CNS), (b) peripheral nervous (PNS)   **7.** (a) somatic, (b) autonomic, (c) sympathetic (d) parasympathetic   **8.** (a) neural tube, (b) brain, (c) spinal cord   **9.** (a) cerebellum, (b) pons, (c) metencephalon, (d) medulla, (e) myelencephalon

10. (a) medulla, (b) pons, (c) midbrain   **11.** (a) muscle tone, (b) posture, (c) equilibrium   **12.** Association

13. (a) superior colliculi, (b) inferior colliculi, (c) red nucleu   **14.** (a) thalamus, (b) hypothalamus, (c) cerebrum   **15.** Hemispheres   **16.** (a) white, (b) myelinated, (c) gray   **17.** neocortex   **18.** (a) convolutions (gyri), (b) sulci, (c) fissures   **19.** (a) dura matter, (b) arachnoid, (c) pia mater, (d)choroid plexus   **20.** second lumbar   **21.** (a) white matter, (b) tracts   **22.** (a) sensory, (b) motor, (c) association   **23.** (a) frontal, (b) parietal, (c) central sulcus   **24.** basal ganglia   **25.** (a) dopamine, (b) inhibition, (c) excitation   **26.** (a) pineal gland, (b) melatonin   **27.** brain stem and thalamus   **28.** (a)electroencephalogram (EEG), (b) alpha, (b) deta, (c) theta, (d) delta   **29.** (a)rapid eye movement (REM), (b)non-REM   **30.** cerebrum   **31.** (a) memory, (b) implicit memory, (c) declarative memory (explicit memory)   **32.** hypocampus   **33.** long-term depression   **34.** temporal lobe   **35.** Broca's area   **36.** (a) sensory receptors, (b) sensory neurons, (c) motor neurons   **37.** (a) cranial, (b) spinal   **38.** (a) sympathetic, (b) parasympathetic   **39.** (a) preganglionic neuron, (b) postganglionic neuron   **40.** psychological dependence

Building Words: **1.** hypothalamus   **2.** postganglionic   **3.** preganglionic   **4.** paravertebral   **5.** sensory   **6.** limbic

Matching: **1.** g   **2.** d   **3.** j   **4.** c   **5.** h   **6.** a   **7.** e   **8.** k   **9.** b   **10.** i   **11.** n   **12.** o

Making Comparisons: **1.** pons   **2.** midbrain   **3.** center for visual and auditory reflexes   **4.** diencephalon   **5.** relay center for motor and sensory information   **6.** hypothalamus   **7.** diencephalon   **8.** metencephalon   **9.** cerebrum   **10.** reticular activating system   **11.** limbic system

Making Choices: **1.** a   **2.** a   **3.** b   **4.** a-e   **5.** a,d,e   **6.** b   **7.** a,b   **8.** d   **9.** a,c   **10.** b   **11.** b   **12.** a,b   **13.** b   **14.** c   **15.** a-e   **16.** a-e   **17.** c,e   **18.** b,e   **19.** b,d   **20.** a,c,d   **21.** a,d,e   **22.** e   **23.** b   **24.** e   **25.** a   **26.** a,d,e

Chapter 43

Reviewing Concepts: **1.** echolocation   **2.** reception   **3.** (a) transduced (converted), (b) electrical   **4.** (a) receptor potential, (b) graded response   **5.** (a) depolarizes, (b) action potential, (c) sensory   **6.** sensory adaptation   **7.** brain   **8.** (a) number and identity, (b) frequency and total number   **9.** (a) selecting, (b) interpreting, (c) organizing   **10.** (a) exteroceptors, (b) interoceptors   **11.** (a) pit vipers, (b) boas   **12.** (a) skin, (b) tongue   **13.** hypothalamus   **14.** electromagnetic receptors   **15.** (a) spinal cord, (b) glutamate, (c) substance P   **16.** change shape   **17.** (a) hair, (b) bristle   **18.** Merkel cells   **19.** (a) Meissner corpuscles, Ruffini endings, (c) Pacinian corpuscles   **20.** (a) muscle spindles, (b)

Golgi tendon organs, (c) joint receptors **21.** (a)hair cells, (b) statoliths **22.** kinocilium **23.** vibrations **24.** (a) canal, (b) sensory hair cells, (c) cupula **25.** (a) saccule, (b) utricle, (c) gelatinous cupula, (d) otoliths, (e) statocyst **26.** (a) angular acceleration, (b) endolymph, (c) cristae, (d) otoliths **27.** (a) cochlea, (b) mechanoreceptors **28.** (a) tympanic membrane, (b) malleus, (c) incus, (d) stapes, (e) oval window **29.** eustachian tube

30. (a) basilar membrane, (b) organ of Corti, (c) sensory neurons, (d) cochlear nerve **31.** (a) frequency, (b) amplitude **32.** (a) gustation (taste), (b) olfaction (smell) **33.** (a) sweet, (b) sour, (c) salty, (d) bitter, (e) umami (glutamate), (f) fatty acid **34.** olfaction **35.** nasal epithelium **36.** limbic system **37.** small volatile molecules **38.** rhodopsins **39.** eyespots (ocelli) **40.** ommatidia **41.** (a) sclera, (b) shape (rigidity) **42.** cornea **43.** vitreous body **44.** aqueous humor **45.** iris **46.** photoreceptors **47.** (a) fovea, (b) cones, (c) sharpest vision **48.** (a) opsin, (b) retinal **49.** dark adaptation **50.** (a) blue, (b) green, (c) red, (d) range of wavelengths **51.** (a) hypothalamus, (b) optic chiasm

Building Words: **1.** proprioceptor **2.** otolith **3.** endolymph **4.** chemoreceptor **5.** thermoreceptor **6.** interoceptor **7.** tactile **8.** statoctyst **9.** olfactory

Matching: **1.** l **2.** b **3.** j **4.** i **5.** n **6.** k **7.** o **8.** d **9.** a **10.** c **11.** g **12.** m

Making Comparisons: **1.** electroreceptor **2.** exteroceptor **3.** pressure waves (sound) **4.** mechanoreceptor **5.** interoceptor **6.** receptor in the human hypothalamus **7.** muscle contraction **8.** interoceptor **9.** thermoreceptor **10.** pit organ of a viper **11.** light energy **12.** photoreceptor **13.** chemoreceptor **14.** mammalian taste buds **15.** chemoreceptor **16.** exteroceptor

Making Choices: **1.** c **2.** d **3.** a **4.** a **5.** c **6.** b **7.** c **8.** b,d **9.** b,d **10.** c **11.** d **12.** e **13.** a,d **14.** b,d,e **15.** a-e **16.** a,c,d,e **17.** c,d **18.** b,d,e **19.** a,d,e **20.** b,e **21.** d,e

Chapter 44

Reviewing Concepts: **1.** cholesterol **2.** (a) sponges, (b) cnidarians, (c) ctenophores, (d) flatworms, (e) nematodes **3.** (a) circulatory, (b) digestive **4.** fluid **5.** (a) blood (circulatory fluid), (b) heart, (c) spaces, (d) blood vessels **6.** (a) open-ended, (b) sinuses (hemocoel) **7.** (a) hemolymph, (b) interstitial fluid

8. (a) arthropods, (b) most mollusks **9.** three **10.** (a) hemocyanin, (b) copper **11.** (a) nutrients, (b) hormones **12.** (a) annelids, (b) some mollusks (cephalopods), (c) echinoderms **13.** (a) dorsal, (b) ventral, (c) five **14.** plasma **15.** ventrally **16.** (a) nutrients, (b) oxygen, (c) metabolic wastes, (d) hormones **17.** (a) plasma, (b) red blood cells, (c) white blood cells, (d) platelets **18.** (a) interstitial, (b) intracellular **19.** (a) fibrinogen, (b) globulins, (c) albumin **20.** (a) red bone marrow, (b) hemoglobin, (c) 120 **21.**)a) anemia, (b) loss of blood (hemorrhage), (c) increased breakdown of RBCs **22.** leukocytes **23.** (a) granular, (b) agranular **24.** (a) monocytes, (b) macrophages **25.** (a) neutrophils, (b) basophils, (c) eosinophiles **26.** (a) lymphocytes, (b) monocytes **27.** thrombocyte **28.** platelets **29.** (a) arteries, (b) veins **30.** capillaries

31. arterioles **32.** (a) ventricles, (b) atria **33.** (a) one, (b) one **34.** (a) three, (b) two, (c) one, (d) sinus venosus **35.** (a) birds, (b) crododillians, (c) mammals **36.** higher **37.** (a) 5, (b) 20 **38.** (a) pericardium, (b) fluid-filled **39.** (a) interventricular, (b) interatrial septum **40.** (a) four, (b) atrioventricular valves, (c) tricuspid valve, (d) mitral valve, (e) semilunar valves **41.** (a) sinoatrial (SA) node, (b) atrioventricular (AV) node, (c) atrioventricular (AV) bundle **42.** (a) systole, (b) diastole **43.** (a) lub, (b) AV valves **44.** (a) dub, (b) semilunar valves **45.** (a) nervous, (b) endocrine **46.** Starling's law of the heart **47.** (a) cardiac output (CO), (b) stroke volume **48.** five liters/minute **49.** (a) hypertension, (b) increase, (c) salt **50.** diameter of arterioles **51.** low-resistance **52.** baroreceptors **52.** (a) pressure changes, (b) cardiac and vasomotor centers **53.** (a) pulmonary circuit, (b) systemic circuit **54.** (a) rich, (b) left **55.** (a) aorta, (b) brain, (c) shoulder area, (d) legs

56. (a) superior vena cava, (b) inferior vena cava **57.** circulatory system **58.** (a) interstitial fluids, (b) an immune response, (c) absorbs fats **59.** lymph **60.** lymph nodes **61.** (a) tonsils, (b) thymus gland, (c) spleen **62.** (a) subclavian veins, (b) thoracic duct, (c) right lymphatic duct **63.** (a) net filtration, (b) hydrostatic, (c) osmotic, (d) osmotic **64.** (a) atherosclerosis, (b) lipid (fat) **65.** (a) ischemia, (b) angina pectoris **66.** thrombosis **67.** (a) angioplasty, (b) stent

Building Words: **1.** hemocoel **2.** hemocyanin **3.** erythrocyte **4.** leukocyte **5.** neutrophil **6.** eosinophil **7.** basophil **8.** leukemia **9.** vasoconstriction **10.** vasodilation **11.** pericardium **12.** semilunar **13.** baroreceptor **14.** sinus **15.** ventricle **16.** leukemia

Matching: **1.** i **2.** j **3.** h **4.** l **5.** d **6.** m **7.** b **8.** o **9.** a **10.** q **11.** k **12.** e

Making Comparisons: **1.** lipoproteins **2.** plasma **3.** cell component **4.** transport of oxygen and carbon dioxide **5.** blood clotting **6.** cell component **7.** defense; differentiate to form phagocytic macrophages in tissue **8.** albumins and globulins **9.** plasma **10.** defense, principle phagocytic cell in blood

Making Choices: **1.** d **2.** a **3.** d **4.** c,e **5.** a **6.** b,c **7.** d **8.** a,b,d **9.** a,c,e **10.** b,d,e **11.** b **12.** a,d,e **13.** c,e **14.** a,b,c,e **15.** c **16.** c **17.** d **18.** a,d,e **19.** a,e **20.** b,d,e **21.** a,b,d,e **22.** b **23.** e **24.** b,c,e **25.** e **26.** a **27.** c **28.** a,d **29.** b,c **30.** b,e **31.** e,c **32.** a

Chapter 45

Reviewing Concepts: **1.** immunology **2.** signal **3.** recognizing foreign or dangerous macromolecules **4.** (a) nonspecific immune responses, (b) specific immune responses **5.** antigen **6.** antibodies **7.** phagocytes **8.** natural killer cells **9.** inflammatory **10.** nonspecific **11.** jawed vertebrates **12.** innate immunity **13.** (a) outer covering, (b) pathogens **14.** defensins **15.** lysozyme **16.** patghogen-associatged molecular patterns (PAMPs) **17.** macrophages **18.** natural killer (NK) **19.** (a) dendritic, (b) interferons **20.** (a) cytokines, (b) complement **21.** tumor necrosis factor (TNF) **22.** interleukins **23.** chemokines **24.** (a) vasodilation, (b) increased capillary permeability, (c) increased phagocytosis **25.** (a) lymphocytes, (b) antigen-presenting cells **26.** (a) T-cells, (b) B-cells **27.** bone marrow **28.** (a)cytotoxic (CD_8), (b)helper (CD_4) **29.** immunocompetent **30.** (a) macrophages, (b) dendritic cells, (c) B-cells, (d) foreign antigens, (e) their own surface proteins **31.** (a) MHC antigens, (b) "fingerprint" **32.** (a) most nucleated cells, (b) antigen presenting cells **33.** properly presented **34.** (a) B, (b) clone **35.** (a) immunoglobulins, (b) Ig, (c) antigenic determinants (epitopes), (d) binding sites **36.** (a) IgG, (b) IgA, (c) IgD, (d) IgE, (e) IgM **37.** (a) IgG, (b) IgA, (c) IgD, (d) IgE **38.** clonal selection **39.** (a) identical, (b) single cell **40.** antigenic determinant **41.** apoptosis **42.** (a) primary immune, (b) IgM **43.** (a) secondary immune, (b) IgG **44.** (a) actively, (b) artificially **45.** temporary **46.** (a) IgG, (b) memory **47.** IgA **48.** (a) molecular structure, (b) source **49.** angiogenesis **50.** protein malnutrition **51.** mucosa **52.** (a) T_H, (b) reveerse-transcribed **53.** self-tolerance **54.** autoreactive **55.** antigen D **56.** erythroblastosis fetalis **57.** IgE **58.** (a) mast, (b) constrict **59.** MHC antigens **34.** (a)allergen, (b)IgE **35.** histamine **36.** autoreactive

Building Words: **1.** antibody **2.** antihistamine **3.** lymphocyte **4.** monoclonal **5.** autoimmune **6.** lysozyme **7.** phagocyte **8.** complement

Matching: **1.** b **2.** r **3.** f **4.** q **5.** g **6.** j **7.** p **8.** o **9.** l **10.** n **11.** e **12.** d **13.** i

Making Comparisons: **1.** plasma cells **2.** memory B cells **3.** B-cells and macrophages **4.** helper T cells **5.** cytotoxic T cells **6.** memory T cells **7.** natural killer cells **8.** macrophages **9.** neutrophils

Making Choices: **1.** a,d **2.** a,c,d **3.** b **4.** b **5.** b,c,e **6.** a-e **7.** a,c,e **8.** b,c,d **9.** a,b,c,e **10.** a-d

11. b,c,e **12.** a-e **13.** c **14.** a-e **15.** a,e **16.** a **17.** c **18.** a,b,c,e **19.** a,b,c,e **20.** d,e **21.** a,b,c,e

22. c,e **23.** a-e **24.** a,c,d,e

Chapter 46

Reviewing Concepts: **1.** organismic respiration **2.** (a) more, (b) faster **3.** 10 **4.** moist **5.** (a) body surface, (b) tracheal tubes, (c) gills, (d) lungs **6.** ventilation **7.** (a) surrounding water, (b) fluid secretions **8.** (a) tracheal tubes, (b) spiracles **9.** tracheoles **10.** (a) dermal, (b) water, (c) coelomic fluid **11.** cilia **12.** filaments **13.** (a) countercurrent exchange system, (b) higher **14.** (a) body surface, (b) body cavity **15.** book lungs **16.** swim bladder **17.** respiratory system **18.** (a) one direction, (b) two, (c) posterior air sacs, (d) into the lungs, (e) anterior air sacs **19.** moist, ciliated **20.** (a) pharynx, (b) larynx, (c) trachea **21.** epiglottis **22.** (a) thoracic, (b) three, (c) two **23.** pleural membrane **24.** pleural cavity **25.** (a) increases, (b) contracting, (c) decreases, (d) relaxing **26.** (a) external intercostal, (b) increasing **27.** pulmonary surfactant **28.** (a) tidal volume, (b) 500 ml, (c) residual volume, (d) vital capacity **29.** pressure (tension) **30.** Dalton's law of partial pressures **31.** (a) partial pressure difference, (b) total surface area **32.** (a) 100 mm Hg, (b) 40 mm Hg **33.** (a) hemoglobin, (b) myoglobin **34.** (a) iron-porphyrin, (b) globin **35.** (a) iron, (b) heme, (c) oxyhemoglobin (HbO_2) **36.** (a) acidic, (b) Bohr effect **37.** (a) carbon dioxide in the plasma, (b) bound to hemoglobin, (c) is carried as HCO_3^- in the plasma, (d) carbonic anhydrase **37.** respiratory acidosis **38.** medulla **40.** (a) medulla, (b) walls of the aorta, (c) walls of the carotid arteries **41.** faster **42.** oxygen **43.** blood pressure **44.** hypoxia **45.** decompression sickness **46.** (a) the metabolic rate decreases 20%, (b) breathing stops, (c) bradycardia occurs **47.** bronchial constriction

Building Words: **1.** hyperventilation **2.** hypoxia **3.** ventilation **4.** oxyhemoglobin **5.** ventilation

6. diffusion **7.** dermal **8.** alveolus

Matching: **1.** n **2.** a **3.** f **4.** l **5.** c **6.** j **7.** b **8.** d **9.** i **10.** g **11.** e **12.** m

Making Comparisons: **1.** gills **2.** some, tracheal tubes; some book lungs **3.** lungs **4.** lungs **5.** body surface **6.** tracheal tubes **7.** dermal gills **8.** gills **9.** lungs

Making Choices: **1.** e **2.** c,d **3.** a-e **4.** b,c,e **5.** a,e **6.** a,b,e **7.** a,b,d **8.** a **9.** b,d,e **10.** e

11. b,c,e **12.** a,d **13.** a,b,d,e **14.** a-e **15.** c **16.** a **17.** a,d,e **18.** d **19.** a **20.** b,d

Chapter 47

Reviewing Concepts: **1.** heterotrophs (consumers) **2.** (a) selection, (b) acquisition, (c) ingestion

3. egestion **4.** (a) herbivores, (b) primary consumers **5.** (a) cellulose, (b) symbiotic **6.** ruminants

7. (a) secondary (b) predators, (c) canine **8.** omnivores **9.** filter feeders **10.** (a) cnidarians, (b) flatworms

11. (a) mouth, (b) anus **12.** peristalsis **13.** (a) pharynx (throat), (b) esophagus, (c) small intestine **14.** (a) mucosa, (b) submucosa, (c) visceral peritoneum **15.** (a)incisors, (b)canines, (c)molars and premolars, (d)enamel, (e)dentin, (f)pulp cavity **16.** salivary amylase **17.** bolus **18.** epiglottis **19.** peristalsis **20.** rugae **21.** (a) gastric pits, (b) parietal, (c) chief, (d) pepsin **22.** (a) duodenum, (b) jejunum, (c) ileum, (d) duodenum **23.** (a) villi, (b) microvilli **24.** mechanical digestion of fats **25.** (a) salts, (b) pigments, (c) cholesterol, (d) lecithin (phosphatidylcholine) **26.** (a) gall bladder, (b) concentrated **27.** (a) trypsin, (b) chymotrypsin **28.** (a)pancreatic lipase, (b)pancreatic amylase **29.** (a) two glucose molecules, (b) amino acids **30.** (a) emulsifies, (b) monoacylglycerols, (c) diacylglycerols, (d) fatty acids, (e) glycerol **31.** endocrine **32.** wall of the microvilli **33.** liver **34.** (a) chylomicrons, (b) lacteal **35.** ileocecal valve **36.** (a) cecum, (b) ascending colon, (c) transverse colon, (d) descending colon, (e) sigmoid colon, (f) rectum, (g) anus **37.** (a) K, (b) B **38.** (a) excretion, (b) elimination **39.** (a) 45 to 65, (b) 20 to 35, (c) 10 to 35 **40.** (a) Calories, (b) kilocalories **41.** (a) starch, (b) cellulose **42.** "complex carbohydrates" **43.** (a) glycogen, (b) fatty acids, (c) glycerol **44.** insulin **45.** triacylglycerols **46.** (a) linoleic acid, (b) linolenic acid, (c) arachidonic acid **47.** 300 **48.** lipoproteins **49.** (a) high-density lipoproteins (HDLs), (b) low-density lipoproteins (LDLs) **50.** (a) saturated fats, (b) cholesterol, (c) unsaturated fats, (d) cholesterol **51.** (a) 9, (b) essential amino acids **52.** (a) deaminate, (b) ammonia **53.** (a) fat-soluble, (b) water-soluble, (c) B, C **54.** fat soluble **55.** (a) sodium, (b) chloride, (c) potassium, (d) calcium, (e) phosphorus, (f) magnesium, (g) manganese, (h) sulfur **56.** trace elements **57.** free radicals **58.** (a) C, (b) E, (c) A **59.** (a) fruits and vegetables, (b) fats **60.** (a) basal metabolic rate (BMR), (b) total metabolic rate **61.** hypothalamus **62.** ghrelin **63.** neuropeptide Y (NPY) **64.** leptin **65.** (a) weight, (b) height **66.** essential amino acids **67.** kwashiorkor

Building Words: **1.** herbivore **2.** carnivore **3.** omnivore **4.** submucosa **5.** peritonitis **6.** epiglottis **7.** microvilli **8.** chylomicron **9.** Intrinsic **10.** Parietal **11.** jejunum **12.** cecum

Matching: **1.** h **2.** e **3.** j **4.** i **5.** o **6.** l **7.** a **8.** c **9.** k **10.** n

Making Comparisons: **1.** splits maltose into 2 glucose molecules **2.** small intestine **3.** pepsin **4.** stomach **5.** trypsin and chymotrypsin **6.** pancreas **7.** RNA to free nucleotides **8.** pancreas **9.** pancreatic lipase **10.** dipeptidase **11.** duodenum **12.** lactose to glucose and galactose

Making Choices: **1.** a,b,d **2.** c,d,e **3.** d **4.** a,d **5.** a **6.** a,b **7.** a,b,c **8.** a,b,c **9.** d,e **10.** a-d **11.** b **12.** c **13.** a **14.** d **15.** c,e **16.** e **17.** c,d **18.** a **19.** a,b,e **20.** b,e **21.** a-e **22.** a,c **23.** c,d **24.** a,c,e **25.** b,c,d,e **26.** c,d,e **27.** c,d **28.** a,d,e

Chapter 48

Reviewing Concepts: 1. water **2.** (a) osmoregulation and excretion **3.** electrolyte **4.** osmol **5.** osmoregulation **6.** (a) water, (b) carbon dioxide, (c) nitrogenous wastes **7.** (a) ammonia, (b) urea, (c) uric acid **8.** (a) uric acid, (b) ammonia, (c) break-down of nucleotides **9.** urea **10.** (a) osmotic equilibrium (isosmotic), (b) osmoconformers **11.** osmoregulators **12.** (a) nephridiopores, (b) protonephridia, (c) metanephridia **13.** (a) gut wall, (b) hemocoel, (c) hemolymph **14.** (a) diffusion, (b) active transport **15.** (a) kidney, (b) skin, (c) lungs or gills **16.** hypertonic **17.** gills **18.** (a) ammonia, (b) urea **19.** dilute urine **20.** hypotonic **21.** drink sea water **22.** (a) urea, (b) hypertonic **23.** concentrated **24.** (a) uric acid, (b) urea **25.** (a) renal cortex (cortex), (b) renal medulla (medulla) **26.** (a) ureters, (b) bladder, (c)urethra **27.** (a) Bowman's capsule, (b) renal tubule, (c) Bowman's capsule, (d) loop of Henle, (e) collecting **28.** (a) cortical, (b) juxtamedullary **29.** (a) filtration, (b) reabsorption, (c) tubular secretion **30.** Bowman's capsule **31.** (a) blood hydrostatic pressure, (b) large surface area for filtration, (c) permeability of the glomerular capillaries **32.** (a) podocytes, (b) filtration slits, (c) filtration membrane **33.** (a) 99, (b) 1.5 **34.** proximal convoluted tubule **35.** tubular transport maximum **36.** (a) becomes irregular,

(b) aldosterone **37.** (a) descending, (b) ascending, (c) counterflow (countercurrent mechanism) **38.** vasa recta **39.** urine **40.** (a) antidiuretic hormone (ADH), (b) posterior lobe of the pituitary gland **41.** hypothalamus **42.** (a) aldosterone, (b) distal tubules, (c) collecting ducts **43.** (a) juxtaglomerular apparatus, (b) angiotensinogen, (c) angiotensin I, (d) angiotensin I, (b) angiotensin II **44.** (a) heart, (b) sodium, (c) blood pressure

Building Words: **1.** protonephridium **2.** podocyte **3.** juxtamedullary **4.** glomerulus

Matching: **1.** e **2.** a **3.** f **4.** n **5.** o **6.** j **7.** m **8.** l **9.** i **10.** b **11.** d

Making Comparisons: **1.** kidney **2.** Malpighian tubules **3.** metanephridia **4.** protonephridia

Making Choices: **1.** a,c,e **2.** b **3.** a,b **4.** a,d,e **5.** c **6.** c,d **7.** b,d **8.** d **9.** c **10.** e **11.** c,e **12.** c **13.** a,c,d **14.** b,d **15.** a,d,e **16.** e **17.** b,c,d **18.** a,c **19.** e **20.** d **21.** a-e **22.** a,c,d **23.** a,b,c **24.** b,d **25.** a **26.** a,b,d **27.** d **28.** a,e **29.** b,c **30.** b,c,e **31.** a,c,d,e

Chapter 49

Reviewing Concepts: **1.** (a) growth and development, (b) metabolism, (c) fluid balance and concentrations of specific ions and chemical compounds, (d) reproduction, (e) response to stress **2.** (a) ducts, (b) into the surrounding interstitial fluid or blood **3.** 10 **4.** bind to a specific receptor on a target cell **5.** slowly **6.** longer **7.** (a) negative feedback systems, (b) homeostasis **8.** (a) fatty acid derivatives, (b) steroids, (c) amino acid derivatives, (d) peptides or proteins **9.** neuropeptides (neurohormones) **10.** bound to plasma proteins **11.** (a) axons, (b) interstitial fluid **12.** (a) autocrine, (b) paracrine **13.** histamine **14.** nitric oxide **15.** (a) receptor up-regulation, (b) receptor down-regulation **16.** (a) in the cytoplasm or in the nucleus, (b) lipid soluble (pass through the membrane) **17.** are not lipid soluble **18.** signal transduction **19.** adenylyl cyclase **20.** protein kinases **21.** (a) inositol trisphosphate (IP$_3$), (b) diacylglyerol (DAG) **22.** receptor tyrosine **23.** neurons **24.** (a) brain hormone (BH), (b) prothoracic, (c) molting hormone (MH), (d) ecdysone **25.** (a) hypersecretion, (b) hyposecretion **26.** pituitary stalk **27.** master gland **28.** antidiuretic hormone (ADH) **29.** oxytocin **30.** (a) releasing, (b) inhibiting **31.** tropic **32.** (a) anterior pituitary, (b) suppresses appetite **33.** somatotropin **34.** (a) gigantism, (b) acromegaly **35.** (a) thyroxine, (b) iodine, (c) triiodothyronine, (d) iodine **36.** (a) anterior pituitary, (b) thyroid-stimulating hormone (TSH) **37.** cretinism **38.** Graves disease **39.** (a) bones, (b) kidney tubules **40.** (a) thyroid, (b) calcitonin **41.** (a) islets of Langerhans, (b) beta, (c) insulin, (d) alpha, (e) glucagon **42.** (a) liver, (b) muscle, (c) fat **43.** diabetes mellitus **44.** (a) epinephrine, (b) norepinephrine **45.** (a)sex hormone precursors, (b) mineralocorticoids, (c) glucocorticoids, (d) testosterone, (e) estradiol **46.** aldosterone **47.** cortisol (hydrocortisone) **48.** (a) corticotropin-releasing factor, (b) adrenocorticotropic hormone **49.** (a) pineal, (b) melatonin **50.** thymosin **51.** atrial natriuretic factor

Building Words: **1.** neurohormone **2.** neurosecretory **3.** hypersecretion **4.** hyposecretion **5.** hyperthyroidism **6.** hypoglycemia **7.** hyperglycemia **8.** acromegaly **9.** homeostasis **10.** steroid **11.** glucogon

Matching: **1.** i **2.** d **3.** l **4.** a **5.** g **6.** c **7.** e **8.** n **9.** m **10.** k

Making Comparisons: **1.** steroid **2.** maintain sodium and potassium balance; increase sodium reabsorption and potassium excretion **3.** amino acid derivative **4.** stimulates metabolic rate; essential to normal growth and development **5.** ADH **6.** glucocorticoids (cortisol) **7.** steroid **8.** prostaglandins **9.** fatty acid derivative **10.** stimulates secretion of adrenal cortical hormones **11.** amino acid derivative **12.** help body cope with stress; increase heart rate, blood pressure, metabolic rate; raise blood sugar (glucose) level **13.** oxytocin **14.** peptide

Making Choices: **1.** c **2.** a,c,e **3.** b **4.** a,d,e **5.** b **6.** a,d **7.** a,d **8.** b,c **9.** b,c **10.** a,c,d **11.** d **12.** a,d **13.** d **14.** b,d **15.** a,b,c,e **16.** a,c,d **17.** a,b.c.d **18.** a,b,c,e **19.** d

Chapter 50

Reviewing Concepts: **1.** (a) male, (b) female **2.** genetically identical **3.** budding **4.** fragmentation **5.** unfertilized egg **6.** zygote **7.** external fertilization **8.** hermaphroditism **9.** (a) harmful mutations, (b) beneficial mutations **10.** (a) spermatogensis, (b) spermatogonium, (c) primary spermatocyte **11.** acrosome **12.** Sertoli **13.** (a) scrotum, (b) seminiferous tubules **14.** inguinal canal **15.** (a) epididymis, (b) vas deferens, (c) ejaculatory duct, (d) urethra **16.** 200 million **17.** (a) seminal vesicles, (b) prostate gland **18.** (a) shaft, (b) glans, (c) prepuce (foreskin), (d) circumcision **19.** (a)cavernous bodies, (b) spongy body (c) blood **20.** (a) androgen, (b) steroids, (c) estradiol **21.** (a) FSH, (b) LH **22.** (a) testosterone, (b) interstitial **23.** (a) oogenesis, (b) oogonia, (c) primary oocytes **24.** (a) follicle, (b) FSH, (c) secondary oocyte **25.** (a) corpus luteum, (b) progesterone **26.** wall of the oviduct **27.** (a) endometrium, (b) menstruation **28.** cervix **29.** receptacle **30.** vulva **31.** clitoris **32.** alveoli **33.** lactation **34.** prolactin **35.** ovulation **36.** (a) estrogen (estradiol), (b) FSH, (c) LH, (d) LH **37.** (a) estrogen, (b) progesterone **38.** (a) hypothalamus, (b) ova, (c) infertile **39.** reabsorbed **40.** conception **41.** (a) 48, (b) 24 **42.** (a) estrogen, (b) prostaglandins **43.** human chorionic gonadotropin or hCG **44.** placenta **45.** parturition **46.** oxytocin **47.** (a) vasocongestion, (b) increased muscle tension (myotonia) **48.** (a) excitement phase, (b) plateau phase, (c) orgasm, (d) resolution **49.** contraception **50.** (a) estrogen, (b) progestin (progesterone) **51.** 90 million **52.** (a) human immunodeficiency viruse (HIV), (b) sexually transmitted diseases (STDs) **53.** progestin pills **54.** (a) vasectomy, (b) vas deferens **55.** (a) tubal sterilization, (b) oviducts **56.** antibodies **57.** binding progesterone receptors in the uterus **58.** (a) human papillomavirus or HPV, (b) 20 million, (c) 50%
Building Words: **1.** parthenogenesis **2.** spermatogenesis **3.** acrosome **4.** circumcision **5.** oogenesis **6.** endometrium **7.** postovulatory **8.** preovulatory **9.** vasectomy **10.** vasocongestion **11.** intrauterine **12.** contraception
Matching: **1.** m **2.** j **3.** k **4.** b **5.** g **6.** c **7.** e **8.** d **9.** i **10.** l **11.** f **12.** n
Making Comparisons: **1.** anterior pituitary **2.** gonad **3.** stimulates development of seminiferous tubules, stimulates spermatogenesis **4.** stimulates follicle development, secretion of estrogen **5.** stimulates interstitial cells to secrete testosterone **6.** stimulates ovulation, corpus luteum development **7.** testes **8.** testosterone **9.** ovaries **10.** estrogen (estradiol)
Making Choices: **1.** c **2.** a,e **3.** a,c **4.** a,d, **5.** a,c **6.** b,d **7.** a-e **8.** a,c,e **9.** c **10.** c,d,e **11.** a **12.** d **13.** a,c,e **14.** e **15.** e **16.** d **17.** a,b **18.** b,d **19.** c,e **20.** c **21.** c **22.** a,b,c,e **23.** e **24.** a,d **25.** a,b,c **26.** a,b,c **27.** d

Chapter 51

Reviewing Concepts: **1.** (a) cell number, (b) cell size **2.** cellular differentiation **3.** morphogenesis **4.** zygote **5.** (a)diploid, (b)sex **6.** (a) zona pellucida, (b) granulosa **7.** bindin **8.** (a)capacitation, (b)acrosome reaction **9.** polyspermy **10.** (a) fast block, (b) calcium, (c) slow block (cortical reaction) **11.** (a)increases, (b) protein **12.** microtubules **13.** totipotent **14.** blastula **15.** (a) invertebrates and simple chordates, (b) holoblastic, (c) radial, (d) spiral **16.** (a) meroblastic, (b) blastodisc **17.** unequal distribution **18.** gray crescent **19.** surface proteins **20.** Blastula **21.** (a) germ layers, (b) ectoderm, (c) mesoderm, (d) endoderm **22.** blastopore **23.** primitive streak **24.**

(a) notochord, (b) neural groove, (c) neural tube, (d) central nervous system **25.** (a) pharyngeal pouches, (b) branchial grooves **26.** (a) chorion, (b) respiration, (c) amnion, (d) allantois, (e) nitrogenous wastes, (f) yolk sac **27.** (a) zona pellucida, (b) blastocyst **28.** trophoblast **29.** secretions from uterine glands **30.** (a) implantation, (b) seventh **31.** (a) chorion, (b) uterine **32.** (a) human chorionic gonadotropin (hCG), (b) corpus luteum **33.** second and third **34.** third **35.** end of first trimester **36.** (a) seventh, (b) convolutions **37.** (a) dizygotic, (b) monozygotic **38.** ultrasound **39.** teratogens **40.** (a) amniocentesis, (b) chorionic villus sampling **41.** carbon dioxide **42.** 5 **43.** produce telomerase

Building Words: **1.** telolecithal **2.** isolecithal **3.** blastomere **4.** blastocyst **5.** blastopore **6.** holoblastic **7.** meroblastic **8.** archenteron **9.** trophoblast **10.** trimester **11.** neonate **12.** acrosome **13.** cleavage **14.** notochord **15.** fetus

Matching: **1.** a **2.** e **3.** j **4.** n **5.** c **6.** f **7.** k **8.** l **9.** i **10.** h **11.** d

Making Comparisons: **1.** gastrulation **2.** 24 hours **3.** fertilization **4.** 0 hour **5.** 7 days **6.** blastocyst begins to implant **7.** fertilization **8.** organogenesis **9.** late cleavage **10.** 3 days

Making Choices: **1.** c **2.** b,c,e **3.** d **4.** c **5.** c,e **6.** a **7.** d **8.** b **9.** c,e **10.** d **11.** c **12.** b,c **13.** b,d **14.** c,d **15.** b **16.** a,b **17.** a,b,e **18.** a,b,d,e **19.** a,b,e

Chapter 52

Reviewing Concepts: **1.** stimuli **2.** natural selection **3.** behavioral ecology **4.** (a) proximate, (b) ultimate **5.** direct fitness **6.** (a) innate, (b) learned **7.** (a) nervous, (b) endocrine **8.** motor programs **9.** sign stimulus or releaser **10.** experience **11.** aposematic or warning coloration **12.** habituation **13.** recognizes **14.** (a) conditioned stimulus, (b) unconditioned **15.** (a) positive reinforcement, (b) negative reinforcement **16.** genetically programmed **17.** insight learning **18.** (a) exercise, (b) learning to coordinate movements, (c) learning social skills **19.** (a) circadian rhythms, (b) diurnal, (c) crepsular **20.** suprachiasmatic nucleus (SCN) **21.** pineal **22.** migration **23.** directional orientation **24.** (a) navigation, (b) map **25.** (a) gathering, (b) capturing **26.** optimal foraging **27.** social behavior **28.** society **29.** Bottlenose dolphins **30.** pheromones **31.** vomeronasal **32.** fighting **33.** testosterone **34.** home range **35.** (a) inherited, (b) preprogrammed **36.** intrasexual selection **37.** intersexual selection **38.** lek **39.** courtship rituals **40.** (a) polygyny, (b) polyandry **41.** parental investment **42.** coefficient of relatedness **43.** kin selection **44.** sentinel behavior **45.** reciprocal altruism **46.** (a) common to a population, (b) transmitted from one generation to the next **47.** inhereited **48.** natural selection

Building Words: **1.** sociobiology **2.** polyandry **3.** polygyny **4.** monogamy **5.** pheromone

Matching: **1.** b **2.** e **3.** i **4.** m **5.** l **6.** g **7.** h **8.** a **9.** c **10.** f **11.** d **12.** n

Making Comparisons: **1.** learned **2.** classical conditioning **3.** operant conditioning **4.** habituation **5.** imprinting **6.** insight learning

Making Choices: **1.** d **2.** e **3.** b **4.** b **5.** a,b **6.** a **7.** a,b,d **8.** a,c,e **9.** d,e **10.** a,b,d,e **11.** a,b,d,e **12.** b,c,e

Chapter 53

Reviewing Concepts: **1.** (a) biotic factors, (b) abiotic factors **2.** (a) population density, (b) population dispersion, (c) birth and death rates, (d) growth rates, (e) survivorship, (f) age structure **3.** population ecology **4.** population dynamics **5.** population density **6.** (a) random dispersion, (b) uniform dispersion, (c) clumped

(aggregated) dispersion   **7.** (a) natality, (b) mortality   **8.** (a) change in number of individuals in the population, (b) change in time, (c) natality, (d) mortality   **9.** (a) immigration, (b) emigration
10. intrinsic rate of increase (r_{max})   **11.** exponential population growth   **12.** carrying capacity   **13.** S-shaped
14. (a) predation, (b) disease, (c) competition   **15.** negative feedback   **16.** high   **17.** (a) death, (b) birth   **18.** (a) intraspecific competition, (b) interspecific competition   **19.** (a) interference competition, (b) exploitation competition   **20.** (a) frost, (b) blizzards, (c) hurricanes, (other weather related events and natural phenomenon)   **21.** (a) semelparous, (b) iteroparous   **22.** fecundity   **23.** (a) *r* strategists, (b) *K* strategists   **24.** survivorship   **25.** (a) type III survivorship, (b) type I survivorship   **26.** (a)source habitats, (b)sink habitats   **27.** (a) birth rate, (b) death rate   **28.** (a) doubling time, (b) less   **29. (a)** replacement-level fertility, (b) 2.1, (c) 2.7   **30.** (a) total fertility rate, (b) lower   **31.** (a) number, (b) proportion   **32.** (a) former, (b) latter
Building Words: **1.** intraspecific competition   **2.** interspecific competition   **3.** biosphere   **4.** ecology   **5.** intrinsic
Matching: **1.** o   **2.** i   **3.** l   **4.** b   **5.** e   **6.** p   **7.** h   **8.** a   **9.** f   **10.** n   **11.** g   **12.** m   **13.** c   **14.** q
Making Comparisons: **1.** density-independent   **2.** environmental factor that affects the size of a population but is not influenced by changes in population density   **3.** density-dependent   **4.** dominant individuals obtain adequate supply of a limited resource at expense of others   **5.** density-dependent   **6.** as density increases, encounters between population members increase as does opportunity to spread disease   **7.** density-independent   **8.** random weather event that reduces vulnerable population
Making Choices: **1.** g   **2.** h   **3.** d   **4.** f   **5.** b   **6.** b   **7.** b   **8.** a,e   **9.** d   **10.** c   **11.** a   **12.** b,c   **13.** a,d   **14.** c   **15.** a   **16.** a,b,c   **17.** b,c,d,e   **18.** c,e

Chapter 54

Reviewing Concepts: **1.** community   **2.** ecosystem   **3.** facilitation   **4.** (a) competition, (b) predation, (c) symbiosis   **5.** niche   **6.** (a) fundamental, (b) realized   **7.** limiting resource   **8.** (a) intraspecific, (b) interspecific   **9.** competitive exclusion principle   **10.** resource partitioning   **11.** character displacement   **12.** coevolution   **13.** (a) aposematic coloration, (b) warning coloration   **14.** cryptic   **15.** Batesian mimicry   **16.** Mullerian   **17.** (a) mutualism, (b) commensalism, (c) parasitism   **18.** pathogen   **19.** (a) ecto, (b) endo   **20.** (a) food, (b) water, (c) other resources   **21.** very common   **22.** food webs   **23.** in a food web   **24.** (a) nitrogen, (b) phosphorous   **25.** (a) species richness, (b) species evenness   **26.** geographic isolation   **27.** less   **28.** ecotone   **29.** edge effect   **30.** complexity   **31.** primary   **32.** secondary   **33.** climax community   **34.** disturbance   **35.** organismic   **36.** individualistic
Building Words: **1.** carnivore   **2.** ecosystem   **3.** pathogen   **4.** intraspecific competition   **5.** herbivore   **6.** interspecific competition   **7.** predation   **8.** cryptic   **9.** parasite   **10.** succession
Matching: **1.** m   **2.** l   **3.** j   **4.** g   **5.** n   **6.** i   **7.** e   **8.** b   **9.** o   **10.** h   **11.** a
Making Comparisons: **1.** commensalisms of species 1 and species 2   **2.** beneficial   **3.** no effect   **4.** predation of species 2 by species 1   **5.** beneficial   **6.** adverse   **7.** parasitism of species 2 by species 1   **8.** beneficial   **9.** adverse   **10.** parasitism of species 2 by species 1   **11.** beneficial   **12.** adverse   **13.** beneficial   **14.** beneficial   **15.** adverse   **16.** adverse   **17.** commensalism of species 1 and species 2   **18.** beneficial   **19.** no effect
Making Choices: **1.** b   **2.** d   **3.** c   **4.** b   **5.** a,c,e   **6.** e   **7.** a   **8.** a,d,e   **9.** e   **10.** a   **11.** c   **12.** b,e   **13.** a-e   **14.** a,b,c,e   **15.** b

Chapter 55

Reviewing Concepts: 1. ecosystem 2. biosphere 3. (a) radiant, (b) chemical, (c) heat 4. food web 5. (a) producers, (b) primary consumers (herbivores), (c) secondary consumers (carnivores) 6. (a) detritus, (b) detritivores 7. (a) numbers, (b) biomass 8. energy 9. gross primary productivity 10. net primary productivity 11. 32, 40 12. bioaccumulation 13. biological magnification 14. A biogeochemical cycle 15. (a) carbon, (b) nitrogen, (c) phosphorus, (d) water 16. (a) photosynthesis, (b) respiration 17. (a) nitrogen fixation, (b) nitrification, (c) assimilation, (d) ammonification, (e) denitrification 18. photochemical smog 19. (a) rocks, (b) soil 20. runoff 21. transpiration 22. groundwater 23. (a) water, (b) temperature 24. long-wave infrared (heat) energy 25. (a) less concentrated, (b) lower temperatures 26. (a) high, (b) low, (c) Coriolis Effect 27. (a) Pacific, (b) Atlantic, (c) Indian, (d) Artic 28. Pacific 29. (a) right, (b) clockwise, (c) left, (d) counterclockwise 30. climate 31. precipitation 32. rain shadows 33. microclimates 34. (a) free nutrient minerals essential for plant growth locked in dry organic matter, (b) remove plant cover and expose the soil, thereby stimulating the germination and establishment of seeds requiring bare soil, as well as encouraging the growth of shade-intolerant plants, (c) increase soil erosion, leaving the soil more vulnerable to wind and water. 35. deforestation

Building Words: 1. biomass 2. atmosphere 3. hydrologic cycle 4. microclimate 5. trophic 6. saprotroph 7. detritus 8. nodules 9. assimilation

Matching: 1. n 2. i 3. e 4. g 5. k 6. l 7. d 8. f 9. c 10. j 11. b 12. m

Making Comparisons: 1. carbon 2. carbon dioxide 3. nitrogen 4. nitrogen 5. hydrologic 6. enters atmosphere by transpiration, evaporation; exits as precipitation

Making Choices: 1. c 2. a,c 3. a,c 4. a-e 5. d 6. b,d,e 7. a,d,e 8. c,e 9. b 10. c 11. e 12. a,c,d,e 13. e 14. e 15. a,b,d

Chapter 56

Reviewing Concepts: 1. (a) terrestrial, (b) climate 2. (a) northern most, (b) permafrost 3. elevations 4. taiga 5. (a) short growing season, (b) mineral-poor soil 6. industrial wood and wood fiber 7. (a) northwestern North America, (b) southeastern Australia, (c) southwestern South America 8. (a) lumber, (b) pulpwood 9. 30-50 inches (75-126 cm) 10. agricultural use 11. 10-30 inches (25-75 cm) 12. tall grass prairies 13. Mediterranean 14. (a) Great Basin Desert, (b) Mojave Desert, (c) Chihuahuan Desert, (d) Sonoran Desert 15. estivation 16. (a) precipitation (b) temperature 17. African 18. over grazing by domestic animals 19. 80-180 inches (200-450 cm) 20. evergreen flowering 21. abiotic factors 22. (a) plankton, (b) nekton, (c) benthos 23. (a) streams, (b) rivers, (c) ponds, (d) lakes, (f) marshes, (g) swamps 24. strength of the current 25. (a) littoral, (b) limnetic, (c) profundal, (d) littoral 26. (a) thermal stratification, (b) thermocline, (c) fall turnover, (d) spring turnover 27. (a) marshes, (b) swamps 28. (a) prairie potholes, (b) peat moss bogs 29. salt marshes 30. mangrove forests 31. intertidal zone 32. (a) distance from land, (b) light availability, (c) depth 33. 4-6 meters (2.5-3.7 miles) 34. neritic province 35. euphotic 36. oceanic province 37. boundries 38. center of origin 39. cosmopolitan 40. (a) Palearctic, (b) Nearctic, (c) Neotropical, (d) Ethiopian, (e) Oriental, (f) Australian

Building Words: 1. phytoplankton 2. zooplankton 3. intertidal 4. thermocline 5. littoral 6. benthic 7. abyssal

Matching: **1.** n **2.** i **3.** j **4.** a **5.** c **6.** l **7.** d **8.** e **9.** h **10.** g **11.** f **12.** k

Making Comparisons: **1.** long, harsh winters, short summers, little rain **2.** permafrost **3.** low-growing mosses, lichens, grasses, and sedges adapted to extreme cold and short growing season; voles, weasels, arctic foxes, grey wolves, snowshoe hares, snowy owls, oxen, lemmings, caribou, etc. (few plant and animal species) **4.** taiga **5.** seasonality, high rainfall **6.** rich topsoil, clay in lower layer **7.** broad-leaf trees that lose leaves seasonally, variety of large mammals **8.** temperate grassland **9.** deep, mineral-rich soil **10.** wet, mild winters; dry summers **11.** thickets of small-leafed shrubs and trees, fire adaptation; mule deer **12.** desert **13.** low in organic material, often high mineral content **14.** savanna **15.** low in essential mineral nutrients **16.** high rainfall evenly distributed throughout year **17.** mineral-poor **18.** high species diversity, distinct stories of vegetation, epiphytes in organic material, often high mineral content

Making Choices
es: **1.** b **2.** d **3.** b,d **4.** a,d **5.** e **6.** a,d **7.** d **8.** b,d **9.** e **10.** b,e **11.** b **12.** c,d **13.** a-e **14.** c **15.** b **16.** b-e **17.** c **18.** b,e **19.** a-e

Chapter 57

Reviewing Concepts: **1.** environmental sustainability **2.** genetic diversity, species diversity, and ecosystem diversity **3.** (a) endangered, (b) threatened **4.** (a) an extremely small range, (b) territory, (c) on islands, (d) low reproductive success, (e) breeding areas, (f) feeding habits **5.** (a) destruction, (b) modification **6.** habitat fragmentation **7.** biotic pollution **8.** nature (the wild) **9.** human controlled settings **10.** animal and plant habitats **11.** landscape ecology **12.** habitat corridors **13.** (a) restoration ecology, (b) *in situ* **14.** (a)artificial insemination, (b)host mothering **15.** protect endangered and threatened species in the U.S. and abroad **16.** endangered animals and plantsconsidered valuable in international wildlife trade **17.** trapping and absorbing precipitation **18.** releasing carbon stored in trees into the atmosphere as carbon dioxide **19.** (a) subsistence agriculture, (b) commercial logging, (c) cattle ranching **20.** industrial wood and wood fiber **21.** CO_2, CH_4, O_3, N_2O, chlorofluorocarbons (CFCs) **22.** greenhouse effect **23.** (a) sea level, (b) precipitation patterns, (c) biological diversity, (d) agriculture **24.** 10 **25.** (a) chlorine, (b) bromine **26.** chlorofluorocarbons (CFCs) **27.** September and November **28.** (a) cataracts, (b) skin cancer, (c) weakened immune system **29.** (a) 50, (b) CFCs

Building Words: **1.** deforestation **2.** stratosphere **3.** endemic **4.** biotic

Matching: **1.** g **2.** o **3.** b **4.** e **5.** h **6.** a **7.** j **8.** n **9.** c **10.** l **11.** d **12.** i

Making Comparisons: **1.** deforestation **2.** permanent removal of forests for agriculture, commercial logging, ranching, use for fuel **3.** reduce current practices **4.** global warming **5.** atmosphere pollutants that trap solar heat **6.** prevent build-up of gases, mitigate effects, adapt to the effects **7.** chemical destruction of stratospheric ozone **8.** reduce use of chlorofluorocarbons and similar chlorine containing compounds